UNDERSTANDING CHIMPANZEES

UNDERSTANDING
CHIMPANZEES

Edited by
Paul G. Heltne
Linda A. Marquardt

with 31 contributors

Published by Harvard University Press
in cooperation with The Chicago Academy of Sciences
Cambridge, Massachusetts, and London, England
1989

Title page photos: Randall L. Susman

This volume was prepared by The Chicago Academy of Sciences
(Special Publication Number 16).

This book is printed on acid-free paper, and its binding materials have been
chosen for strength and durability.

Library of Congress Cataloging-in-Publication Data

Understanding chimpanzees / Paul G. Heltne and Linda Marquardt, editors.

    p.   cm.
    ISBN 0–674–92091–0
    1. Chimpanzees.   I. Heltne, Paul G.   II. Marquardt, Linda.
QL737.P96U56   1989   599.88'44–dc20
89–15379   CIP

The genus *Pan* consists of two species that are endemic to equatorial Africa: *Pan troglodytes* (traditionally known as the common chimpanzee, although "common" is a misnomer now that the implied abundance no longer exists) and *Pan paniscus* (also known as the pygmy chimpanzee, now more often referred to as the bonobo).

Taxonomists generally agree that *Pan troglodytes* can be further divided into three main populations that exhibit mutually exclusive geographical ranges: *P. t. verus* in western Africa, *P. t. troglodytes* in central Africa, and *P. t. schweinfurthii* in eastern Africa.

— *Geza Teleki*, this volume.

CONTENTS

Additional Sites

2 / CURRENT FIELDWORK: *Pan paniscus*

3 / THE CHIMPANZEE MIND

4 / CHIMPANZEE CONSERVATION

5 / EVOLUTION and EPILOGUE

FOREWORD

Jane Goodall

The papers in this volume represent a wide range of different aspects of chimpanzee and bonobo behavior: social behavior and ecology in the field; the rich variety of cultural traditions between one population and another in various parts of Africa; behavior in captive colonies; the incredible cognitive abilities of chimpanzees in the language acquisition labs. All of this tremendously increases our understanding of chimpanzees.

Two things most particularly impress me. Both are aspects which I have thought about before, but they are driven home by paper after paper – by the Japanese, by the Dutch, by the Americans, by the British. The first is the variability of chimpanzee behavior, the richness of individuality, and the way in which individual chimpanzees have influenced not only the history of their own communities in the wild but also scientific thinking. Washoe was Washoe; if the Gardners had started off with another chimp, a less gifted chimp, a less practical chimp, that project might have died in infancy – they might have picked a chimpanzee dunce. The chimpanzees working in enriched environments (such as those provided by the Gardners, the Fouts, and Sue Savage-Rumbaugh and Duane Rumbaugh) are just a few individuals representing whole species. (It's like taking a handful of people to be looked at on another planet. We could all say, "My goodness, if they picked so-and-so they would have a pretty dim view of humans.") If a particular chimpanzee fails to accomplish tasks that other chimpanzees have mastered, the psychologist working with that individual cannot assume that Roger Fouts, Duane Rumbaugh, and Sue Savage-Rumbaugh have made exaggerated claims regarding the intelligence of chimpanzees in general – the psychologist may have picked a dimwit.

The second thing that impresses me is how much more there is to learn. The attempt to understand the nature of the chimpanzee is really very recent. It began with the pioneering work of Wolfgang Köhler, who has long been a hero of mine, and the work of Robert Yerkes in the United States. Yerkes gathered together a group of scientists eager to work with chimpanzees; but then, when the war broke out, the interest declined and possibilities for research became fewer. It did not pick up again until the early '60s when, once more, there came a surge of interest in chimpanzee behavior. This time investigators tramped out into the African forests – Adriaan Kortlandt to Zaire (the Congo then), Vernon Reynolds to Uganda, Junichero Itani and I to Tanzania. And at the same time, the Gardners initiated the ASL and other language-acquisition projects with Washoe. Yet even now we have a long way to go in our attempts to understand the true nature of the chimpanzee.

The difference in cultural tradition between the chimpanzees of Gombe and those of the Mahale Mountains, only 100 miles away, has already proved to be very great. Once we have increased collaboration, more visiting of each other's study sites, we shall find even more differences between the traditions of one group and the other. Let me give a couple of examples: (1) At Mahale, when two chimpanzees groom together, they very often sit facing each other, and each, with one arm in the air, holds hands with his partner. This does not happen at Gombe, where chimps hold an overhead branch. (2) At Gombe the chimps fish for vicious, biting, driver ants, using a long stick. The stick is pushed down into the nest, withdrawn (coated with ants), swept through the chimp's free hand, and the mass of insects thus collected are then chewed. There are many of these ants in Mahale, but the chimps there don't eat them. Chimps in Mahale, however, use twigs, which they push into holes in a tree trunk to fish for the tree-living carpenter ants. There are plenty of carpenter ants at Gombe, but the Gombe chimps have not learned how to eat them. These are just a few examples.

The differences in cultural tradition are even greater between the chimps at Gombe, in the eastern part of the range, and the chimps far west in the Ivory Coast, the Tai Forest, where Christophe and Hedwige Boesch have been working for nine years. We have only just begun to understand the variability of behavior in chimpanzees across Africa.

As Duane Rumbaugh points out in his paper, there is also a great deal more to be learned about the cognitive abilities of the chimp. Clearly, an understanding of behavior and cognitive capability will be, to a large extent, determined by the imagination and sensitivity of the people who are working with individual chimps. As Roger Fouts states, there must be opportunity for the animals themselves to tell the experimenters new facts about their personalities, about their intelligence, about their emotions.

This brings me to the urgent future of chimpanzee research. What is the urgency? The urgency is made very plain in Part Four, the chapter on chimpanzee conservation. The tropical forests in Africa are disappearing at a most horrifying rate. People in the United States are probably more familiar with the destruction of the rain forest in South America because it is nearer to home. The same sort of thing is happening in the great tropical rain forests worldwide, and Africa is no exception.

I have lived in Africa for years, but even I am horrified by Geza Teleki's information regarding the situation of chimpanzee populations across the continent. He paints a picture of the future which is dismal, grim, and dark. The habitat of the chimpanzee in its native home is dwindling at a terrifying rate and, as yet, there are very few areas where the chimps are totally protected.

The most we can hope for in the future is to set up a series of preserves and national parks where the chimps, to some extent, can live out their natural lives. But even if we manage to do this – and unless we act soon, it will be too late – we can hope at best for a series of islands within which the chimps will be imprisoned. Gombe is such an island. Cultivation has come right up around the boundaries of the park, whereas when I arrived in 1960, I could climb up to the peaks of the rift escarpment and look out to the east over chimpanzee country. Now it is tilled fields and almost all the forest is gone. There is hunting around the boundary. Gombe is only 30 square miles and it only supports three different social groups of chimps – only about 150–160 individuals. This is a very tiny population, and its future is jeopardized.

The wild population has been depleted across Africa not only by the destruction of the habitat and the explosive growth of the human population, but also by the demand from zoos, the entertainment industry, and biomedical labs. Even in captivity the possibility of establishing a self-sustaining population is grim. The best breeding mothers are those who were caught as infants in the wild. To capture the baby, hunters shoot the mother. You can imagine the trail of destruction that is left behind: females who died, infants who died in the forest, and infants who died during the transportation, having been wrenched from their mothers' bodies.

Despite the shocking trauma endured by infants captured in this inhumane way, the female infants who survived have turned out to be the best breeders. They gained invaluable experience during their early years and so are able to care for their babies. If a chimp born in captivity is taken from the mother at a very early age (as is practiced in most of the labs), she is unlikely to be a good mother when she grows up. In one experiment, a female chimpanzee was taught, successfully, how to look after her baby: she was shown films and given a toy infant so that she could practice maternal behavior while she was still pregnant. On the whole, however, these captive-born females are not good breeders, and the chimpanzee population in this country, and other countries, is not increasing as a result of births in captivity.

Unless something is done about captive breeding, the scientists engaged in medical research, particularly hepatitis and AIDS research, will undoubtedly want to obtain more chimpanzees from their natural habitat. This is going on today, even between countries where it is illegal both to export and import chimps. There are unscrupulous dealers who manage, often by means of bribes, to get chimpanzees out, leaving a trail of dead, dying, and wounded individuals behind them. Methods of importing can be equally destructive.

This, then, is why chimpanzee research is urgent. If we do not take action – those of us who care – if we do not do something now, it is going to be too late. In another 100 years there won't be any more chimpanzees. Future generations of humans will never be able to learn first hand of the wonder of chimpanzees. If any chimpanzees survive, they will be remnants of a dying species.

A promising plan came out of discussions held in 1986 between scientists and others concerned for the survival and care of chimps. A chimpanzee action committee has been formed (the Committee for Conservation and Care of Chimpanzees) to try to

(a) raise the status of chimps in the United States from threatened to endangered, (b) conduct surveys and establish protected areas in Africa, (c) monitor international trade, and (d) promulgate humane conditions for all chimps in captivity.

The picture of the chimpanzee that is portrayed in this volume is that of a creature, a *being*, with a very high level of intellectual sophistication, much greater than was thought even ten years ago; a being whose emotional states are quite similar to our own and who is capable of feeling pain, sorrow, happiness; a being who can trust and whose trust it is very easy to betray. This volume raises an ethical question: "Is it right for those in laboratories, who are seeking to further human knowledge and alleviate human suffering, to use beings of this sort?" Maybe sometimes this can be justified. If a tremendous amount of human suffering can be alleviated by performing experiments on a handful of chimps, maybe this is as justifiable as it would be to use a handful of humans for that same purpose. If we used the humans – if humans volunteered themselves, as many of those dying with AIDS have done – we would treat them well. We would not take a human volunteer and put him in a tiny cage, alone and with no stimulation. But that is exactly what is so often the lot of the chimps. Are we justified in this? I think if we really ask ourselves, deep within our hearts, the answer will be "no." The chimpanzee, because he is what he is, deserves far better treatment at our hands.

A story from Lion Country Safari, in Florida, makes the point clearly. There are five islands at Lion Country and a total of 30 chimps from all kinds of backgrounds. One island has an alpha male, called "Old Man," and three females. When a young man named Marc Cusano was employed at Lion Country as a keeper, he was told, "Don't go onto the islands. Those chimps are dangerous. They hate people." Marc didn't see how he could properly care for them if he couldn't go onto the islands. He began a very slow process whereby he gradually got closer and closer and gradually formed friendly relationships with each of those 30 chimps. He went to the hospital a number of times during that process, which took about two years, and this says a good deal about his dedication. During that time, one of Old Man's three females gave birth.

One day, after Marc had become friends with Old Man, he accidentally slipped while he was putting food out on the island, and he fell close to the infant, who screamed. Instantly the mother, her protective responses aroused, flew at Marc and bit him. The other two females rushed in to help, as chimpanzee females will. Marc was bitten on his arm: he felt one hand go dead. He was also bitten on his back and his leg. The females hung on and Marc thought: "This is it."

And then Old Man – Old Man who had clearly been badly mistreated in his youth by humans, who may have been rescued from a lab – charged up to rescue Marc. Marc was probably the first person in many years with whom he had had friendly contact. He lifted those females off one by one and literally, physically, hurled them away. He stayed with Marc, displaying around him with his hair bristling, keeping the females off. They were highly aroused and screaming. Marc managed to drag himself to his little boat and paddle away.

He said to me afterwards that there was absolutely no question in his mind but that Old Man saved his life. I think there is something in this for us. Old Man is a chimpanzee being who has suffered at human hands, reaching out to help a friend of another species. If a chimpanzee can do that, we can too. And we can each help in different ways. Some of us can help by giving our time. Some of us can help by talking about chimps; by spreading the word, particularly to children; by going to the zoo and watching chimps, trying to understand that they are intelligent beings and that we humans are not the only ones who matter in this world. Some of us can help by making donations to the "Committee for Conservation and Care of Chimpanzees." [For further information and address, see the Epilogue.] Any money raised will go to aid conservation in the wild and toward improving conditions for the captive chimpanzees in labs. I would urge each of you to help, as you can, because I think the chimps deserve it. I hope that the information in this volume will have a big impact, not only on our knowledge and understanding but also in helping the conservation and care of these very remarkable chimpanzee beings.

It is particularly exciting for me to be contributing to this volume along with people working out in nature, people working in labs, people working with groups of chimps in captivity. There is the great contingent of hardworking field, zoo, and laboratory researchers from Great Britain, from Holland, from the United States, and from Japan. And there are those who, although not actively involved in research at present, are totally dedicated to the welfare of chimpanzees – caring people such as Geza Teleki, sick with river blindness from his work in Sierra Leone, and Shirley McGreal of the International Primate Protection League. For these people, we and the chimpanzees must be really grateful.

It has only been possible for me to continue the research at Gombe, and it was only possible to pull together the tremendous volume of information contained in the book *The Chimpanzees of Gombe*, because of the work of so many other people. It tends to be forgotten that but for the foresight of Louis Leakey way back in 1957, when he first thought of sending young women out to study the great apes, Gombe research would never have been begun, at least not at that time and certainly not by me.

There has been a whole succession of students and researchers at Gombe who have done so much to collect careful and precise data. It is really delightful that some of them are represented in this volume: Chris Boehm, Bill McGrew, Geza Teleki, and Richard Wrangham. These people and many others have made great contributions to the growing body of research at Gombe.

Also, let me pay great tribute to the Tanzanian field assistants, because they are there day in and day out. These are men without higher education, but they have been trained, and trained very rigorously, in the natural school, in the laboratories of Gombe. Their observations are precise and painstaking. They know the chimpanzees and the terrain as well or better than anyone else. They are absolute experts in keeping up with fast-moving chimpanzees in very difficult conditions. They are far better at it than I am. In fact, the last time I was at Gombe with a man I consider to be one of my best current researchers, Yahaya Alamasi, we came to a certain place and he said, "I don't want you to come here, you are not here often enough. You go 'round that way and I will go this difficult way and we will meet at the other end." These men really are tremendous. They have now attained a very sophisticated level of collecting data, which Chris Boehm describes in Part One of this volume. It was his idea to introduce complicated tape recorders into the recording equipment of the field staff, and this led to the use of video cameras. This opens up a whole new area of research possibilities at Gombe.

There has been a tremendous input into the interpretation of the Gombe results because of the work psychologists and others are doing with captive chimpanzees on problems of cognition. (There is a lot of talk about psychologists who exploit their chimpanzee subjects. Sadly, this is sometimes true. But it is absolutely not true of all of them. The scientists and caretakers who work with chimps in the laboratory or the zoo and who are represented in this volume are, without exception, people who are very dedicated to the future of their animals – animals who have given us so much knowledge, taught us so much about chimpanzee beings.) It has been very meaningful to be able to write about intelligent performances of chimpanzees in the wild against the background of knowledge that has been provided by this painstaking research in the lab, where cognitive ability and intelligence can be rigorously tested and where chimps can be pushed, as it were, toward the upper limits of their mental abilities. So often, in the wild, an intelligent behavior is discounted as a mere anecdote. In fact, as I have always firmly believed, a collection of such anecdotes becomes a very powerful tool when trying to understand some of the more complex aspects of the behavior of a creature who is as undoubtedly intelligent as the chimpanzee.

All of these people, all of these studies, help us to draw a more perfect picture of the creature who, along with the bonobo and the gorilla, is closest to human. This picture, I sincerely hope, will prove to the scientific community and to other people in the world today that we must consider, much more carefully than we do now, the kinds of things that are ethical to do with these creatures when we bring them out of the forest into our own countries.

PREFACE

This volume represents several firsts: it gathers together current field research on the three subspecies of chimpanzee *(Pan troglodytes)* in the eastern, central, and western regions of Africa; it joins this research with field and zoo observations on bonobos *(Pan paniscus)* to further the comparison of the two species; it adds to this rich mix laboratory studies of chimpanzee and bonobo behavior and cognition; and it reviews the endangered status of chimpanzees and bonobos and gives high priority to their conservation. The chapters herein represent the work of 31 researchers with diverse backgrounds and concerns. Only by establishing such an interplay of ideas and research can *Homo sapiens* hope to come to an understanding of chimpanzees, bonobos, and humans.

The publication of this volume rests on an extraordinary amount of support from many people. We are deeply indebted to Jane Goodall for inspiring this project. She played an invaluable role in structuring the topics, selecting the contributors, and looking always to the big picture. We also offer a very special note of thanks to Toshisada Nishida for his advice and for his help in coordinating the many Japanese contributors. During the planning stages, valuable advice was also provided by Frans de Waal, Duane Rumbaugh, and Richard Wrangham.

This book was prepared at The Chicago Academy of Sciences. A number of people have been absolutely essential during its production. We owe a particular note of gratitude to Bettie Leslie, whose perseverance has been extraordinary and whose support has been unflagging during the overall preparation of the manuscripts. We are especially grateful to Suzanne W. Brown for her creative energy and personal commitment to the preparation of this volume, and to Sandra Lancaster for her comprehensive and skillful help with editing and manuscript preparation.

Very special recognition and gratitude are due Christopher Dunn, who designed and maintained the technical and computer systems which made this publication possible and who was unfailingly on site to attack the gremlins. We especially thank Nicholas Howell, who provided expert and generous assistance with the many visuals accompanying the papers and with the final camera-ready copy. Special thanks go to

Elizabeth Thompson for copy-proofing and for cheerful support even in the most mundane areas and to Sheila Ary for preparing the index. We also thank Joanne Softcheck and Bonnie Needham, both of whom provided extensive, experienced assistance with preliminary copy editing. Diane Hutchinson served as consultant on the book design, with assistance from Vita Jay on layout.

We join the authors in recognizing the institutions that support their work and the public and private foundations and institutions who contributed to the development of material in this volume and who continue to support chimpanzee and bonobo research. In particular, we thank the trustees of The Chicago Academy of Sciences for their continuing support, and we recognize the L.S.B. Leakey Foundation and the John D. and Catherine T. MacArthur Foundation for initial grants to this project.

Finally, to the authors of this volume, we express heartfelt thanks. They are among the pioneers in the field of chimpanzee and bonobo studies, following scouts such as Wolfgang Köhler and Robert Yerkes. They have worked the fields with care and creativity, and they have made discoveries unpredicted and unimagined. Their cooperation, patience, and enthusiasm during the preparation of the book made possible this significant contribution to the understanding of chimpanzees and bonobos.

Paul G. Heltne
Linda A. Marquardt

1

CURRENT FIELDWORK
Pan troglodytes

Gombe

Mahale Mountains

Additional Sites

GOMBE:
HIGHLIGHTS AND CURRENT RESEARCH

Jane Goodall

TOOL USING, HUNTING, AND SOCIAL RANK

I want to review some of the highlights of 26 years of research at Gombe and discuss some of the current work going on at Gombe and elsewhere.

In the sphere of tool using and tool making, we have wonderful examples of some of the cultural variations among chimpanzees in different geographical areas (Kortlandt, this volume).

The fact that chimpanzees hunt, and hunt cooperatively, is well known today – although this was very startling when first reported. When I first published a description of hunting and meat eating, there was a spate of papers suggesting that this was totally abnormal behavior and undoubtedly caused by the fact that the chimps had been given bananas. Now, however, other people have moved into the field, and we know that hunting is typical of chimps throughout their range.

One of the fascinating aspects over the years at Gombe has been to look at the different techniques employed by adult male chimpanzees in their attempts to gain supremacy, in their constant struggle to maintain or better their position in the dominance hierarchy.

Dominance as a concept will surely always have its ups and downs in the behavioral literature and in discussions between scientists, but there is absolutely no question that the chimpanzee does have an inherent, powerful, and compelling desire to work his way up through the dominance hierarchy. So much so that when we have the odd individual, as we do at Gombe, who does not seem particularly interested in his social rank, we regard him as distinctly unusual and want to burrow into his childhood to see if we can find clues there as to why he shows this surprising lack of the dominance drive.

CANNIBALISM AND WARFARE

The first observation of cannibalism came from Dr. Suzuki in Uganda, back in the early '70s. This observation was very startling then. I have to admit that when I first read the report, I thought: "He can't have been near enough. It must have been a black infant baboon that the chimpanzees were eating." Then, about a year later, we had the first example coming from Gombe, observed by David Bygott. These observations point to the value of long-term research: during the first ten years at Gombe we clearly had not seen the full extent of the repertoire of creatures who are as long-lived and have such variable and flexible behavior as the chimpanzee.

The division of the main study community at Gombe into two groups (1970–72) led to the most fascinating and, in a way, disturbing series of events that we have seen. The two communities established themselves in two different parts of what had originally been the common range, and a few years later, we saw the first of the brutal attacks by males of the Kasakela community on isolated individuals of the smaller Kahama community. Within the next seven years, that splinter group in the south, the Kahama community, had been exterminated – or at least all its members had disappeared and were never seen again.

These attacks, as well as other aspects of inter-community aggression, are probably similar to behaviors in our own ancestors which eventually led to human warfare. Dr. Eibl-Eibesfeldt, in his long discussion of the emergence of warfare in our own species, has said that in some primitive tribes, warfare takes the form of a series of hunting raids, and that is precisely what we saw at Gombe.

After the Kahama community had vanished, the Kasakela victors and their females moved into the newly vacated area in the south. They fed and nested there without concern. But it was only just over a year later that they began to be harassed on their southern boundary by the very powerful Kalande community from further south. In a period of just over a year, they had not only been driven out of the territory they had acquired as a result of their warlike behavior, they had been driven even further north than their "pre-war" boundary. At the same time, another large community in the north, with between six and ten fully adult males, was pushing southward.

There was a period in the early '80s when I feared that all the chimps of my study group, chimps I had been with for 20 years, might disappear. Three adult males vanished rather like the males of Mahale's C-group had vanished, one by one, and we shall never know what happened to them. By 1981 there were only four fully adult males in the community. The range of the community had shrunk from around 15 square kilometers, which it had been at the end of the "war," to something between eight and nine square kilometers in 1981 – an area that certainly wasn't big enough to support the community's 15 females and their young.

Fortunately, there were many adolescent males growing up. They began to spend longer and longer periods with the adult males and to accompany them on some of their patrols. Had it come to a fight, they would have been too inexperienced to be very helpful; but I suspect that their pant-hoots and drumming displays, serving to increase the volume of sound, enabled our Kasakela community, with its four adult males, to seem stronger than in fact it was. Whether or not for this reason, gradually the boundaries expanded again, and at the time of writing, our community is in a very healthy situation, with seven adult males and a range expanded to about 12 square kilometers.

As far as we know, there are no more of these violent inter-community interactions occurring. No longer do we see Kalande groups chasing the Kasakela males back to the center of their range, as we saw in the early '80s. Rather, we have the more typical situation: when neighboring males encounter one another near the periphery, both sides give terrific calls and displays and then discreetly retreat, each back toward the center of its home range. So, on that front, all is fairly peaceful at Gombe today.

THE IMPACT OF THE INDIVIDUAL

I cannot resist stressing the tremendous impact that individual chimpanzees can make on the history of their community. We tend to think of this as a human prerogative. We think of the Hitlers and Napoleons, and so on, who did indeed alter the course of the world. We find the same sort of thing happening in chimpanzee society too.

The first example concerns the mother-daughter pair, Passion and Pom, who developed extraordinary cannibalistic tendencies. This had an immense impact on the whole future of the Kasakela community. For a period of something over four years, all but one of the infants born in the central part of the community range were known either to have been killed and eaten by Passion and Pom, or to have disappeared during the first few weeks of life (they were suspected to have met a similar fate).

This meant that at the end of that five years, because the typical inter-birth spacing is about five years, any adult females who had *not* produced infants during that period were ready to give birth again. Also, because females who have lost infants can conceive again within a few months, the victimized females were ready to give birth again too. So when Passion and Pom stopped killing infants, we suddenly had more infants born in one year than had ever been the case before.

It was, in a way, a baby boom. It began in 1977 when Passion herself became pregnant. After this she was seen to make only two further attempts on other females' infants; both were unsuccessful. The "baby boom" went on through the first part of 1978. Thirteen babies were born between July 1977 and the end of 1978 (seven of these were in 1978, whereas usually only one or two are born during a given year).

All these infants were growing up together within the community. And, because mothers of similarly sized infants very often tend to spend quite a bit of time traveling about together, there were more nursery parties than usual, and they were larger.

The other significant thing was that, because all the females were suddenly lactating and looking after infants, there was a marked sparsity of cycling females available to the Kasakela males. I think this may have led to a certain amount of increased patrolling behavior (when the males have an opportunity to recruit adolescent females from neighboring communities) in the south of the range. Certainly at that time (although at Gombe, unlike Mahale, we don't know too much about transfer of females), many "stranger" females were sighted traveling about with the Kasakela males. It is possible that this lack of Kasakela cycling females may have had something to do with the outbreak of inter-community violence between the Kasakela and the Kalande community in the far south.

Thus the impact made by Passion and Pom was very considerable. We even had, five years later, another clustering of births, when many of the females who gave birth in 1977–78 had babies again. However, this "baby boom" was less dramatic than the first. In the interim, some of the 1977–78 babies had died (so that, after about a year, their mothers had given birth again), and some adolescent females had matured and also produced infants. Thus the birth pattern had already become more spaced out, more similar to the pre-cannibalism era.

The second individual I want to mention is Gigi (Fig. 1). Gigi shows many male-type behaviors. She has been cycling fairly regularly since 1965; she is apparently sterile. We don't know that she has ever been pregnant. She certainly has not had an infant. Just recently her cycles have become somewhat less regular, but even so, her sexual skin still swells and goes down month after month. As a result of this, she has enormously expanded sexual skin; she has the biggest swelling of any of the Gombe chimps. For years Gigi, when swollen, has attracted many of the community males. As Richard Wrangham has pointed out, when males gather together in parties, they are more likely to go out on patrols around the periphery of the home range. Thus Gigi, by her constantly recurring swellings, by serving as a magnet to attract males who then patrol, has probably served to increase the number of contacts between neighboring social groups. She has *certainly* served to create, on many occasions, large gatherings and outbursts of excitement and a great deal of the social activity which is always stimulated when chimps get together in large numbers.

Figure 1. Gigi.

Gigi has another fascinating characteristic, although this has not had much influence on the community as a whole. About 1975, she became unusually fascinated not by infants in general, but by one particular infant. This was a two-year-old male, Michaelmas, son of Miff. Gigi would follow Miff around; she would reach out to Michaelmas and embrace him; she would carry him; he would sometimes run to her when he was frightened, rather than to his mother. Then, when he was about three years old and a little more independent, Gigi lost interest in him. She turned to Athena, who had an infant 1½ years old, and she began to spend time with Athena and carry, play with, and groom *her* infant.

She has, since that time, traveled with a succession of mothers when their infants were between about 1½ and 2½ years of age, acting as a real "auntie" (using the term in its primatological sense).

This was highly significant in the case of one mother. The behavior of this female, Patti, is one of Gombe's mysteries which I hope continued research will help to illuminate. Patti was a transfer female. She moved into the Kasakela community as an adolescent. She became quite clearly pregnant, but the baby disappeared. She may have been subjected to an attack such as described elsewhere in this volume, although I don't think so, as she never appeared with wounds; we simply did not see the baby. Around 14 months later she gave birth again. To our amazement, she showed the kind of maternal behavior which I had previously associated only with lab or zoo chimps that have had a very disturbed, abnormal upbringing. She did not know how to handle her infant. She threw him on her back as a mother might carry a dead baby; she carried him, supporting him ventrally, but sometimes supported his rump rather than his back, so that his head bumped along the ground. She held him, as she rested, in what we call the "trouser pocket" between the thigh and groin. Sometimes, as she reached forward for food, the baby tucked firmly into her groin pocket, she all but squashed the breath out of him: he gave little high-pitched pathetic sounds. She then looked down at him, but she did not do anything about it. She even carried him once by one leg, with his head and back bumping along the ground. It was not surprising that the baby was dead within a week.

Then came the second infant, Tapit. Patti seemed to have learned a little, whether from watching other females, or whether from that brief experience with the first infant, we shall never know. But she supported this infant more securely in the ventral position – although very often back-to-front, with his head facing backwards. She still showed many bizarre maternal behaviors. I will give a couple of examples:

1. When a tiny infant is climbing near its mother, and she is ready to move on, she usually reaches up, puts one hand behind the infant's back, and with the second hand she draws the infant to her. Not Patti. She would just reach up, grab an ankle, and pull; then there was a terrible tug of war, with Tapit screaming and hanging on for dear life. Eventually he would be pulled off, would land upside down on Patti's tummy, and off she would go with Tapit back-to-front, still screaming.

2. When Tapit first began to toddle away from her, at a much younger age than most mothers will tolerate, he would sometimes become confused. He clearly was trying to get back to his mother, but would go to different individuals. His whimper-

ing would get louder and louder until he was screaming. Patti would sit there and watch him. To be anthropomorphic, it was as though she was thinking, "What a cute little thing he is." Sometimes it would be another female who would rescue him. Occasionally, when he was a little older, Patti would actually leave him altogether. Once, for example, I heard loud screaming as I followed Melissa and daughter Gremlin. Gremlin left Melissa and moved toward the screaming. She found Tapit, completely by himself. She gathered him up and carried him until they encountered Patti.

This is where Gigi, "auntie" Gigi, came in. She became a very frequent companion to Patti. In fact, only when she was going through periods of swelling and traveling with males were she and Patti apart. This continued until Tapit was about four years old. He spent as much time, sometimes more time, in contact with or near Gigi than with his own mother during "follows." When Patti had another infant, Gigi behaved in precisely the same way. The interesting thing, as far as Patti is concerned, is that by this time she was an excellent mother.

We know from studies on wild Japanese monkeys that females who lost their mothers during infancy, but survived, have turned out to be abusive mothers. Their own first-born infants have a much higher percentage of deaths than do infants born to females who did not lose their mothers while they were young. I suspect that Patti may have been an orphan.

FAMILY HISTORIES AND AFFECTIONATE BONDS

Another of the fascinating aspects of the long-term research at Gombe has been learning more and more about the strength of the affectionate bonds between mothers and their families. This is so well illustrated by the behavior of infants or youngsters after the deaths of their mothers.

Merlin was the first orphan we knew about. He was somewhere between four and five years of age when his mother, Marina, died. He developed so many behaviors typical of socially deprived young primates, whether human or nonhuman. He pulled hair on his legs. His thighs were completely bare. His belly was almost bare, and he also plucked hairs from the inner part of his arm (Fig. 2). He would sometimes hang upside down from his feet for over five minutes at a time. His social behavior deteriorated. Sometimes, for example, he seemed not to realize he should get out of the way of the displays of charging males.

He was adopted by his elder sister, Miff. She was around eight or nine at the time, and she took him everywhere with her. Initially she allowed him to ride on her back, but he was a bit heavy for her, so she discouraged that. But she allowed him to share her nest at night. However, it seemed that the psychological blow caused by the death of his mother was such that he never recovered; he became more and more physically emaciated and, in this state, more susceptible to any disease that happened to be going through the group. Merlin died 18 months after his mother, during an epidemic of polio.

Figure 2. Orphan Merlin, about one year after his mother's death. He has plucked hair from his belly, legs, and arms.

The second family I want to discuss very briefly, Flo's, spans 26 years of research. When Flo grew old and died, her son Flint was unable to get over the psychological loss, the psychological blow caused by the death of his mother, even though he was 8½ years old. We don't know how old Flo was; she was probably closer to 50 than 40. She looked very old and emaciated. When she died, Flint stayed by her body for a day and then fell into a state of grief and depression (Fig. 3). He died 3½ weeks after Flo.

Flo's daughter Fifi, like other mothers, spent much time on her own with Freud, her firstborn son, after the death of old Flo. This is sometimes boring for the infant, especially if his mother is not very attentive or playful. But for Freud it was not too bad. Fifi, like Flo before her, was a playful female who spent a lot of time playing with Freud. And, in the wild, he was able to find many ways to occupy himself when his mother was busy grooming others, feeding, or resting.

Freud went through the typical weaning depression. Weaning starts around the age of $3\frac{1}{2}$ years, and it lasts for at least a year, peaking during the fourth year. The mother increasingly denies the infant access to the nipple and prevents him (or her) from riding on her back during travel. Often the infant will throw violent tantrums; some even hit and bite the mother. What is so important and significant, at this difficult time in the life of a young chimpanzee, is the attitude of the mother. She may prevent the child from suckling, she may actually bite or even hit the child when it flies into a tantrum, but she will immediately afterwards embrace it. It is as though her message is, "You can't have milk, you can't ride on my back, but I still love you anyway."

Fifi's next infant, her second son, Frodo, was born after a five-year interval. Just as Fifi had been fascinated with her own infant brother Flint, Freud was fascinated by his new infant brother Frodo. He continually wanted to touch and play with Frodo. Fifi, like Flo before her, did not punish these attempts. Instead, she diverted Freud's attention by briefly tickling him and play-biting his fingers, so that for a while he forgot his fascination with the baby.

As Frodo grew older, Freud was allowed to pull him away from Fifi (as long as he was not too rough with the youngster) and even allowed to carry him, take him away, and play with or groom him.

As Frodo grew up, he found he had a built-in playmate in Freud, who continued to travel about with his mother and young sibling. There is a big difference in the early experience of a second-born child since he (or she) will almost always have an older

Figure 3. Flint grieving after Flo's death.

Figure 4. Early experience is different for the second sibling. The older sibling is a built-in playmate, and this helps the mother too.

sibling available for play (Fig. 4), for grooming, or for help in any difficult or dangerous situation. The first-born, for hours at a time, only has his mother.

It is different for the mother as well. She, who had to give so much attention to the firstborn child, can now sit and, if she thinks at all, think her own thoughts while her youngsters play around her.

In addition, the elder sibling will serve as a role model (Fig. 5). The close attention that is paid to the behavior of others is, of course, a characteristic of chimp behavior in the wild. And it is because they can learn from one another through observation, imitation, and practice, that the different cultural traditions can be passed from one generation to the next (Fig. 6).

Frodo began to leave his mother, to move away, traveling with groups of adults for two or three days at a time, when he was only five years old. Usually a young male does not start to do this until he is at least eight. Frodo was able to leave early and become precociously independent because he went with big brother Freud, who was there to keep an eye on him.

Fifi became pregnant again and, when Frodo was five years old, gave birth to her first daughter, Fanni. In the tradition of the family, Frodo was very fascinated by his little sister Fanni. He carried Fanni around to play with and groom, just as Freud used to carry him.

Figure 5. The elder sibling serves as a role model for the younger.

Figure 6. Social behavior can be passed from one generation to the next by observational learning.

Figure 7. Freud, a young adult male, still spends time with his mother, Fifi. So a bond develops between him and infant Fanni.

Freud, at ten years old, was a young adolescent by this time; nevertheless, he still spent a good deal of his time with Fifi, and so a bond developed between Freud and his sister Fanni (Fig. 7).

When Fanni was five years old, Fifi gave birth to her second daughter, Flossi. It shouldn't surprise us that Fanni was fascinated by Flossi (Fig. 8).

Fifi and her family speak as ambassadors for the chimpanzees of Gombe. We want to know what happens to Freud, to Fanni, and to Flossi as they grow up. We want to study their life histories. We also want to collect this kind of information from other places.

The third family I want to discuss is that of the infant-killer, Passion (Fig. 9). Figure 10 shows (left to right) Passion's family in 1978: Prof, juvenile son; Passion herself; adult daughter Pom and her newborn baby; and Passion's one-year-old son Pax. (He was so named because I hoped that with an infant of her own to cope with, Passion would be unable to attack other mothers and seize their infants. And this was, indeed, the case.)

Passion was a somewhat harsh mother when Pom was born in 1965. We had long wondered how Pom would treat her own first infant. She was not the most playful of mothers, but she cared for her own baby better than she had been cared for herself (Fig. 11).

Figure 8. Fanni is fascinated by her young sister Flossi.

Figure 9. Passion.

I have already mentioned that I have a high level of emotional involvement with many of the chimps. At the time when Passion (with Pom's help) was killing and eating infants, I think I hated her. But three years later she became sick. She had some abdominal problem and was sometimes quite clearly in considerable pain, doubling over and crouching to the ground. Figure 12 shows Passion just a few days before her death. She had become very emaciated and her ribs and hips protruded. How could I go on hating her? After all, she was not a human, she was a chimpanzee and could scarcely be held responsible for her behavior.

Figure 10. Left to right - Prof, Passion, Pom and Pan, and Pax. Pax is interested in the new member of the family, Pan.

Figure 11. Pom was a better mother than Passion had been to her.

Figure 12. Passion, just before her death. Pax trying to nurse. Prof grooms his mother. Pom, not in the picture, is nearby.

In Figure 12, four-year-old Pax, not yet weaned, is trying to nurse, though I doubt if his mother was producing more than a few drops of milk (if any) by then. Prof, shown grooming his mother, was nine years old when Passion died. He was still very emotionally dependent on her. Pom (not shown in the photograph) was almost always with the rest of the family.

Following Passion's death, Pax became depressed and lethargic (Fig. 13). Gradually, however, after the first few weeks, he began to show signs of improved psychological well-being. He no longer looked depressed. He began once more to play with other youngsters. He traveled with Pom and Prof. The three were hardly ever apart (Fig. 14).

After about six months, Prof began to leave Pom quite often and travel with the adult males. Pax remained closely attached to Pom. They were almost like a mother-infant pair. This close relationship between Pax and his sister lasted for about a year. Then, of his own volition, Pax began to spend more time with his elder brother, Prof.

Figure 13. Pax, depressed after Passion's death.

Figure 14. Prof and Pax (left) watch Pom termite fishing. The three were almost inseparable for months after Passion's death.

Figure 15. Prof and Pax.

It was about six months after this that Pom emigrated. She left the group and moved to the Mitumba community in the north.

By this time Pax and Prof had become all but inseparable (Fig. 15), and Prof showed a great deal of "maternal" concern and affection for his young brother. If the pair became separated, then Pax, being the younger, was more distressed than Prof; but if Prof suddenly noticed his young brother was not there, he would immediately leave any group he happened to have been traveling with and move off in the direction they had come from, to search for Pax.

On one occasion when he lost Prof, Pax nested very close to one of the adult males, but this was clearly no substitution for proximity to his absent brother. He cried on and off all night and continued to do so until he was reunited with Prof the next morning.

Six years after Passion's death (at the time of writing), Prof and Pax are still almost always together. Often they travel about on their own, away from other chimps – although, of course, they frequently join social groups as well.

Next let me review certain aspects of Melissa's family. Goblin, her first surviving infant, born in 1964, is alpha male today. In Figure 16, he sits beside his adult sister Gremlin, who has just given birth to *her* first baby, Getty. Getty became one of the most innovative youngsters Gombe has known. In Figure 17, 3½-year-old Getty has invented a new game – the only Gombe infant ever seen to play with sand. Melissa developed a very close and meaningful relationship with her grandson.

Figure 16. Goblin (alpha male), sister Gremlin cradling newborn, Getty.

Melissa herself produced Gombe's only known twins, Gyre and Gimble. It was hard for her to cope with two babies, and Gyre, the weaker, died when he was ten months old. Figure 18 (left to right) shows Melissa, Gremlin and Getty, and Gimble. When Gimble was eight years old, Melissa gave birth to tiny, underdeveloped Groucho, who died at 13 months (Fig. 19). Ten days later (October 1986), old Melissa died.

After her death, Gimble, Melissa's surviving twin, nine years old but not much bigger than a normal five-year-old, began traveling around with his adult brother Goblin. Goblin was born in 1964, and Gimble was born 13 years later. But despite this, despite the fact that Gimble had not spent as much time with Goblin as he would have with a sibling nearer his own age, the bond between them was strong. Goblin waited for Gimble during travel and was very quick to protect him in any kind of trouble.

Finally, let me mention two other case histories. Cindy was orphaned in 1966. She was three years old when her mother died. She had no older sibling. She traveled around alone or with one or another of the adult males for the two months that she survived her mother.

Figure 17. Getty playing with sand.

Figure 18. Left to right – Melissa, Gremlin and Getty, Gimble.

Figure 19. Melissa and the tiny, under-developed Groucho.

Figure 20. Skosha and Kristal stayed close together after Pallas' death.

Five-year-old Skosha was not completely weaned when her mother died, but she was quite able to survive independently as far as nutrition was concerned. She may represent our first example of an orphan being adopted by a nonrelative – although, in fact, Pallas, with whom she formed a very close bond and who became her foster mother, might have been a biological sibling of Skosha's mother.

Pallas had lost an infant about six months before she adopted Skosha. She waited for Skosha, she allowed the child to ride on her back, she provided reassurance on the many occasions when Skosha became frightened or upset during social interactions. Then, about two years after she had adopted Skosha, Pallas gave birth to Kristal. A naive observer would certainly have assumed that theirs was a biological family group.

When Kristal was five years old, the age at which Skosha herself had been orphaned, the mother Pallas died. It was very touching to see how Skosha, the foster daughter, then adopted her young foster sister (Fig. 20). Unfortunately, Kristal, like so many of the orphans, became more and more depressed, and she only lived for about nine months after her mother's death.

We have an important lesson to learn from some of these case histories, a lesson which is directly relevant to the way chimpanzees are so often treated in labs. They are put in stressful situations where any disease is far more likely to affect them than it would if they were in large cages in social groups and psychologically healthy and happy.

CURRENT RESEARCH

I want to discuss a collaboration planned between the research being done at Gombe and the research in the very opposite extreme of the chimpanzee range, in the

Tai Forest of the Ivory Coast. I want to share just one or two of the differences between the Tai chimps and the Gombe chimps, with Christophe and Hedwige Boesch's permission.

One of the fascinating differences concerns hunting behavior. In both Gombe and Tai, the most frequent prey is the red colobus monkey. Tai chimpanzees seem frequently to hunt in a very close, cooperative way. It is normal in a hunt at Tai for at least six, and sometimes more, chimpanzees to all hunt the same prey. At Gombe it is very often the case during colobus hunts that there are two or three different hunts going on at the same time – chimps up in the trees, chimps on the ground, chimps and monkeys all over the place – sometimes very confusing.

Perhaps even more fascinating is the fact that in Tai the chimpanzees hunt adult monkeys almost exclusively. They don't quite go to the extent of discarding an infant, although Christophe has said that sometimes it seems that males just drop infants in order to chase an adult. Younger chimps or females are then able to catch and eat the discarded infants.

At Gombe, however, it is quite clear that it is the *infant* that is normally the target when a hunting chimpanzee catches a mother. Almost always the mother is released after the infant has been torn from her. Females are sometimes killed and so, too, are adult males. But by and large, the victims are infants seized from their mothers, or infants who have just begun to travel independently, or juveniles. This is a very interesting difference.

Another difference is the fact that at Gombe it is typical for the male colobus monkeys, if they have a chance, to charge down and harass the chimpanzee hunters. When they do this, it is extremely likely that they will be successful, that the chimpanzees will retreat in the face of such aggression, will leave the hunt and climb precipitously to the ground. They may try again later, but for the moment they are repulsed. The Gombe records provide many examples of chimpanzees actually releasing captured colobus youngsters when faced with very persistent, aggressive attacks by the male monkeys, including biting, especially around the scrotum. When Christophe and Hedwige heard about this, they found it difficult to believe. This does not happen in Tai.

There is also a tremendous difference in food sharing. The Tai chimps hunt more cooperatively, and they also share with less aggression than is the case at Gombe. When we sat and talked about this, it was quite clear that to make a proper comparison, Christophe had to come and visit Gombe (or I had to go and visit Tai). We hope that he will be able to spend about four months at Gombe one summer, studying hunting behavior. We are making very careful reports of the hunting at Gombe and sending them to Tai. We hope to end up with a really good comparative study of hunting behavior – and of many other behaviors as well.

Finally, the study of chimp vocalizations (Boehm, this volume) is another very exciting study that is going on at Gombe. I hope that we can collaborate with the scientists in the Mahale Mountains and other study areas and eventually learn whether there are geographic differences in vocal communication similar to those now described in nonverbal communication patterns and feeding traditions.

CHIMPANZEE USE OF MEDICINAL LEAVES

Richard W. Wrangham and Jane Goodall

Primate diets are known to include a wide variety of secondary compounds that are damaging or neutral when eaten (Glander, 1982; Waterman, 1984). Bioactive secondary compounds that are beneficial to wild primates, on the other hand, have not been investigated in detail. This is surprising because a number of primates are known to eat plant items containing physiologically significant levels of secondary compounds which could affect parasites in the gut or elsewhere (Hamilton et al., 1978; Janzen, 1978; Peters and O'Brien, 1984; Philips-Conroy, 1982). Detailed investigations are required to test such hypotheses, however, partly because the primates could have detoxification mechanisms that interfere with potential medicinal properties (Janzen, 1978). Furthermore, even if the toxins remain biologically active, they may be eaten only incidentally, because they happen to be components of nutritionally valuable foods.

Ingestion of foods that lead to pharmacological control of internal parasites has apparently not yet been demonstrated in wild animals. However, there is good evidence that certain species of animals select plant parts to control external parasites. For example, starlings *(Sturnus vulgaris),* like many other birds, tend to add fresh green vegetation when they reuse an old nest. Selected plants are more effective than non-preferred vegetation as biocides against ectoparasites and pathogens. Starlings have been shown to be capable of distinguishing preferred plants on the basis of their volatile chemicals. This suggests that the basis for nest-material selection is the concentration of compounds controlling ectoparasites or pathogens (Clark and Mason, 1985, 1987).

If animals can select plants for external medicinal use, they may be able to do so for internal problems. Here we review evidence that chimpanzees *(Pan troglodytes schweinfurthii)* select and ingest leaves of three species of *Aspilia* (Heliantheae, Asteraceae = Compositae), leaves that do not contribute important nutritional benefits. This report synthesizes previous observations of feeding behavior (Wrangham and Nishida, 1983) and plant chemistry (Rodriguez et al., 1985) and presents new data on the uses of *Aspilia* spp. leaves by chimpanzees and people.

Aspilia spp. are herbaceous straight-stemmed plants up to 2 m tall, occurring in grassland or deciduous woodland. Five species have been recorded at Gombe: *A. congoensis* S. Moore, *A. kotschyi* (Hochst.) Oliv., *A. mossambicensis* (Oliv.) Wild, *A. pluriseta* Schweinf. ssp. *gondensis* (O. Hoffm.) Wild, and *A. rudis* Oliv. and Hiern ssp. *rudis* (Clutton-Brock and Gillett, 1979). The species are not easily distinguished in the field; unless specimens were collected for botanical identification, they are referred to as *Aspilia* spp. Only *A. congoensis*, *A. kotschyi*, and *A. mossambicensis* have been recorded in Mahale. Chimpanzees have been observed eating *A. pluriseta* and *A. rudis* at Gombe, and *A. mossambicensis* in Mahale (Wrangham and Nishida, 1983).

METHODS

Methods of observation of chimpanzees selecting *Aspilia* spp. in Gombe National Park in Tanzania were described by Wrangham and Nishida (1983). Adult male chimpanzees in the Kasakela and Kahama communities were observed as focal individuals for up to 13 hours continuously by R. Wrangham in 1972–73. Data for 1976, 1979, and 1983 were collected by field assistants as "B-Records" under the supervision of J. Goodall. Only all-day observations were analyzed. Details of data collection and checks on reliability were described by Goodall (1986).

RESULTS

Feeding Behavior

The feeding behavior of chimpanzees selecting *Aspilia* spp. leaves is unusual in several respects. First, individual leaves are selected more slowly and carefully than normal. Only young leaves are eaten, between about 2 cm and 10 cm long, and up to about 4 cm broad. Although occasionally several leaves are stripped together from the stem, the common pattern is for each leaf to be selected (either as a whole or by clipping the distal half) and swallowed individually before the chimpanzee seeks the next one. There is no evidence that the leaves are ever chewed. Selection is carried out both visually and by touch or taste: chimpanzees sometimes close their lips over a leaf, remain still for a few seconds, then abandon the leaf without detaching it from its stem. Each selected leaf is held in the mouth for several seconds while it is rolled around by the tongue (Wrangham and Nishida, 1983).

It is not surprising that the leaves are eaten slowly, because they are not easy to swallow. The leaves are very hispid, covered in bristly hairs that give them a rough surface. The difficulty of swallowing a sharp, rough, dry object may explain why chimpanzees roll the leaves around in the mouth for about five seconds, perhaps moistening and lubricating them.

It is also possible that there may be unpleasant chemical cues, though RWW could taste nothing unusual by mouthing *Aspilia* spp. leaves. The possibility of noxious chemicals is suggested by two observations at Gombe in 1972–73 when chimpanzees selecting *Aspilia* spp. leaves showed behavior that was not seen during feeding on other foods. First, a past-prime adult male, Hugo, vomited four minutes after starting to eat young leaves of *Aspilia pluriseta*. He had previously been eating *Garcinia huillensis* fruit. Hugo resumed eating *A. pluriseta* leaves within four minutes. At least five other individuals in the same party also ate *A. pluriseta* leaves, but without eating fruit first. They showed no discomfort. Hugo's general condition at the time seemed excellent, and he subsequently participated fully in an active day of traveling with a large excited party (December 21, 1972). Second, an adult male, Jomeo, three times lifted and wrinkled his nose while swallowing *Aspilia pluriseta* leaves. Both observations suggested that eating *Aspilia* spp. leaves may be unpleasant to chimpanzees.

Ingestion rates vary with the size of the leaf. The size of leaf selected varied little within days, but on different days preferences varied from very small (less than 2 cm long and not yet fully open) to full-sized but still young (ca. 10 cm long, fresh green). The median rate was five per minute (range 2.4–15.4, n = 7 bouts), compared to 37.3 ± 6.5 leaves per minute (n = 12 bouts) when chimpanzees ate leaves of *Mellera lobulata* (Acanthaceae), a commonly selected shrub with leaves of similar size which are collected and eaten in the typical manner for most leaves. The ingestion rate for *Aspilia* spp. leaves is thus as low as 15% of the rate when eating ordinary leaves (i.e., those that are chewed). The low intake rate is particularly striking given that chewing takes up much of the time when ordinary leaves are eaten. For ten accurately timed occasions in Gombe, adult males spent a mean of 13.9 minutes eating *Aspilia* spp. leaves (range 1–25).

The second unusual aspect of chimpanzee feeding on *Aspilia* is that the leaves are not chewed. At Gombe and Mahale, unchewed leaves of *Aspilia* spp. have been found in chimpanzee feces in all months of the year, whenever they are eaten. Fecal leaves (after washing) have a similar color to fresh material. The principal damage is that many leaves have been folded, though often only once or twice. Scanning electron microscopy reveals sufficient rupture of surface cells to release significant amounts of potent chemicals (E. Rodriguez, pers. comm.).

Third, Gombe chimpanzees tend to select *Aspilia rudis* and *A. pluriseta* leaves within an hour of leaving their sleeping nests, before their first big meal. In 1972–73 the median time for selecting *Aspilia* was 0715 h, and *Aspilia* spp. leaves were swallowed as the first "food" of the day on 68% of days (n = 22). This pattern was repeated in 1976–83: *Aspilia* spp. leaves were recorded to be the first "food" on 83% of the days they were swallowed (n = 18), and they were not later than the third "food" of the day, even on one occasion when they were eaten as late as 1230 h. *A. mossambicensis* leaves, by contrast, are selected at all times of day (Wrangham and Nishida, 1983).

Fourth, although all individuals from at least two years of age occasionally select *Aspilia* spp. leaves, the frequency differs by sex. This is shown in Table 1, which records frequencies of selection by all Gombe focal individuals for whom there were at least five complete days of observation in 1979 and 1983. Females selected *Aspilia* spp. leaves significantly more often than males (median 11.4% of days for females, 2.6% of days for males; Mann-Whitney U = 11.5, n = 11, 7; p < 0.02). These data

Table 1. Sex differences in frequency of selecting *Aspilia* spp. leaves.

	Females			Males	
Individual	%	n	Individual	%	n
PI	27.3	11	EV	5.9	17
LB	20.0	15	ST	5.3	19
MF	14.3	7	FG	4.5	22
PS	12.5	8	GB	2.6	38
GG	12.5	8	JJ	0.0	20
WK	11.4	35	SH	0.0	9
PM	7.1	14	AL	0.0	8
GM	6.7	15			
ML	3.3	30			
FF	2.9	34			
AT	0.0	8			

Notes: % = percent of days (when individual was observed as target all day) on which the target selected *Aspilia* spp. leaves. n = number of days observed. Data are from Gombe B-Records, 1979 and 1983.

come from all times of year, with sexes sampled approximately equally at different times. Evidence that the frequency of leaf selection varied in parallel for females and males is shown in Table 2, Part II: both sexes selected *Aspilia* spp. leaves often in 1979 and 1983. These data suggest that females at Gombe prefer *Aspilia* spp. leaves more than males do. Sex differences have not been investigated at Mahale. It is also possible that females spend more time selecting *Aspilia* spp. leaves than males do. Data from 1972, 1973, 1979, and 1983 show that for females, 47.1% of timed bouts were longer than 15 minutes (n = 17), compared to only 18.2% for males (n = 11).

Fifth, temporal variation in selecting *Aspilia* spp. leaves does not follow a consistent annual pattern. Within years for which there are data on all months, there is a trend toward peak frequencies in January–February at Mahale, and either January or July at Gombe (Table 3). However, selection of *Aspilia* spp. leaves has been recorded in all months; and within years, peak frequencies have occurred in January (4 years), February (2), March, May (2) and July. This variation raises the possibility that the frequency of selection is related less closely to abundance than it is for most foods.

While there are no direct data on *Aspilia* spp. leaf abundance, two additional observations support the suggestion above. First, chimpanzees often make special journeys, which may be short or long, to obtain *Aspilia* spp. leaves. Yet on other days they do not forage for *Aspilia* even when fresh leaves are abundant close to the nest site. For instance, on January 4, 1973, an adult male, Mike, found *A. pluriseta* leaves within one minute's walk of his nest site. Having eaten just the distal half of only two leaves, he walked back past his nest site to a large fruit patch, where he fed. On that

Table 2. Interannual variation in the frequency of *Aspilia* spp. leaf selection.

	(I) Fecal analysis		
Data source:		%	n
Gombe feces	1964	8.2	455
	1965	2.6	455
	1966	2.1	511
	1967	1.1	525
Mahale feces	1975	0.5	295
	1976	2.5	831
	1977	3.1	510
	1978	0.5	413
	1979	3.6	260

	(II) Direct observation		
Data source	% (Females)	% (Males)	n
1976	1.4	0.0	109
1978	1.0	0.0	168
1979	8.9	3.7	185
1983	9.0	4.2	162

(I) Fecal analysis % = fecal samples containing *Aspilia* spp. leaves.

(II) Direct observation % = proportion of observed population selecting *Aspilia* spp. leaves.

occasion the *Aspilia* were close. On January 7, 1973, by contrast, another adult male, Jomeo, left his nest (in forest) at 0639 h and walked, quickly and directly, to the nearest patch of wooded grassland, which he reached after 13 minutes. He then spent 19 minutes eating 45 *A. pluriseta* leaves before changing direction and walking for an additional 11 minutes to reach a ripe fruit tree. These observations contrast with data on adult males who left their sleeping nests in the same period without selecting *Aspilia* spp. leaves. Thus, on January 4, 12, and 13, 1973, males encountered patches of *A. pluriseta* within five minutes of leaving their nests, but paid no attention to them.

Second, interannual variation in the recorded frequency of *Aspilia* spp. leaf selection is substantial. Table 2 presents three data sets; in each set the frequency of *Aspilia* spp. leaf selection was recorded consistently for four or more years. In each case there were significant differences in frequency between years (χ^2, $p < 0.01$). Because *Aspilia* bushes are perennials, variation between years in the amount of new leaves available seems unlikely to be high. Furthermore, whenever chimpanzees stopped eating *Aspilia* spp. leaves, there were normally abundant young leaves remaining.

Finally, an earlier hint of a further unusual aspect of feeding behavior has not been consistently verified. In a previous analysis from eight matched pairs of days in 1972–73, there was a slightly significant trend for the total feeding time on all foods to

Table 3. Monthly variation in *Aspilia*-swallowing.

Month	Mahale Feces 1975–79		Gombe Feces 1964–67		Gombe Observation 1976, 1978–79, 1983	
	%	n	%	n	%	n
Jan.	10.5	114	1.0	101	15.1	53
Feb.	12.2	115	0.0	106	2.6	39
Mar.	1.7	235	3.8	132	5.0	40
Apr.	0.0	199	0.0	121	2.7	37
May	1.0	195	3.6	166	4.8	63
June	0.0	235	3.5	170	0.0	70
July	1.6	320	13.2	227	2.8	71
Aug.	2.6	274	6.1	379	0.0	49
Sept.	0.9	218	3.0	198	2.1	47
Oct.	1.0	105	0.0	110	0.0	49
Nov.	0.5	204	0.0	113	8.5	47
Dec.	1.1	95	2.4	123	3.8	52
Mean	2.8		3.1		4.0	

Sources: Mahale feces, Wrangham and Nishida, 1983, Table 1; Gombe feces and Gombe observations, Gombe Research Center data files.
Notes: n = number of feces examined or number of days of observation.

be higher on mornings when *Aspilia* was eaten (Wrangham and Nishida, 1983). This has not been found in other years. Thus, in 1979 there was no difference in the recorded feeding time between 15 days when *Aspilia* spp. leaves were eaten and 15 control days (matched for sex of focal individual and date to within one week).

There is currently no indication that swallowing *Aspilia* spp. leaves is associated with any other changes in behavior for the rest of the day. Swallowers were sometimes alone and sometimes with others; they sometimes had active days and sometimes had quiet days. Nor is there evidence that individuals who selected *Aspilia* spp. leaves were in poor condition. In ten observations of Gombe focal individuals in 1972–73, males were in parties seven times when they ate *Aspilia* spp. leaves. Twice their companions did not eat, but on at least four occasions, others also ate *Aspilia* leaves.

Medicinal Uses of *Aspilia* spp. in Africa

Wrangham and Nishida (1983) and Rodriguez et al. (1985) reported that two African species of *Aspilia* are used for medicinal purposes by indigenous peoples. However, the species cited by these papers *(A. africana, A. holstii)* are not known to be selected by chimpanzees. It is now clear that at least some of the *Aspilia* species

Table 4. Uses of *Aspilia* as medicine in Africa.

Species of *Aspilia*	Number of records					Sources
	Stomach	Skin	Eye	Other	Total	
africana	2	2	1	2	7	1,3,5
kotschyi	0	1	0	1	2	3
holstii	0	0	0	1	1	4
mossambicensis	3	2	2	5	1 2	2,3
pluriseta	0	2	1	0	3	4,6
rudis	1	0	0	1	2	6
spenceriana	0	0	0	2	2	7

Sources: 1. Ayensu (1979); 2. Kokwaro (1976); 3. N. K. Mubiru (pers. comm. 1984: data files of the Natural Chemotherapeutics Research Laboratory, Entebbe, Uganda); 4. Watt and Gerdina (1962); 5. Notes from plant collections in the Makerere University Herbarium, Entebbe, Uganda (extracted by R. Wrangham, July, 1984); 6. E. Mpongo (pers. comm.); 7. Bouquet and Debray (1974).

Note: Each record represents a report of a specific part of the plant being used by a particular human population for a particular purpose.

Table 5. African species of *Aspilia* and their use by chimpanzees and humans.

Species of *Aspilia*	Chimp use at:[1]		Human use as medicine [2]				Location
	Gombe	Mahale	Leaf	Root	Bark	Unknown	
africana	A	A	4	1	–	2	Ghana to Uganda
asperifolia	A	A	–	–	–	–	Southern Uganda
chrysops	A	A	–	–	–	–	Central Tanzania
congoensis	P (NS)	P (NS)	–	–	–	–	Tanzania
helianthoides	A	A	–	–	–	–	Uganda
holstii	A	A	–	1	–	–	Tanzania to S. Afr.
kotschyi	P (NS)	P (NS)	2	–	–	–	Uganda to Tanzania
mossambicensis	P (NS)	P (S)	7	4	1	–	Uganda to Tanzania
pluriseta	P (S)	A	3	–	–	–	Uganda to Tanzania
rudis	P (S)	A	2	–	–	–	Tanzania
spenceriana	A	A	–	–	–	2	Ivory Coast
subpandurata	A	A	–	–	–	–	Uganda
Total			18	6	1	4	

Sources: Ayensu (1979), Bouquet and Debray (1974), Kokwaro (1976), Watt and Gerdina (1962), Wrangham and Nishida (1983) and records of Natural Chemotherapy Research Laboratory (Kampala, Uganda), Makerere University Herbarium, and Gombe Research Center Herbarium.

Notes: The genus *Aspilia* includes approximately 90 species in Central and South America, Southern Europe, Africa and Madagascar. The table shows a selection of species recorded in East Africa. Also included are species from West Africa with known medical uses. The distributions shown are not complete.

1. Chimpanzee use: P = present, A = absent; S = swallowed, NS = not observed to be swallowed or mouthed.
2. Human use: figures show number of recorded uses.

selected by chimpanzees are used for medicinal purposes by Africans, and that the leaf is the preferred part (Tables 4, 5). Thus, 12 reports for *A. mossambicensis* included seven uses of leaves, four of roots, and one of bark (Kokwaro, 1976; Mubiru, pers. comm.); the only three reports for *A. pluriseta* were of leaf use (Table 5). There is some indication that *Aspilia* spp. are used more frequently than most plants. For instance, the number of reports of medicinal use of *Aspilia mossambicensis* cited by Kokwaro (1976) is greater than the number of uses reported for 99% of the ethnobotanical species in his compilation.

The *Aspilia* parts which are recommended for medicinal use (leaves and roots) differ according to ailment. Leaves are reported to be used primarily for topical or stomach conditions: nine out of 18 reports of leaf use are for topical application (wounds, burns, rashes, ringworm, conjunctivitis, etc.); six for stomach problems (pain or worms); and three for other complaints (cough, fever). This contrasts with six uses of roots for snakebite, whooping cough, back pain, cystitis, gonorrhea, and milk flow (with no reference to topical or stomach treatments).

Of the seven species used by people, four have been recorded in chimpanzee habitats *(A. kotschyi, A. mossambicensis, A. pluriseta, A. rudis;* Table 5). It will be interesting to find whether *A. kotschyi* is also used by chimpanzees. Gombe and Mahale also contain a species, *A. congoensis,* for which there are no recorded uses by either chimpanzees or people.

Phytochemistry and Pharmacology

Chemical and pharmacological analysis of the leaves of *A. mossambicensis* from Mahale and *A. pluriseta* from Kenya established that they contain a high concentration of a sulfur-containing red oil. The major constituents are thiarubrine-A and -B, dithiacyclohexadiene polyines previously identified in various species of Compositae, including *Chaenactis douglasii,* a plant used by native Canadians to treat skin sores (Rodriguez et al., 1985). Thiarubrine-A and related thiophenes are also present in the roots of *Aspilia.* Tests to date indicate that thiarubrine-A has numerous biological effects which suggest that it "holds some promise as a new type of antibiotic and phototoxic agent" (Towers et al., 1985). Thiarubrine-A exhibits antifungal, anthelmintic, antibacterial, and antiviral properties.

Thiarubrine-A shows antifungal activity against *Aspergillus fumigatus* and *Candida albicans.* Thiarubrine-A is about as effective as equivalent concentrations of amphotericin-B (Fungizone™), a standard antifungal. The relative efficacy of the two compounds varies with experimental conditions. *Aspilia* compounds are also effective against *Saccharomyces cereviseae.* Anthelmintic activity has been shown against *Caenorhabditis elegans,* a free-living nematode: nematodes were killed in all experimental conditions with concentrations of at least 5 ppm thiarubrine-A, suggesting that parasitic gut nematodes may also be vulnerable to thiarubrine-A. Effective antibiotic activity has been recorded against *Bacillus subtilis, Escherichia coli,* and *Mycobacterium phlei.* Growth of *Staphylococcus albus* and *Streptococcus fecalis,* by contrast, is inhibited only in the light (presence of ultraviolet-A), but not in the dark, while *Pseudomonas fluorescens* and *P. aeruginosa* growth is not affected in light or dark (Towers et al., 1985).

Antiviral properties have so far been found only against viruses that have membranes. Two mammalian viruses, murine cytomegalovirus and Sindbis virus, are extremely sensitive to thiarubrine-A, but only in the presence of UV-A radiation. The bacteriophage T4 is slightly affected (again, only in UV-A), whereas the bacteriophage M13 is not affected in the dark or light (Hudson et al., 1986).

The toxicity of thiarubrine-A to mammalian cells has not been investigated extensively. Towers et al. (1985) found that in the presence of UV-A, mitosis was inhibited in Chinese Hamster ovary cells (CHO) at concentrations as low as 0.12 ppm; in the dark, a dose of 2 ppm was required to inhibit mitosis. However, experimental manipulation shows that under certain conditions CHO cells readily detoxify the drug and flourish even in concentrations of 5 ppm. The mechanism of detoxification is unknown. Studying effects on mouse fibroblast cells, Hudson et al. (1986) also found that toxicity was increased in the presence of light. Significant mortality occurred among cells exposed to 1.0 µg/ml thiarubrine-A. Chemical studies of *Aspilia* spp. are continuing at the University of California, Irvine, since other metabolites, such as flavonoids and diterpenes, are present with thiarubrines and thiophenes and may have biological significance (Rodriguez, pers. comm.).

DISCUSSION

Few data are currently available on the chemistry of chimpanzee diets. Hladik (1973, 1977) analyzed the nutritive value of 23 plant items eaten by captive chimpanzees released onto an island in Gabon. He found that the chimpanzees, which were heavily provisioned with agricultural foods, selected wild fruits to maximize caloric intake, whereas they ate leaves to compensate for protein shortage in fruits. They did not appear to select plants in relation to their mineral content. Hladik noted that high levels of toxins were sometimes avoided, but found that chimpanzees are not always deterred by secondary compounds and therefore suggested that toxins are not as effective at deterring feeding as is often believed. The only other survey of the food chemistry of free-living chimpanzees was by Wrangham and Waterman (1983), who found that unripe fruits rich in condensed tannins are avoided, as expected. In general, the foods of Gombe chimpanzees appear not to have significant levels of toxins, because unlike some of the foods of sympatric primates such as baboons *(Papio anubis),* they are relatively palatable to humans (RWW, pers. obs.).

Aspilia spp. leaves therefore appear to be the first chimpanzee foods known to contain high concentrations of a powerful bioactive drug. Three hypotheses may be suggested to account for chimpanzees in Gombe and Mahale eating these leaves. First, selection of *Aspilia* spp. leaves may have no positive adaptive significance. If so, it could be a local tradition maintained by cultural drift (and therefore likely to become extinct as a result of unpredictable changes in fashion). Alternatively, differences between chimpanzee populations in the expression of this behavior could be genetically based, maintained by genetic drift.

The drift hypotheses cannot be firmly rejected at present. However, the observations of directed travel to *Aspilia* spp. patches and the fact that chimpanzees spend a long time selecting leaves clearly indicate that the behavior can be costly in energy and time. Accordingly, if it provides no benefit, we should expect cultural or natural selection to act rapidly to reduce its frequency. The fact that the behavior is expressed in two separate populations suggests that selection is acting against it slowly, if at all.

Second, *Aspilia* spp. leaves may be selected because of their unusual physical properties. The most obvious candidate here is the roughness of the leaf surface, since this is the most striking physical feature of *Aspilia* plants. However, although *Aspilia* spp. leaves certainly have rougher surfaces than most plants, they are not uniquely rough. For instance, *Ficus exasperata* has such rough-surfaced leaves that it is known as the "sandpaper fig." Yet chimpanzees pick and chew them rapidly, in the manner typical of most leaves. An alternative possibility is that *Aspilia* spp. leaves influence the digestibility of other foods (e.g., by helping to form a bolus). However, because *Aspilia* spp. leaves emerge almost undamaged in the feces, it seems unlikely that they function to bind digesta.

Third, *Aspilia* spp. leaves may provide specific chemical properties for the chimpanzees. The low rate of eating, the lack of chewing, and the minimal damage suffered by each leaf during passage through the gut all suggest that nutritional benefits are too small to be significant. It remains possible that vitamins or minerals are released in effective amounts; chemical analyses of the nutrients in *Aspilia* spp. are still needed. As a comparable example, Oates (1978) concluded that occasional selection of uncommon species of leaves by black-and-white colobus monkeys *(Colobus guereza)* was most easily explained as a means of obtaining sodium. Unlike the selection of *Aspilia*, however, the leaves taken by the colobus monkeys came from a variety of unrelated species sharing a particular habitat (muddy pools), and were eaten in a normal way.

Aspilia leaf selection by chimpanzees, therefore, does not appear to fit any previously described pattern of feeding behavior. Pharmacological significance was suspected prior to chemical analysis both because nutritional benefits appeared minimal and because the selection of each leaf was elaborate. Furthermore, the tendency for two of the species to be selected at dawn, but for the third to be selected at any time of day, suggests that it is the quality of the plant which varies according to time of day rather than the effect on the chimpanzee. A variable likely to be affected by diurnal rhythms is the concentration of particular secondary compounds (Robinson, 1974). The discovery of a potent bioactive compound at high concentration in the leaves of *A. mossambicensis* and *A. pluriseta* clearly supports the idea that *Aspilia* spp. leaf selection has pharmacological effects. Indeed, recent studies by E. Rodriguez at the University of California, Irvine, indicate that older leaves of *Aspilia* spp. lack the thiarubrines, with only small amounts detected in younger tissues. This suggests that the concentration of thiarubrine-A may fluctuate with time and season.

Effects could be euphoric, painkilling, or curative. There is at present no direct evidence in favor of any of these. Medicinal properties are suggested, however, first by the fact that thiarubrine-A is an effective antibiotic, killing certain fungi, nematodes, viruses, and microorganisms at low concentrations; and second by the widespread human use of *Aspilia* spp. leaves for ailments that seem likely to be improved

Table 6. Leaves in the diet of chimpanzees at Gombe and Mahale.

PTERIDOPHYTA
Dennstaedtiaceae
    G, M   *Pteridium aquilinum* (L.) Kuhn   C

ANGIOSPERMAE: DICOTYLEDONS
Acanthaceae
    G, M   *Asystasia gangetica* (L.) T. And.   A
    M   *Blepharis buchneri* Lindau   A (M), E (G)
    G   *Mellera lobulata* S. Moore   A
Anacardiaceae
    G   *Rhus vulgaris* Meikle   D
Annonaceae
    G, M   *Annona senegalensis* Pers.   D
    G, M   *Uvaria angolensis* Oliv.   D
Apocynaceae
    G, M   *Diplorhyncus condylocarpon* (Muell. Arg.)   D
    G, M   *Saba florida* (Benth.) Bullock   D
    G, M   *Tabernaemontana holstii* K. Schum.   D
Araliaceae
    G   *Cussonia kirkii* Seem.   D
Asteraceae
    G, M   *A. mossambicensis* (Oliv.) Wild   C
    G   *A. pluriseta* Schweinf.   C
    G   *A. rudis* Oliv. and Hiern   C
    G, M   *Bidens grantii* (Oliv.) Scherff.   D
    M   *Vernonia amygdalina* Del.   D
Begoniaceae
    G   *Begonia princeae* Gilg.   B
Convolvulaceae
    G, M   *Ipomoea cairica* (L.) Sweet   A
    G   *I. eriocarpa* R. Br.   A
    G   *I. obscura* Ker-Gawl   A
    M   *I. rubens* Choisy   A
    G   *Merremia pterygocaulos* (Choisy) Hall   A
Combretaceae
    G, M   *Combretum molle* G. Don   D
Cucurbitaceae
    M   *Mukia maderaspatana* (L.) R. J. Roem.   D
Euphorbiaceae
    G, M   *Acalypha ornata* A. Rich.   D
    G   *Hymenocardia acida* Tul.   A
Flacourtiaceae
    G   *Lindackeria kivuensis* Bamps   B
Hypericaceae
    G   *Harungana madagascariensis* Poir.   D
Malvaceae
    G, M   *Hibiscus aponeurus* Sprague and Hutch   D
    M   *H. cannabinus* L.   D
    M   *H. nyikensis* Sprague   D
    M   *H. rostellatus* Guill. and Perr.   D
    G, M   *H. surratensis* L.   D

(continued)

Table 6. (continued)

Menispermaceae		
M	*Dioscoreophyllum volkensii* Engl.	D
G, M	*Stephania abyssinica* (Dillon and A. Rich)Walp.	D
G, M	*Tinospora caffra* (Miers.) Troupin	A
Mimosaceae		
G, M	*Albizia glaberrima* (Schumach. and Thonn.)	D
G	*Newtonia buchananii* (Bak.) Gilb. and Bout.	D
Moraceae		
G, M	*Chlorophora excelsa* (Welw.) Benth. and Hook.	A (Galls only)
G	*Ficus capensis* Thunb.	B
G, M	*F. congensis* Engl.	A
G, M	*F. exasperata* Vahl.	A
G	*F. gnaphalocarpa* (Miq.) A. Rich	A
G, M	*F. ingens* Miq.	A
G	*F. kitubalu* Hutch.	B
G	*F. lukanda* Ficalho	A
G	*F. polita* (Miq.) Vahl.	A
G, M	*F. urceolaris* Hiern	A
G, M	*F. vallis-choudae* Del.	A
Myristicaceae		
G, M	*Pycnanthus angolensis* (Welw.) Warb.	B
Papilionaceae		
G	*Aeschynomene baumii* Harms.	B
G, M	*Baphia capparidifolia* Bak.	A
G, M	*Crotalaria lachnophora* A. Rich.	D
G, M	*Dalbergia malangensis* E. P. Souza	B
M	*Erythrina abyssinica* DC.	A
G	*Pterocarpus angolensis* DC.	D (Baboons often)
G, M	*P. tinctorius* Bak.	A
Proteaceae		
G	*Protea welwitschii* Engl.	D
Rosaceae		
G, M	*Parinari curatellifolia* Benth.	A
Rubiaceae		
G	*Canthium hispidum* Benth.	D
G	*Sabicea orientalis* Wernham	D
Sterculiaceae		
G, M	*Sterculia quinqueloba* (Garcke) K. Schum.	B
G, M	*S. tragacantha* Lindl.	A

ANGIOSPERMAE: MONOCOTYLEDONS
Smilacaceae

G, M	*Smilax kraussiana* Meisn.	A

Sources: Data shown for Gombe (G) and Mahale (M)
(Wrangham, 1975; Nishida and Uehara, 1983).

Notes: A = eaten frequently in high quantity   D = eaten rarely and in low
B = eaten rarely, but in high quantity   quantity despite abundance
C = eaten frequently in low quantity   E = not eaten
Items eaten "rarely" are those with fewer than five feeding records.
Items eaten "frequently" are those with five or more feeding records.

by an antibiotic or anthelmintic (Tables 4, 5). Medicinal action is consistent with the observation that chimpanzees select *Aspilia* leaves at different frequencies from year to year (Table 2). Medicinal properties do not account for the sex differences in the rate of leaf selection. This is the first report of sex differences in plant food selection among Gombe and Mahale chimpanzees, and it remains a puzzling observation. However, possibly females are more physiologically stressed than males. For instance, Nishida (pers. comm.) notes that both at Gombe (Goodall, 1986) and Mahale the only adults that have been recorded eating feces have been females, suggesting that they may be more susceptible than males to nematode infestations.

The fact that several individuals within a party often eat *Aspilia* spp. leaves simultaneously is unexpected if the function of selecting them is medicinal. However, it is possible that several individuals feel ill at the same time. The intensity of nematode infestations, for example, may vary seasonally (File et al., 1976). Use of *Aspilia* spp. to control gut parasites is therefore a possible explanation for the observations.

It is striking that baboons *(Papio anubis)* in Gombe National Park have not been seen eating *Aspilia* spp. leaves in the chimpanzee style; the only reports are of one or two occasions when baboons may have eaten *Aspilia* spp. leaves in the typical leaf-eating manner (A. Sindimwo, pers. comm.). The diet of Gombe baboons is well known as a result of direct observations since 1969. In general, the diet of baboons and chimpanzees is similar, with differences between them due partly to differences in their ability to open difficult fruits. Where there are differences associated with taste, baboons are more willing than chimpanzees to eat items containing bitter, hot, or otherwise noxious chemicals. For example, both species eat large quantities of the young leaf of *Pterocarpus tinctorius* (Papilionaceae) which has a pleasant, nutty taste for humans. But the closely related *P. angolensis* is eaten in large quantities only by baboons; *P. angolensis* has an unpleasant, bitter taste to humans and perhaps to chimpanzees also. Observations such as this suggest that baboons are able to detoxify a wider range of chemicals than chimpanzees and that they are therefore able to sustain a more varied diet. The fact that baboons do not select *Aspilia* spp. leaves is therefore worth further investigation. More generally, comparisons of apes and other species could test the hypothesis that medicinal use of plants is associated with higher cognitive abilities.

Aspilia spp. leaves have drawn attention because of the odd way they are eaten, but it is possible that many other plant items are selected by chimpanzees partly for the plants' bioactive secondary compounds. The two most detailed studies of chimpanzee feeding behavior have been at Gombe and Mahale, where diets and feeding times have been recorded daily for over ten years at each site. These studies have recorded 201 plant-food types from 146 species at Gombe (Wrangham, 1975) and 328 plant-food types from 198 species at Mahale (Nishida and Uehara, 1983). The majority of food items appear to be chosen for their nutritional value. For instance, similarities and differences in the foods eaten at the two sites are generally explicable in terms of their abundance (Nishida et al., 1983), and seasonal changes in diets are strongly correlated with food availability (Wrangham, 1977). In many cases, however, even abundant food items are eaten only occasionally. Thus, Table 6 lists 64 species whose leaves

Figure 1. Subadult female (Jolly) in a *Celtis durandi* at Kanyawara, Kibale Forest. Whole leaves resembling *C. durandi* have been found in chimpanzee feces in the Kanyawara area, but identification is still not certain. Chimpanzees commonly eat *C. durandi* fruit, which can be seen here.

are eaten in Gombe and/or Mahale, and shows that 27 of these (42%) are eaten rarely and in small quantities even though they are abundant. Most cases are probably accounted for by individuals occasionally trying out new foods. However, E. Mpongo (pers. comm.) reported that the leaves of *Hibiscus aponeurus* are eaten like those of *Aspilia* – singly, slowly, without being chewed, and only in the early morning. He also reported that a few seeds of *Strychnos* spp. (Loganiaceae) are occasionally extracted individually from unripe fruits and slowly chewed. Such observations offer the opportunity to identify more precisely the chemicals responsible for apparently non-nutritional feeding by chimpanzees and are currently being investigated by the Phyto-chemical Laboratory of the University of California at Irvine. Further investigation of plants eaten in unusual ways is required throughout the range of chimpanzees.

POSTSCRIPT

Recent evidence (Wrangham et al., in prep.) shows that Ugandan chimpanzees also swallow medicinal leaves. Between December 1987 and January 1989, 401 chimpanzee fecal samples were collected from three sites in Kibale Forest, Uganda (Kanyawara, Ngogo, Kingo). Whole leaves were found in 16 samples (4.0%), representing at least four species (Fig. 1). The only leaf definitely identified is *Rubia cordifolia* (Rubiaceae): one dung sample contained 16 leaves of *R. cordifolia*, all whole and without toothmarks. This species is well-known to people living near

Kibale, who report that they find its leaves so effective for curing stomach ailments that they plant it in their gardens in order to have it readily available. Its biochemistry is currently being investigated.

ACKNOWLEDGMENTS

This report is part of a cooperative investigation which includes work by T. Nishida, E. Rodriguez, and G. H. N. Towers. We thank T. Nishida for sharing field data, and T. Nishida and E. Rodriguez for valuable comments. The field studies were made possible by the Government of Tanzania and funded from numerous sources cited by Goodall (1986). The Gombe B-Records were collected by a team of Tanzanian field assistants led by H. Matama. E. Mpongo has been particularly helpful in the collection of data on *Aspilia* spp. Recent investigations by R. Wrangham of chimpanzee and human use of *Aspilia* spp. were made possible by a grant from the L.S.B. Leakey Foundation.

For the 1987-1989 work in Uganda, we wish to thank J. Basigara, K. Clement, A. Clark, G. Isabirye-Basuta, and M. Hauser for help in collecting dung samples, J. Byaruhanga and E. Tinkasimire for dung analysis, the Government of Uganda for permission to work in Kibale Forest, and the National Science Foundation and National Geographic Society for financial support.

REFERENCES

Ayensu, E. S. 1979. *Medicinal Plants of East Africa.* Michigan: Reference Public Press.

Bouquet, A., and M. Debray. 1974. *Plantes Medicinales de la Cote D'Ivoire.* Paris: O.R.S.T.O.M.

Clark, L., and J. R. Mason. 1985. Use of nest material as insecticidal and anti-pathogenic agents by the European starling. *Oecologia* 67:169–176.

_____. 1987. Olfactory discrimination of plant volatiles by the European starling. *Animal Behaviour* 35:227–235.

Clutton-Brock, T. H., and J. B. Gillett. 1979. A survey of forest composition in the Gombe National Park, Tanzania. *African Journal of Ecology* 17:131–158.

File, S. K., W. C. McGrew, and C. E. G. Tutin. 1976. The intestinal parasites of a community of feral chimpanzees. *Journal of Parasitology* 62:259–261.

Glander, K. E. 1982. The impact of plant secondary compounds on primate feeding behavior. *Yearbook of Physical Anthropology* 25:1–18.

Goodall, J. 1986. *The Chimpanzees of Gombe: Patterns of Behavior.* Cambridge: Harvard University Press.

Hamilton, W. J., III, R. E. Buskirk, and W. J. Buskirk. 1978. Omnivory and utilization of food resources by chacma baboons, *Papio ursinus. American Naturalist* 112:911–924.

Hladik, C. M. 1973. Alimentation et activite d'un groupe de chimpanzes reintroduits en foret gabonaise. *La Terre et la Vie* 27:343–413.

_____. 1977. Chimpanzees of Gabon and chimpanzees of Gombe: some comparative data on the diet. In T. H. Clutton-Brock, ed., *Primate Ecology,* pp. 481–502. London: Academic Press.

Hudson, J. B., E. A. Graham, R. Fong, A. J. Finlayson, and G. H. N. Towers. 1986. Antiviral properties of Thiarubrine-A, a naturally occurring polyine. *Planta Medica* 1:51–55.

Janzen, D. H. 1978. Complications in interpreting the chemical defenses of trees against tropical arboreal plant-eating vertebrates. In G. G. Montgomery, ed., *The Ecology of Arboreal Folivores,* pp. 73–84. Washington, D.C.: Smithsonian Institution Press.

Kokwaro, J. O. 1976. *Medicinal Plants of East Africa.* Nairobi: East African Literature Bureau.

Nishida, T., and S. Uehara. 1983. Natural diet of chimpanzees *(Pan troglodytes schweinfurthii)* long-term record from the Mahale Mountains, Tanzania. *African Study Monographs* 3:109–130.

Nishida, T., R. W. Wrangham, J. Goodall, and S. Uehara. 1983. Local differences in plant-feeding habits of chimpanzees between the Mahale Mountains and Gombe National Park, Tanzania. *Journal of Human Evolution* 12:467–480.

Oates, J. F. 1978. Water-plant and soil consumption by guereza monkeys *(Colobus guereza):* a relationship with minerals and toxins in the diet? *Biotropica* 10(4):241–253.

Peters, C. R., and E. M. O'Brien. 1984. On hominid diet before fire. *Current Anthropology* 25:358–360.

Philips-Conroy, J. E. 1982. Baboons, diet and disease: food plant selection and schistosomiasis. *International Journal of Primatology* 3:280.

Robinson, T. 1974. Metabolism and function of alkaloids in plants. *Science* 184:430–435.

Rodriguez E., M. Aregullin, T. Nishida, S. Uehara, R. W. Wrangham, Z. Abramowski, A. Finlayson, and G. H. N. Towers. 1985. Thiarubrine-A, a bioactive constituent of *Aspilia* (Asteraceae) consumed by wild chimpanzees. *Experientia* 41:419–420.

Towers, G. H. N., Z. Abramowski, A. J. Finlayson, and A. Zucconi. 1985. Antibiotic properties of Thiarubrine-A, a naturally occurring dithiacyclohexadiene polyine. *Planta Medica* 3:225–229.

Waterman, P. G. 1984. Food acquisition and processing as a function of plant chemistry. In D. J. Chivers, B. A. Wood, and A. Bilsborough, eds., *Food Acquisition and Processing in Primates,* pp. 177–211. New York: Plenum Press.

Watt, J. M., and M. Gerdina. 1962. *The Medicinal and Poisonous Plants of Southern and Eastern Africa.* Edinburgh: E. and S. Livingston.

Wrangham, R. W. 1975. Behavioural ecology of chimpanzees in Gombe National Park, Tanzania. Ph.D. thesis, Cambridge University.

———. 1977. Feeding behaviour of chimpanzees in Gombe National Park, Tanzania. In T. H. Clutton-Brock, ed., *Primate Ecology,* pp. 503–538. London: Academic Press.

Wrangham, R. W., and T. Nishida. 1983. *Aspilia* spp. leaves: a puzzle in the feeding behavior of wild chimpanzees. *Primates* 24:276–282.

Wrangham, R. W., and P. G. Waterman. 1983. Condensed tannins in fruits eaten by chimpanzees. *Biotropica* 15:217–222.

METHODS FOR ISOLATING CHIMPANZEE VOCAL COMMUNICATION

Christopher Boehm

IMPORTANCE OF ISOLATING THE VOCAL CHANNEL

Postural, gestural, and facial-expression channels of communication would seem to be fundamental for nonhuman primates (Washburn, 1982; Lancaster, 1975). Vocal communication, the subject of this paper, may be equally fundamental. But it is difficult to sort out its independent functions since, in contexts of friendly communication within the group, vocal communication so often is combined with non-vocal communication.

Even *Pan troglodytes* – a relatively voluble primate – relies substantially upon non-vocal communication. Under free-ranging conditions, at close quarters, the rich repertoire of visual, tactile and olfactory signals of chimpanzees enables them to communicate quite effectively without vocalizing under many social circumstances. They easily get one another's attention and communicate non-vocally by approaching, touching, changing posture, changing body orientation, gesturing, erecting hair, changing gait, changing facial expression, scratching, self-grooming, drumming a tree, branching, or, at Gombe, by grooming a leaf (Goodall, 1986).

When working with very small groups on twelve hour follows at Gombe Stream National Park, I sometimes have recorded not a single vocal sound. And when the chimpanzees go out on patrol, they often maintain a strict silence for hours, during which they manage, nevertheless, to behave very much as a coordinated group.

It is easy to think of routine circumstances under which chimpanzees regularly combine vocalized signals with these other modes of communication in a way which appears to be merely optional. But sometimes the vocal signal can be of primary importance for communication purposes. For example, when a crouching subordinate male also pant-grunts (this always signals social apprehension; see Goodall, 1986) as he approaches the alpha male, he is combining a postural signal with a vocal one in a situation in which either signal might suffice. However, if the alpha male happens to be displaying nearby in thick brush, then pant-grunting loudly may be a better signal of submission by the subordinate than crouching, since a submissive stance by itself might be visually misinterpreted through the foliage by the alpha male, and an attack might result.

In most circumstances, then, when chimpanzees are in visual contact, it is difficult to disentangle and assess the separate functions of these two styles of communication. One way to better evaluate the functional importance of the vocal channel of communication would be to factor it out definitively from the gestural, postural, facial, and olfactory channels with which it is generally combined.

Fortunately, it is not necessary to move to experimental contexts in order to do this, since wild chimpanzees rely almost exclusively on vocal communication in their long-distance exchanges. Long-distance exchanges are defined as any vocalizations that take place between any animals of the same species who are out of direct visual contact with each other and whose movements cannot be seen. In these exchanges, chimpanzees often use vocal signals to resolve problems of territorial boundary competition with conspecifics. Within the community, they give predator alarm calls. Vocalizations help locate lost individuals and also enable individuals under attack to receive help from allies (Goodall, 1986).

The above situations either are "unfriendly" or have an "emergency" nature. Similar vocalization behaviors are to be found in many primate species. However, due to the unusually large amount of fission and fusion that routinely takes place within a chimpanzee territorial community, the chimps also do an unusually large amount of calling at long distance which might be classified quite differently: it is both "friendly" and "not of an emergency nature." In this respect, chimpanzees, along with bonobos, may be somewhat distinctive among the nonhuman primates.[1]

The evolution of language in humans surely saw the vocal channel of communication becoming more and more ascendant over the olfactory, gestural, postural, and facial-expression channels, and it can be plausibly hypothesized that this took place largely in a context of friendly cooperation within the group. For this reason, the distant vocal communication of chimpanzees is of special linguistic and anthropological interest, aside from the need to study such communication behavior in its own right as a contribution to primatology.

UNITS AND THEIR COMBINATION

This paper reports research that is about to leave the pilot stage, a study of long-distance calls made by free-ranging chimpanzees at Gombe Stream National Park in Tanzania. Presently, I shall describe the study and its methods, and afterwards discuss some theoretical issues which are relevant to any attempt to understand fully the vocal signals emitted by chimpanzees and other primates whose calls are finely graded. But first, it will be useful to consider the units of communication that become readily apparent to human observers who work with chimpanzees, and the extent to which various units seem to be combined. In dealing with linguistic aspects of the study, I shall avoid unnecessary use of technical terms or jargon.

Chimpanzees vocalize by making several types of calls:

1. *Discrete unpanted calls.* These are minimal vocalizations, usually monosyllabic, which can be emitted alone, that is, without being in some way combined with some other call. They always involve a unidirectional air stream expelled through the mouth and, presumably, the nose. A few of these unpanted discrete calls are always emitted alone (e.g., the soft cough of admonition, possibly *hoo* calls made by adults on seeing strange beings or objects, or eagle *wraa's* of hostility toward foreign beings). Others are emitted either alone or in combination with some other vocal production (e.g., hoots given in excitement, defiant *waa* barks, anxious screams, excited barks, a variety of grunts, and hedonistic food calls).

2. *Discrete panted calls.* These are multi-syllabic vocalizations which can be emitted alone, that is, without being in some way combined with another call. They always are produced by a bidirectional, alternating *panted* flow of air in and out of the vocal cavity creating the effect of a series of similar-sounding vowel-syllables. Two panted calls which are always emitted alone are the unvoiced grooming pant and the copulation pant. Panted calls which are emitted either alone or in combination with some other vocal production include the voiced or unvoiced "laughing" which accompanies play.

3. *Discrete unpanted calls which also can be panted.* Many, but not all, of the calls listed in the first category can be panted (e.g., distressed pant-screams, apprehensive pant-grunts or pant-barks, and a variety of different calls presently lumped together as pant-hoots). However, the *wraa* and *waa* calls (described above) are discrete unpanted calls which never become panted.

4. *Discrete unpanted calls that can be combined.* In category 1, it was specified that certain unpanted discrete calls may also be combined with other calls. This seems to involve, on rare occasions, a single call which sounds like a blend of two discrete calls (e.g., a compromise between a scream and a *waa*). Far more frequently, one discrete call gradually *grades* (changes along a continuum) into another (e.g., a scream into a *waa* or vice versa, a *waa* into a *wraa*, a whimper into a scream, or a food call into a pant-hoot).

5. *Discrete panted calls that can be combined.* Many panted calls can be combined in the same way as unpanted calls, along a graded continuum. Examples are pant-hoots which, due to a sudden threat, grade into pant-screams; or pant-grunts, given by a subordinate to a higher ranking individual who is approaching, which grade into excited pant-barks and then possibly into pant-screams as the latter begins to display. Pant-hoots also can grade into food calls, which are unpanted.

6. *Other possible categories.* It is conceivable that certain chimpanzee calls have no discrete form, that is, they are always combined with some other call. A great deal of further research is needed to know if there are other viable categories.

This classification of call types[2] is preliminary: at present, one cannot state with confidence that an unpanted hoot, for example, is the same vocal production or signal as that used in a panted hoot. Snowdon (1982) has successfully analyzed the units used by pygmy marmosets, and this can serve as a partial model for working with chimpanzee calls. But the grading inherent in much of chimpanzee vocalization will present special problems for discerning "units."

Probably the majority of calls emitted by chimpanzees out in nature are ones that are either discrete, always panted, or discrete and panted, rather than ones that are blended by grading or *combined* (i.e., in an ungraded series). We are not yet in a position to treat the matter statistically, but when wild chimpanzees are followed everywhere they go, pant-grunts, pant-hoots, food calls, whimpers, screams, and "laughing" are very common in their ungraded forms. However, calls combined by grading are also encountered very frequently, especially in association with pant-hooting and pant-grunting, and with food calls.

An important analytical issue, discussed by Marler (1976), is whether what human ears hear as intermediate forms of graded calls are being heard by chimpanzees as merely a continuous series of transition points between one discrete call and another, or whether chimpanzees may be decoding calls that are intermediate on the grading continuum as being different calls, on a par semantically with what I have identified above as discrete calls. In this connection, recent experimental research by Petersen (1982) and Snowdon (1982) suggests that other nonhuman primates can break up a sound continuum on the basis of fine acoustical distinctions, in a way similar to what humans do when they discriminate one familiar phoneme from another which is acoustically similar. It is likely, then, that chimpanzees can regularly emit (and recognize) calls that are acoustically quite similar. But how this ability relates to grading, and to possible perception of intermediate forms, remains a matter for future research.

The above discussion of units and their combination has been a simplified one; it follows linguistic intuitions of my own, and of the other field workers who have named the calls, in assuming that if a call can occur in isolation from any other call, it somehow is "basic." The conclusions were derived largely from Goodall (1968), Marler (1976), and Goodall (1986), but also from more than fifteen months of my own field research at the Gombe Stream Research Centre in Tanzania.[3] One important purpose of the current research project at Gombe, conducted jointly by Dr. Jane Goodall and myself, is to explore the question of what kinds of units chimpanzees are using in vocal communication and how they combine these units. A first step will be to accumulate a very large corpus of vocalizations, recorded under maximally natural circumstances, so as to make an acoustical and behavioral analysis of all uncombined calls, that is, those which are emitted discretely on a basis that is acoustically and temporally independent of other vocalizations. (Temporal independence will need further definition; the idea is that two calls by the same individual, occurring with only a few seconds intervening, are not likely to be independent; obviously, the real issue is not temporal but whether the first call stimulates the second.)

A question of profound theoretical interest is whether calls emitted in combinations may be communicating significantly more than the mere sum of the messages normally communicated by the constituent single calls. While chimpanzees in captivity have been shown to have what might be called "syntactic" capacities – the ability to combine different symbols in such a way that additional levels of meaning may emerge from the patterned combinations – the very frequent use of graded combinations by wild chimpanzees remains to be assessed in this respect. A minimal assumption about graded calls is that these broadcast a change of emotional state (Marler, 1976). But much more is possible, and new research methods must be devised to explore this important matter.

Thus, when an observer hears a wild chimpanzee who is out of sight first giving a pant-grunt that grades to a pant-bark, then suddenly a series of loud screams which diminish and grade into a series of squeaks, he assumes that a subordinate animal has felt threatened or has been attacked briefly in spite of having given submissive signals. The observer may assume, also, that the dominant animal was merely making a side attack in the process of carrying out an intimidation display, rather than making a serious attack. The observer may also be able to surmise who the vocalizer was from hearing the animal's voice, although this capacity seems to vary greatly from one experienced observer to the next. When another chimpanzee of the same community hears this same sequence, it seems quite certain that it knows exactly who is vocalizing (Bauer and Philip, 1983) and understands at least as much as I have stated above about what was taking place; it probably understands far more. It would not be unreasonable to suggest that some of his comprehension results from reading adjacent calls as special combinations: for example, if the subordinate individual's vocalization chain had graded from screams into *waa's* instead of squeaks, a logical conclusion would be that the subordinate became emboldened after the dominant individual "displayed away." But such hypotheses must be substantiated.

A basic taxonomy of wild chimpanzee calls was published by Goodall in 1968; this taxonomy consisted of about two dozen different acoustical productions, almost all vocalizations, that served some purpose of expression or communication. Her more recent assessment suggests that in some cases a number of calls which seemed acoustically similar to human ears had been lumped together, and that the number of different communication signals that wild chimpanzees can differentiate is much greater than 24 (Goodall, 1986). In the present expanded list of 34 calls, she still leaves the "pant-hoot" as a "catchall" category, so far as acoustical patterns are concerned, and also suggests that screams need to be further differentiated; I would propose also the "*waa*-bark" and the "pant-grunt" as prime candidates for further "splitting."

By contrast, Marler's analysis of carefully recorded chimpanzee calls, based on sound-spectrographic techniques, resulted in a taxonomy that included just over a dozen different calls.[4] He collapsed many of the categories previously set up by Goodall (1968) and arrived at a little more than half as many vocalization types. The majority of Marler's taxonomic cuts do coincide with the finer demarcations made by Goodall, but a considerable amount of information was lost by "lumping."

I emphasize that none of these prior efforts addresses *substantively* the question of whether combined or graded calls of wild chimpanzees may involve something like a proto-syntax. It is hoped that the vocal communication project described in the pages that follow will make direct contributions to resolving this important question.

TAXONOMIC CONSIDERATIONS: Splitting and Lumping

Research design is necessarily affected by the kinds of scientific expectations that the researchers hold. I shall specify these at four levels:

1. A conservative working hypothesis is that wild chimpanzee vocalizations are purely expressive and do nothing more, communicatively, than to advertise emotional states. By this view (Smith, 1977), graded calls of nonhuman primates merely express changes in emotions.

2. A less conservative working hypothesis is that calls of chimpanzees also communicate information about the social and physical environments, as has been demonstrated definitively for other primates (Seyfarth and Cheney, 1982). There is no doubt in my mind that when a chimpanzee hears a large group that is out of sight, first hunting and then fighting over meat, that quite a bit of information is being communicated, probably quite inadvertently, about the presence of game, about the availability of meat, about the social situation, and also about the emotions of the actors (Goodall, 1986). When *hoo's* and *wraa's* are given over discovery of a large python on the ground, and then other individuals quickly travel to that place and show cautious interest or fear and hostility, more seems to be understandable from the calls than pure and simple emotional states.

3. A still less conservative hypothesis, supported by Goodall (1986), is that, in many instances, there is an *intention* to communicate on an immediate basis. By this I mean that the communicator wishes for the information he transmits to have an effect upon the hearer. Any human observer in the field who has received an admonitory cough from a high-ranking male after inadvertently coming too close to him, and receives yet another cough for not backing away fast enough, is likely to arrive at such an intuitive working hypothesis. This is true, regardless of his or her philosophical position on chimpanzee mental abilities or on the limits of reliable scientific description. It would appear that such intentions also may be directed at an individual who is out of sight, but hopefully not out of hearing, as when a juvenile who is lost alternates between soft whimpering and occasional loud outbursts of whining[5] that are combined with rapid visual scanning. Again, when a subordinate is attacked, its screams do not necessarily signify an intention of alerting an ally to help; but in many cases, an intensely screaming victim does appear to be actively looking around for assistance as it calls, and it may even solicit assistance directly through a gesture (Goodall, 1986). In both of these cases, acoustically very similar vocalizations apparently can be either "expressive" or intentionally communicative, depending on whether they are used manipulatively.

4. A still less conservative hypothesis is that wild chimpanzees are capable of communicating intentionally in matters that involve displacement: that is, they can deliberately communicate vocally to another individual about something that either (a) is out of sight for one or both of them or (b) resides in the past or the future. If this is so, then wild chimpanzees can be said to be operating at a symbolic level of communication – that is, at a symbolic level as many humans have defined this for humans. A food call, for example, means something like "food worth eating"; the call is used both

expressively and, if emitted loudly, as a social invitation. Along this line, deceptive behavior similar to that studied experimentally by Menzel (1971, 1973) possibly has been observed in the wild, in a context of vocal communication being used to deceive others about a food source. The two-point recording technique discussed in the next section may make it possible to support or disconfirm such hypotheses.[6]

Long-term study of long-distance vocal communication may well offer some substantive clues for evaluating which of the above levels of communication take place in the wild. In designing the vocalization project at Gombe, none have been ruled out. Some more specific working hypotheses are that purely vocal communication can transmit (1) information about the age, sex, identity, and emotional state of the vocalizer relative to other individuals in hearing range; (2) information about the size and makeup of a party; (3) information about the nature and intensity of agonistic situations; (4) information about interest in being sociable; (5) information about feeding conditions; (6) information about specific activities such as hunting and nesting, and, possibly, about readiness to go on patrol; (7) information about the presence of members of a neighboring community; (8) information about other dangers, or novel items, in the environment; (9) information about location and distance of the vocalizer from the hearer; and (10) if several calls are given, information about direction of travel. But to firmly establish such claims, it would appear that new methods of study must be devised.

When Marler worked at Gombe, he used excellent but cumbersome Kudelski (Nagra) high-fidelity audio recording equipment (Marler, 1976). This made it feasible to operate only in camp, where a relatively high level of banana provisioning was used as a method of attracting subjects. While Marler did manage to reduce several dozen vocalization types to just over a dozen, it would appear that his analysis was prematurely parsimonious. Almost certainly he "lumped" together calls which seem acoustically rather similar to a human ear but have discrete meanings for chimpanzees. Some of this lumping would appear to have been an artifact of relying upon the sound spectrograph, which in some cases cannot make discriminations which are possible for an experienced human ear.[7]

An alternative research strategy is to further "split" the almost three dozen categories in Goodall's current list wherever multiple call types can reasonably be hypothesized within an existing category. The next step would be to use a combination of different field recording techniques and different methods of analysis in order to test these "splitting" hypotheses. In the present project, the idea is to concentrate upon "catchall" categories like the "pant-hoot," "scream," "pant-grunt," or "*waa*," so as to differentiate multiple calls within such categories. This will be done by identifying varying behavioral patterns which appear to have acoustical correlates – and then by following this up with intensive spectrographic and human-ear analysis of any acoustic patterns that seem to co-vary with the behaviors. A second strategy is first to identify acoustical sub-patterns and then to try to identify their behavioral correlates. These two standard sorting strategies of primatologists can be usefully combined to achieve within-category "splitting."

In the case of very complex calls, such as pant-hoots, this acoustical analysis will be particularly difficult, even for a human ear that is quite experienced with wild chimpanzees. The problem is that we do not fully understand what the chimpanzees are using as distinctive features. Is it pitch, tenseness, phrasing, duration, volume, frequencies of formants, vowel quality, intonation pattern, number of repetitions, or other features? It seems likely that many of these could be relevant. It also is quite possible that chimpanzees may be relying upon other distinctive features that are not suggested by interpolation from human phonemics.

The present methodological strategy is to further extend Goodall's (1986) expanded list by using a combination of techniques, including several that would appear to be innovative. The research methods to be outlined here include the use of both tape recorders and highly portable video cameras in combination with written notes, and working with two-point recording techniques. One hope, in sharing these methods early, while research is just getting underway, is that other researchers who study *Pan troglodytes* may become interested in obtaining similar data. Ultimately, this could lead to comparison of finer points of vocal communication systems of different groups, and hopefully it could lead to the identification of differences that are cultural, that is, differences of dialect (see Goodall, 1986). Comparisons with *Pan paniscus* will also be of great interest.

FIELD TECHNIQUES AND EQUIPMENT

The initial pilot research strategy was simply to incorporate tape recording into the basic daily "follows" conducted at the Gombe Stream Research Centre. These follows involve the intensive observation of a target animal from nest to nest, an effective, basic research routine which has prevailed for over a decade at Gombe, with two-man teams of carefully trained Tanzanian field assistants keeping a written record of the behaviors observed (Goodall, 1986).

During field work in the summer of 1984, I began to explore the use of low-fidelity, miniaturized Sony M-5 and M-205A mini-cassette recorders having built-in microphones and automatic recording level controls; these were small enough to carry in a pocket on all-day follows. A few field assistants were trained to record vocalizations, along with spoken comments sufficient to tie the calls to the detailed written record. Pilot results were promising, so in the late summer of 1985 all 15 of the Gombe field assistants who specialize in observing chimpanzees were trained to do field recording. However, the technical quality of the recordings remained very limited, especially in terms of signal-to-noise ratio and frequency response. Later, at the end of the 1985 dry season, high-fidelity professional Sony "Walkman" (WM-D6Cs) were introduced and tested. In spite of their larger size and external microphones,[8] the WM-D6Cs were found to be feasible for the sometimes trying field conditions at Gombe, under which the size, weight, or shape of carried objects quickly becomes critical. Once I had trained the field assistants to use the equipment, the recording of vocal behavior became a routine part of data collection during the dry season of 1986. Because these machines have manual recording-level controls, they can record faintly heard, distant vocalizations quite clearly.[9]

In the summer of 1986, I made two further pilot innovations under the auspices of a small pilot grant from the L.S.B. Leakey Foundation. One was the introduction of reliable super-lightweight Sony 8mm video cameras to document more fully the social context of vocalization and to provide clues about the articulation of vocal productions. After five weeks of experimentation with two different kinds of video cameras, only the Sony version of a video "brownie camera," the handy-cam, proved compact enough to be carried on the average Gombe follow. Not only can a Gombe follow involve making one's way quickly over rough terrain, but it often requires the rapid negotiation of thickets or large brush patches laced with vines and brambles. The second innovation was what might be called two-point recording. My intention was to record long-distance vocal exchanges from both ends of the exchange, with the field assistants using the professional audio recorders while I used the handy-cam video recorder, which makes excellent audio recordings. With its standard microphone, this machine makes audio recordings superior to those obtained on conventional half-inch VHS or Beta cam-corders.[10]

The two-point recording technique alluded to above must be further described. It records individuals who are exchanging long-distance calls at both ends of the exchange, so as to study the entire exchange as a macro-unit of behavior. Methodologically, this requires that observers be with both parties so as to obtain (1) accurate, nearby recordings of the vocalizations; (2) full records of what the animals are doing during and between vocalizations; and (3) precise location of the two sets of vocalizations in time, so that their exact sequence can be reliably determined. One way of obtaining the fullest possible record will be to use an 8mm video camera-recorder unit, since the wide-angle lens can record visually the behaviors of other individuals in the group who provide an important part of the vocalizer's social context.

The ideal situation is to have such a recorder at each end of the long-distance exchange and to record with both machines simultaneously throughout the exchange. With respect to describing precisely the sequence of a series of exchanged calls, recording of times using synchronized watches is one method which is useful; however, having a sensitive microphone and a recorder with a good signal-to-noise ratio serves the same purpose, since distant signals will be recorded clearly enough to compare them with nearby recordings of the same signals (except when both groups call simultaneously, which is not usually the case). A high quality audio recorder and a directional professional microphone such as a relatively compact Sennheiser ME-80 are ideal for this task, but the Sony 8mm cam-corders with standard, very compact omnidirectional microphones also make reasonably clear recordings of distant calls if there are no competing sounds nearby. By establishing a protocol by which the observer at each end leaves the recorder on continuously once a long-range exchange seems likely or has begun, the two recordings will make it possible to plot the exchange accurately over time and to determine unambiguously the source and sequence of the calls.

POSSIBILITIES FOR PLAYBACK EXPERIMENTS

Playback experiments, as used by Seyfarth and Cheney (1982), are useful insofar as they establish still more firmly that a previously well-observed and statistically significant behavioral response pattern goes with a given type of vocalization. The stimulus is a previously recorded call which is played back in high fidelity from a hidden location at a time when the actual caller from whom it was recorded is out of sight and hearing.

In a sense, a conversational sequence of long-distance calls and return-calls can be taken to be a series of natural "playback experiments." That is, where a call follows a highly distinct acoustical pattern (e.g., a short series of pant-hoots having a brief high climax, accompanied by drumming a tree buttress or other object) and where the call has been observed a number of times to elicit a response or a predictable range of responses, each further incidence of a similar call can be treated as a stimulus, and the response can be observed and recorded to see if the same behavioral pattern continues to appear.

Of course, a classical playback experiment (e.g., Seyfarth and Cheney, 1982) uses stimulus calls known to be identical to prior natural calls because the natural ones have been accurately reproduced electronically. At Gombe, one possibility for a playback experiment is to initiate long-distance calling (or some other behavioral response such as scanning or traveling) with a tape-recorded stimulus from a previous natural "initiation" (e.g., a group emitting loud food calls upon discovering an unusually ample food patch). Another possibility is to play a "return call" recording when a natural initiating call has not been returned by other animals within the normal response period. Walkie-talkie contact between the person activating the playback and the person observing the group would facilitate such experimentation.

With distance calls, the classical playback experiment technique presents several problems. One is that very powerful, high-fidelity portable sound-reproduction equipment is expensive and also cumbersome and therefore difficult to place. A more serious problem with a species whose social groupings and patterns of travel are highly unpredictable is ascertaining that the individual whose recorded voice was used is nowhere within hearing. This problem is mentioned by Byrne and Byrne (1986), who have attempted to rely upon playback experiments in studying distance calls made in Mahale, in terrain presenting more serious problems than Gombe. Because playbacks would have to be broadcast at very high volume with only partial control over directionality, and because chimpanzee groupings and daily paths of travel are extremely unpredictable, it would be difficult – even with walkie-talkie contact – to ascertain that voices used for stimulus tapes were not those of individuals who might be within hearing and in the company of other individuals. The latter obviously would become confused; there is also the unknown variable of an individual hearing his own voice reproduced. If recognized, the effect would be unpredictable in an animal that is capable of visual self-recognition. If unrecognized, it would appear – to that individual alone – that a *stranger* of the same sex and developmental status had entered the community's territory.

In the summer of 1988, when I had spent about fifteen months observing the Gombe study group and had been reproducing chimpanzee vocalizations for several years in my lectures, a "playback experiment" took place which had unpredictable but suggestive consequences. I was trying to contact and follow a group of males who were frequently vocalizing at a distance with two other smaller groups as they all travelled, when suddenly all vocalization ceased. I reached the top of the ridge just south of the provisioning camp, in the center of the Kasakela community's territory, and waited for further sounds, but none were emitted. After half an hour's wait, I thought that all three groups might have passed silently into the next valley and out of my hearing range. More likely they had stopped to feed nearby, without making any food calls.

Observers trying to locate chimpanzees at Gombe frequently wish that they could emit an authentic-sounding locational pant-hoot so as to stimulate a return call. And while the researchers' policy at Gombe is one of non-intervention and contact avoidance, tourists, when not strictly supervised, occasionally try to stimulate the animals with crude call-imitations. There is never any response by the animals beyond an occasional quick glance. Thus, the chimpanzees of the study group have a long but limited experience with humans improvising their calls. It was on the basis of this knowledge that I had been planning for several years to try a stimulus call of my own, if I could find the right circumstances.

I decided that the situation was optimal to try a locational pant-hoot, and also decided that I would emit only the beginning of the call, which is in a low register. My assumption was that human vocalizing inadequacies would be magnified in trying to reproduce the more complicated, high-frequency portion of the locational pant-hoot sequence. I assumed that the longer I vocalized, the greater the chance that I would deviate seriously from the chimpanzee acoustical pattern. I also thought that perhaps the low-register portion of the pant-hoot would carry less of a voice-signature. Chimpanzees rather frequently break off their pant-hoots at the point where I terminated mine.

My intention was to reproduce the incipient pant-hoot of an adult male, but not to imitate any particular adult male, in the hope that one of the groups might answer even though they were not certain who was calling. Due to dense vegetation, I was totally out of sight up on the ridge top. I emitted the first several pairs of "syllables" of an incipient pant-hoot, keying inspiration and expiration to the chimpanzee pattern. Immediately, from a location several hundred yards below the top of the ridge, at least four or five male voices emitted a variety of short, unusual, excited-sounding unpanted calls which I was unable to classify. These ceased abruptly. After a minute, I switched on the Sony handy-cam videorecorder and repeated the abridged call. This time I recorded only two responding voices, those of juveniles or early adolescent males, and I wondered whether my deception had failed with the older males the second time.

A few minutes later, the alpha male Goblin appeared, walked toward me, alertly sniffed foliage, sat down for a few moments about fifteen feet away and looked in my direction, then erected his hair and displayed, running in a straight line away from me down the ridge top to the west, out of sight; a tree was then drummed several hundred feet away. A minute later I heard a second drumming, quite distant and from the same direction, and I inferred that Goblin or another male had made a second running display.

Goblin's approach and foliage sniffing were unprecedented. There are many possibilities in interpreting his response to such a poorly controlled stimulus. Minimally, one can state that Goblin's curiosity was aroused and that he was aroused in such a way that when he detected no trace of another chimpanzee with me in the location where he had heard the call, he displayed in an unusual manner over a long distance. For students of wild chimpanzee vocal communication, this anecdote points to the potentially intrusive nature of playback experiments.

Territorial patrolling of wild chimpanzees is an area of research which deserves intensive study. A controlled experiment, which would go far beyond the verification-by-playback techniques of Seyfarth and Cheney, would be to record the actual voice of a single stranger-male emitting a pant-hoot and to play this back in or near a territorial overlap location where such a solo marauder might penetrate but normally would not vocalize. This would obviate the problem of a familiar, friendly voice confusing members of the study group.

Such an experiment would test for aversive versus aggressive responses to single strangers. The experimental design could be varied to test different territorial locations, the size of the defending group, and the sex of the stranger. The experimenter could also measure subsequent changes in patrolling behavior. But such experiments, which could add significantly to our understanding of chimpanzee territoriality, would have to be scrutinized with great care before being done. They could alter a study group forever in that area of behavior, and the territorial equilibrium between two communities could be disrupted. It is unclear at what point the findings would be so important as to justify such an experiment when "natural experiments" are available for study.

I believe that long-distance call sequences must be studied as ongoing "conversations" rather than as a series of independent stimulus-response episodes; it would be difficult to adapt a field playback strategy to include the continuance of exchanges after the first pair of vocalizations. Given the limitations and problems discussed above, the initial study will emphasize full behavioral descriptions of a sizeable corpus of natural exchanges through a two-point recording technique. Later, playback experiments will be considered for testing specific behavioral hypotheses under controlled conditions.

PROSPECTS FOR RESEARCH IN THE IMMEDIATE FUTURE

During the summer of 1986, there were few estrous females at Gombe, and coincidentally, many of the dry-season fruit trees were bearing poorly. Thus, at the outset of the vocalization project, food patches tended to be too small for large groups to share, and the usual large groups of chimpanzees were absent. Consequently, very little long-distance communication had taken place by early August, when I had to leave the field. However, by that time I had trained one of the field assistants to use a Sony M-8 handy-cam 8mm video camera on daily follows. We also worked out the logistics of setting up two observers for different community sub-groups which were within vocalizing range. Hopefully these efforts can be continued in the dry season of 1989.

In 1987–1988, the training in videotaping was extended, and as a side effect of a different research project, a substantial corpus of one-point recordings of vocalizations was collected. These recordings await spectrographic analysis and interpretation. Field assistants are now trained to produce two-point video recordings. Over the next several years, I hope to obtain recordings of long-distance vocal exchanges and analyze the behaviors that precede, accompany, and follow from such "conversations."

These intensive techniques should produce a corpus of recordings, with contextual behavioral data, which will permit observers to discern some of the more subtle phonetic distinctions that are being made by chimpanzee vocalizers and listeners when they communicate. It is possible, by relying upon sonograms and also upon some newer, computerized techniques of sound spectrographic analysis, that a sharper analytical instrument will assist the only-too-human ears of the observers. But another methodological innovation will be to use ethnographic techniques to systematically debrief the more experienced paraprofessional field assistants, who have been listening to chimpanzees vocalize on a weekly basis for up to 15 years. It may also prove to be an advantage that their African phonemic and cultural points of reference are different from those of the professional researchers. Preliminary work in this direction has already begun (Goodall, 1986).

In the matter of methodology, many of the long-distance vocalizations may prove to be elusive, especially with respect to obtaining and reliably interpreting vocal or behavioral responses to playback experiments. For example, experienced observers are often at a loss to predict when calls will be emitted or answered. This is because groups of chimpanzees do not respond – vocally or in any other appreciable way – to many of the distant calls that they hear. However, carefully selected experiments might produce some more predictable results. For example, it can be strongly predicted that long-distance pant-hoots which grade into food calls serve sometimes as a stimulus to travel. There might also be a social factor. This could be narrowed down experimentally: a food call/pant-hoot from a particular individual might precipitate travel by another individual in the direction of the caller, while a similar call from a different individual might elicit no response. Another hypothesis to be tested is that travel is more likely to take place with SOS screams than with food calls, particularly where the vocalizer and the hearer in both cases are well bonded. Another hypothesized strong basis for travel would be purely social interest, as when two closely bonded individuals are known to have been traveling separately for a time, and in pant-hooting they establish contact.

As mentioned before, long-distance pant-hoots in combination with calls associated with active hunting seem to announce hunting activities and consequently to bring other chimpanzees running to the scene of the hunt (Goodall, 1986). A few playback experiments might be possible in this area without causing undue confusion; this is difficult to say. But more subtle experiments are possible: I presently have a video recording of pant-hoots emitted while a group of three males had stopped traveling to gaze at a group of colobus monkeys in trees not far away; they made these calls in response to distant pant-hoots. It would have been very interesting to have videotaped the distant group's behavior, to have determined whether the monkeys were also within their sight, and, in any event, to have recorded the behavioral reaction of the distant group when their pant-hoot was returned. As a corpus of such natural materials is accumulated, it will be possible to devise other playback experiments. One of the field assistants has suggested that males who are about to begin a patrol may pant-hoot in a special way that brings other males to join them. There are, of course, many other behavioral contexts that will offer the opportunity for testing semantic hypotheses with playback experiments; but these examples at least illustrate some well-controlled possibilities.

The methods discussed above surely are labor-intensive, since a large number of attempts will be necessary simply in order to obtain a substantial sample of successful two-point recordings of vocal exchanges, and an even larger number of attempts will be needed to successfully conclude a series of playback experiments. An innovation at Gombe, one that has been in existence for over a decade at present, is the local training and employment of an excellent paraprofessional field staff. Goodall (1986) has discussed the training and supervision, and the quality of the work. At Gombe, having such a staff brings down the costs of intensive research, such as that outlined above, to a level that is reasonable. Hopefully, this innovative approach to staffing might be diffused to other research sites where intensive research is undertaken, including studies of long-distance vocal communication.

PHONETICS AND SEMANTICS: Some Theoretical Considerations

The acoustical characteristics of chimpanzee vocalizations have been discussed descriptively by Goodall (1968, 1986) and from a more technical linguistic standpoint by Marler (1976), and both authors have dealt with the behavioral contexts of vocalization. Marler also has discussed problems that may be encountered by researchers who attempt to apply what is known about syntax in human languages when hypothesizing possible combinatorial properties of wild chimpanzee vocalizations (Marler, 1977). However, any current substantive understanding of "chimpanzee semantics" and the capacities of these animals to combine their signals syntactically have come out of captive experiments exclusively (e.g., Fouts, 1975; Savage-Rumbaugh et al., 1978; Premack and Premack, 1972).

Being merely human, our starting point in thinking further about the meanings encoded and decoded in wild chimpanzees' vocalizations surely will be human. We humans, when informally learning first or second languages, determine the meanings of words either by (1) common-sense observation of the contexts in which the words are used, banking on our capacity to generalize, or (2) by asking native speakers for definitions of words – a technique which has been exploited in ethnosemantic investigations of cultural anthropologists (e.g., Casagrande and Hale, 1967; Werner, 1967; Boehm, 1980). It is reasonable to assume that chimpanzees learn both chimpanzee calls and their behavioral applications by observation and generalization, just as human children do. It also is reasonable to assume that human beings can try to learn the meanings of wild chimpanzees' vocalizations on the same basis. But as observers, we face the disadvantage of being outside of an exotic system of communication, rather than being able to participate in that system as anthropologists do in their field research. (The playback experiments discussed above would provide a vicarious, highly limited form of "participation.")

A preliminary approach to understanding vocal communication of chimpanzees may well benefit from what is known of the semantics of human language, in that our understandings about a familiar form of communication can help us to generate hypotheses to be tested in the field with chimpanzees. For example, in a human language the stock of lexical items (words) is always smaller than the stock of meanings that are communicated. This can be verified by reading any dictionary. Such economy of semantic transmission is accomplished through polysemy and homonymy.[11] In homonymy, a single acoustical unit that a linguist isolates as a "word" may have several totally different meanings, for example, bear (the animal); bear (carry); bare; and possibly a dog who is a bayer, depending upon dialect. In polysemy, the same word has several "senses" that can be quite divergent in meaning, but which are not totally unrelated as in the case of homonymy, for example, dim (light condition); dim (unintelligent); dim (minimal, as with hopes); or, to return to the "bear" example, bear (carry); bear (have a child); bear (as in bear down), etc. Since both polysemic units and homonyms are acoustically identical, hearers must rely exclusively upon context to decode the precise meaning. Whether vocal signals which are heard by chimpanzees as being acoustically identical can have multiple meanings – as chimpanzees decode them in the light of various different contexts – remains an open question.

Of course, present categorizations by Goodall (1986) and Marler (1976) do lump many different behavior contexts under the label of single calls, such as pant-hoots. If such categories were based on calls which chimpanzees truly perceived as being acoustically identical, this would suggest a rather extensive use of homonymy or polysemy. And this would suggest, in turn, that the chimpanzees' vocal repertoire was far smaller than their set of communication needs. However, limitations in human abilities to hear or to electronically identify and isolate phonetic features that are distinctive to chimpanzees are likely to be a major problem here: there may be literally dozens of acoustically different pant-hoots, so far as chimpanzees are concerned, which to them have different meanings. To determine the number, we will have to identify the distinctive features by which chimpanzees differentiate the different pant-hoots. In this respect, Goodall has gone much further than Marler in suggesting possibilities for behavioral variants, but these hypotheses need to be tested systematically by acoustical analysis. When this work is done, it may be possible to resolve the separate question of whether chimpanzees' vocal communication involves anything like homonymy or polysemy. Playback experiments could be useful for final verification.

Another peculiarity of human communication units is synonymy: certain pairs of acoustically very disparate words may have identical or, at any rate, very similar meanings. Whether chimpanzee communication exhibits such subtleties can be determined only after a great deal of further research, but the possibility must not be ruled out. For example, I have heard a subordinate male scream "thinly" when he approached a dominant male in a situational context that was likely to provoke apprehension or anxiety, where my intuitive prediction would have been that he would either pant-grunt loudly or else pant-bark. In this case, two acoustically dissimilar calls might be employed to convey the same message – one of apprehensive acknowledgment of rank.

It also is quite conceivable that chimpanzee units of communication, like words in English, are inflected. When I listen to calls that are presently categorized as defiant *waa's,* three acoustical pattern points stand out to my ear: I sometimes hear "waa," sometimes "wow," and sometimes "woe."[12] One possibility, of course, is that I am making distinctions based on acoustical features which, for chimpanzees, are merely in free variation. Another is that the phonetic differences merely reflect different degrees of emotional intensity rather than more radically demarcated messages. But, since the three calls that I hear take place in a variety of contexts (most of them clearly agonistic), it is possible that the messages they carry are being modified systematically, in some patterned way: that is, *waa, wow,* and *woe* are acoustically similar calls which regularly occur in similar yet somewhat different behavioral contexts. If so, something similar to grammatical inflection could be taking place; that is, the semi-vowel "w" sound that humans hear is comparable to a root-morpheme, and the vowel-like sounds that follow it are acting like suffixes. If other calls were found to be transformed on a similar basis, a stronger case could be built in this direction. Obviously, videotaping will help to sort out the behavioral contexts, if such differences exist and they are subtle.

From the standpoint of data analysis, it will make sense to concentrate on understanding the frequently emitted discrete calls of chimpanzees before trying to understand combinations of calls, and, in particular, the graded combined calls. In analyzing discrete calls, it makes perfectly good sense to test hypotheses by a method that takes into consideration the nature of single *words* in our own language, keeping in mind that Indo-European languages tend to distill relatively small amounts of meaning into relatively simple semantic units, ones that can be intuitively identified as "words." Such languages then combine words into sentences: this, therefore, is our intuitive model for the combining of semantic elements.

However, many other languages take a different approach (e.g., agglutinative languages such as Eskimo, Navajo, or other highly inflected languages). Models like English, Japanese, or American Sign Language may be appropriate at the stage of analysis where discrete calls are being treated. But when it comes to understanding the more complex graded calls of chimpanzees, then human languages other than English may serve as better models for generating working hypotheses about how the chimpanzee combinations are being put together. This means that if chimpanzees' more complicated calls are to be evaluated for possible "syntactic" properties, the better model may well be an agglutinative language. With a language such as English as a model, a grammarian tends to focus on sentence structure; with an agglutinative language, the focus is more on the morphological makeup of the words, which are complex structures built largely out of affixes. Such an agglutinative model may serve as a better model for understanding calls like pant-hoots, pant-grunts, pant-barks, screams, squeaks, and food calls when these are combined by grading. But more generally, the key to analysis will be the ability to make subtle distinctions among behavior contexts, a process which will be enhanced by use of videotapes, and the ability to make subtle acoustical distinctions keyed to learning more about chimpanzee perceptions of distinctive features.

Another important aspect of the research will be to understand chimpanzee vocal exchanges as *conversations,* in which the context becomes increasingly complex as an exchange continues; simultaneous, two-point recording techniques focused on long-distance vocal exchanges will provide a major advantage in this pursuit. Also, much probably can be learned at short range, where, for example, grunts in a series sometimes have the appearance of being used as queries and replies.

SUMMARY

It will be difficult to further advance the study of wild chimpanzees' vocal communication in the wild so long as vocal, postural, gestural, and facial-expression channels of communication are studied simultaneously. Of course, nonvocal communication can help us to understand the concomitant vocalizations. But when the other channels are in operation, this makes it difficult to factor out the specific communicative effects of the vocal channel. Given the methods outlined in this chapter, it is anticipated that the vocal channel of communication can be studied quite effectively, in isolation from the other major channels.

In studying chimpanzees' vocal communication, the fundamental approach remains the one practiced by primatologists over several decades: calls which sound different are carefully assessed for differences in the accompanying behaviors (Snowdon, 1982; Goodall, 1968). If such a pattern is found, then the vocalization can be reliably correlated with the behavior and its meaning for communication can be hypothesized and, in some instances, tested. This is very similar to the "context of situation" approach to semantic studies of human language suggested by Malinowski (Malinowski, 1935; Ogden and Richards, 1970; see also Boehm, 1972), an approach also employed by sociolinguists.

This similarity to semantic approaches developed by cultural anthropologists (who, like primatologists, work with exotic semantic systems) is worth considering further. Malinowski's (1935) technique in *Coral Gardens and Their Magic, Vol. II,* was to seek definitions for words and then to conduct an exhaustive ethnographic analysis of the total situational context of an utterance containing the words (e.g., a magical incantation employed while planting yams). By normal ethnographic standards, this technique was highly labor-intensive.[13] In working with nonhuman primates, the costs of obtaining information obviously will be high, since our subjects cannot be asked questions in the wild – except, in a sense, by means of playback experiments. Thus, Malinowski's original approach to the investigation of semantics, which intensively studies the situational context of an utterance and then defines its meaning in terms of its function, is the appropriate method for analyzing vocal communication of nonhuman primates.

The ethnosemantic approach is one of deep contextual analysis: if you fully understand the physical, social, and cultural contexts of an utterance, you can describe its semantic load and determine what makes it distinctive in comparison with other utterances in different contexts. The assumption, with human language, is that you already know the language and that your subjects (usually) will be able to answer your questions. In applying similar approaches within primatology, needed refinements have come in the form of quantitative sorting, spectrographic analysis, and use of playback experiments. These partially compensate for unfamiliarity with the linguistic forms and for the inability to query informants.

The long-term vocalization study which is resuming at Gombe will focus on long-distance vocal communication. However, it will concentrate as well on data collection of vocal communications made at short range, particularly in order to obtain a fully representative corpus of communicatively significant calls that are emitted discretely. A relatively large corpus will be needed to permit quantitative sorting necessary to set up subcategories within those calls which are similar acoustically and which at present have been "lumped." This phase of research will not isolate the vocal channel from the other channels of communication; rather, gestures, postures, facial expressions, and so on, will be used as contextual factors to aid in sorting vocalizations within categories. The analysis of long-distance vocalizations will be built upon this base and eventually may be tested, wherever feasible and safe, through use of playback experiments. This analysis of purely vocal communication should also shed further light on combined calls which are used in close-quarters situations, where these also are being combined with gestures, postures, and so on.

It is emphasized that, wherever possible, the vocal exchanges of chimpanzees will be studied not just as a series of independent vocalized signals with their subsequent behavioral reactions, but potentially as *conversations.* By adding two-point recording of long-distance vocalization episodes to the arsenal of techniques already developed by primatologists, something paralleling discourse analysis in human language can be attempted. What has been learned already in captive chimpanzee experiments about the chimpanzee potential for complex communicative interactions will be a valuable resource in going forward with this project. But an open approach is needed. The long-distance vocal exchanges of free-ranging chimpanzees may well follow rules of their own, and for this reason a reasonably inductive initial stance is in order.

With captive chimpanzees, a great deal has been learned about the interactive aspects of primate communication, as chimpanzees have held conversations with human beings and now hold them with other chimpanzees, using "manual" languages that humans have devised. The present project, if successful, will tell us something about how chimpanzees in the wild hold natural two-way (and sometimes three-way or four-way) conversations, using the vocal channel alone. These findings should be of interest to ethologists who specialize in chimpanzee behavior, to psychologists who study linguistic capacities of captive chimpanzees, to the anthropologists, archaeologists, linguists, and evolutionary biologists who develop hypotheses on the origins of human language, and to primatologists in general.

ENDNOTES

1. Hamadryas baboons have a somewhat similar social organization (Kummer, 1968), but due to the harem structure and a far looser "territorial" structure, hamadryas social behavior exhibits less of the "ad hoc" quality found with chimpanzees.

2. I follow Goodall (1986) in associating different chimpanzee calls with the presumptive emotions and observed behavioral orientations which accompany the calls; except where noted otherwise, calls used for exemplification are derived from Goodall (1986).

3. I am grateful to the government of Tanzania and its agencies for help in facilitating research at Gombe, particularly the National Research Council, the Serengeti Research Institute, and Tanzanian National Parks. I also thank Jane Goodall, who, as collaborator, has made the resources and records of the Gombe Stream Research Centre available and has provided other logistical support and intellectual stimulation in this joint effort. I also have profited from discussion with Philip Lieberman and John D. Newman. The L.S.B. Leakey Foundation provided a small pilot grant which made possible the introduction of video recorders for purposes of vocalization research, and more recently, the Sony Corporation has made similar contributions in this area. Northern Kentucky University awarded a Faculty Summer Research Fellowship and a sabbatical leave which aided my training in ethological research methods in 1984. The H. F. Guggenheim Foundation (1987–1989) has supported further intensive videotaping of behaviors relevant to conflict resolution, a side effect of which has been a significant augmentation of the corpus of recorded vocalizations. All this support is gratefully acknowledged.

4. Marler's field recording took place entirely in the camp where provisioning was done, so his findings did not represent the entire range of normal chimpanzee vocal behavior. One likely sampling problem in his study was that extensive provisioning led to unnatural levels of agonism.

5. This call is listed by neither Goodall nor Marler; it sounds very much like a

dog's continuous modulated whining and presumably is included in Goodall's (1986) whimper-to-crying graded productions. I have recorded it once and have heard it one other time, in both cases from juvenile males.

6. A less definitive, similar example is reported by Goodall (1986): adolescent Figan, long before becoming alpha, learned to suppress his own food calls when he was given bananas just after the senior males had left camp (since the latter had returned on several previous occasions to take his bananas away after hearing him vocalize). In this case, he manipulatively avoided communicating about something the others could not see.

7. One major problem with the sonograms, identified by Lieberman (pers. comm.; see also Lieberman, 1972), is that they do not provide the full, accurate, and detailed representations that are needed to help an analyst to identify distinctive features that may be significant to chimpanzees but are unfamiliar to humans. Fortunately, sonograms can be augmented with computerized techniques, and both methods will be relied upon in the described study if funding is adequate.

8. Compact Sony microphones, designed for the WM-D6C, having a fair degree of directional function, were of a suitable size. Unfortunately, the larger professional "shotgun" microphones which will sharply isolate the voice of a particular individual when a group is vocalizing would pose major problems at Gombe, insofar as they would either take a very long time to put away for travel or else they would tend to tangle in vines and brambles. The result of such entanglement is the loss of one's target. The ideal solution to the microphone problem would be to have one stereo channel recorded with a shotgun microphone, and the other with an omnidirectional microphone that would pick up observer comments and also the vocalizations of individuals other than the target.

9. A specially devised recording technique was to fix permanently the recording-level control of the WM-D6C on the highest position. This means that the instant the recorder is activated, it will pick up faintly heard, long-distance calls without the necessity of reaching for a second control. Over-modulation of very loud, near calls is controlled by angling the cardioid (moderately directional) microphone above the target in order to effectively reduce the recording level. Use of metal tape, which we could not afford, makes such a strategy even more feasible since metal tape accepts a stronger signal without distorting. The technique has two advantages. One is that for paraprofessional field assistants, it simplifies the recording task, at little cost, in a direction that enhances the recording of distant calls. The other is that it is easier to use. Even for someone who is very familiar with recording techniques, it can be difficult in the bush, where physical constraints can easily become critical, to operate a recording-level control at the same time that one is directing a microphone and trying to follow and assess the behaviors that accompany vocalization.

10. One problem with such cameras in their very compact original form is that the microphones are not sufficiently directional to isolate a single individual's voice if several individuals are vocalizing.

11. Green and Marler (1979) have discussed the possibilities for animal calls.

12. For simplicity's sake, I have retained standard English language orthography here.

13. In my doctoral dissertation (Boehm, 1972), the approach was streamlined: I collected and analyzed 10,000 definitions collected from 40 informants, using a standard cue-word list. As a way of quickly assessing the situational context, I asked each informant to provide a concrete example of a behavior or situation appropriate to the word being defined. This also facilitated quantitative analysis.

REFERENCES

Bauer, H. R., and M. Philip. 1983. Facial and vocal individual recognition in the common chimpanzee. *Psychological Record* 33:161–170.

Boehm, C. 1972. Montenegrin ethical values: an experiment in anthropological method. Ph.D. diss., Harvard University.

_____. 1980. Exposing the moral self in Montenegro: the use of natural definitions to keep ethnography descriptive. *American Ethnologist* 7:1–26.

Byrne, R. W., and J. M. Byrne. 1986. Wild chimpanzee vocalization. *Anthroquest* 36:12–13.

Casagrande, J. B., and K. L. Hale. 1967. Semantic relationships in Papago folk-definitions. In D. Hymes, ed., *Studies in Southwestern Linguistics,* pp. 164–193. The Hague: Mouton.

Fouts, R. S. 1975. Communication with chimpanzees. In G. Kurth and I. Eibl-Eibesfeldt, eds., *Hominisation and Behavior,* pp. 137–158. Stuttgart: Fischer Verlag.

Goodall, J. 1968. Expressive movements and communication in free-ranging chimpanzees: a preliminary report. In P. C. Jay, ed., *Primates: Studies in Adaptation and Variability,* pp. 313–374. New York: Holt, Rinehart and Winston.

_____. 1986. *The Chimpanzees of Gombe.* Cambridge: Belknap Press.

Green, S., and P. Marler. 1979. The analysis of animal communication. In P. Marler, and J. G. Vanderbergh, eds., *Social Behavior and Communication,* pp. 73–158. New York: Plenum.

Kummer, H. 1968. *Social Organization of Hamadryas Baboons: A Field Study.* Chicago: University of Chicago Press.

Lancaster, J. B. 1975. *Primate Behavior and the Emergence of Human Culture.* New York: Holt, Rinehart and Winston.

Lieberman, P. 1972. *The Speech of Primates.* The Hague: Mouton.

Malinowski, B. 1935. *Coral Gardens and Their Magic, Vol II.* London: George Allen and Unwin.

Marler, P. 1976. Social organization, communication and graded signals: the chimpanzee and the gorilla. In P. P. G. Bateson and R. A. Hinde, eds., *Growing Points in Ethology,* pp. 239–280. New York: Cambridge University Press.

_____. 1977. The structure of animal communication sounds. In T. H. Bullock, ed., *Recognition of Complex Acoustic Signals,* pp. 17–35. Berlin: Dahlem Konferenzen.

Menzel, E. W. 1971. Communication about the environment in a group of young chimpanzees. *Folia Primatologica* 15:220–232.

_____. 1973. Leadership and communication in a chimpanzee community. In E. W. Menzel, ed., *Precultural Primate Behavior,* pp. 192–225. Basel: Karger.

Ogden, C. K., and I. A. Richards. 1970. *The Meaning of Meaning: A Study of the Influence of Language Upon Thought and of the Science of Symbolism.* New York: Harcourt, Brace and World.

Petersen, M. R. 1982. The perception of species-specific vocalizations by primates: a conceptual framework. In C. T. Snowdon, C. H. Brown, and M. R. Petersen, eds., *Primate Communication*, pp. 171–211. Cambridge: Cambridge University Press.

Premack, A. J., and D. Premack. 1972. Teaching language to an ape. *Scientific American* 227:92–99.

Savage-Rumbaugh, E. S., D. Rumbaugh, and S. Boysen. 1978. Symbolic communication between two chimpanzees *(Pan troglodytes). Science* 201:641–644.

Seyfarth, R. M., and D. L. Cheney. 1982. How monkeys see the world: a review of recent research on Ververt monkeys. In C. T. Snowdon, C. H. Brown, and M. R. Peterson, eds., *Primate Communication*, pp. 239–252. Cambridge: Cambridge University Press.

Smith, W. J. 1977. *The Behavior of Communicating: An Ethnological Approach.* Harvard University Press.

Snowdon, C. T. 1982. Linguistic and psycholinguistics: approaches to primate communications. In C. T. Snowdon, C. H. Brown, and M. R. Peterson, eds., *Primate Communication*, pp. 212–238. Cambridge: Cambridge University Press.

Washburn, S. L. 1982. Human behavior and the behavior of other animals. In T. C. Wiegel, ed., *Biology and The Social Sciences*, pp. 95–117. Boulder, CO: Westview.

Werner, O. 1967. Problems of Navajo lexicography. In D. Hymes, ed., *Studies in Southwestern Ethnolinguistics*, pp. 145–164. The Hague: Mouton.

THE RESEARCH AT GOMBE:
ITS INFLUENCE ON HUMAN KNOWLEDGE

Roger S. Fouts

When I first began to work on this paper, I was overwhelmed with the size of the task. My goal was an examination of the research from Gombe Stream and its influence on human knowledge, which is certainly something that several scholars could spend several lifetimes studying as compared to the few months I had available to devote to this article.

First of all, I found that the notion of human knowledge was extraordinarily broad in the sense of requiring the inclusion of the influence of the research from Gombe Stream on science, academic thought, philosophy, our conception of our species' place in nature, and so on. This indeed is a formidable task, not because Gombe's influence is rare and fleeting, but rather because it is so very pervasive.

I decided initially to examine the direct influence of the research from Gombe Stream on science and to limit my focus to Jane Goodall's contributions to that research in order to make things simpler for me, but in no way intending to make less of the prolific and worthwhile contributions of others who have worked at Gombe.

With my two self-imposed limitations of area and researcher, I went to the Science Citation Index (SCI), which lists scientific works that refer to a particular individual's research papers, and began counting the number of times "Jane Goodall" was listed. This task is, in some small empirical way, a measure of a person's research contribution to his or her discipline. Working back from 1986 to 1979, I found that Jane was being referenced in over 100 scientific articles per year, in a wide variety of scientific disciplines. As one would expect, the usual primatological, zoological, and animal behavior journals certainly had their share of references. Over these seven years I found over 50 different journals that had referenced Jane's research. The journals ranged over a diversity of disciplines: *American Journal of Obstetrics and Gynecology, American Naturalist, American Journal of Disabled Children, Clinical Pediatrics, American Journal of Physical Anthropology, Placenta, Mammalia, Psychological Record, Hormones and Behavior, Journal of Human Evolution, Developmental Psychology, British Journal of Hospital Medicine, Behavior, Advanced Veterinarian*

Science, Journal of Endocrinology, African Journal of Ecology, Human Biology, Recherche, Applied Animal Ethology, Developmental Medicine, Journal of Tropical Pediatrics, Journal of Child Psychiatry, Practitioner, Lancet, and *Florida Entomologist,* just to name a few.

This may seem to be a rather superficial measure of Jane's contribution to science and human knowledge. But the SCI did tell me that her research has had wide influence, with effects on medicine, child development, mother/infant behavior, anthropology, and even entomology. But it didn't really tell me the "why." What is it about her approach to understanding the chimpanzee that results in such profound influence on science and human knowledge? I began my second search by going directly to Jane's original works.

We as academicians, scientists, and scholars value knowledge. Our knowledge of the world is influenced by a great deal more than just objective scientific fact. It is influenced by the habits of thought of our times and our cultures. These biases of thought and theory, of scientific view and world view, are not readily amenable to change, and they may represent the darkness of ignorance. Indeed, knowledge is limited by ignorance in the sense that each discovery waits upon the last. We were ignorant of microbes until the microscope was discovered. Discoveries that are accepted by culture help to remove ignorance and, in this manner, change the nature of our knowledge. But it is often the very ideas that are considered to be knowledge today that must be overcome if new discoveries are to be accepted.

It is indeed amazing when we stop and consider what it must have been like, that night in 1960, when a diminutive English girl, her mother, and their African cook lit that first campfire on the shores of Lake Tanganyika. That flame was eventually to light new areas of knowledge. Jane's discoveries, innocent and unpretentious, have made a generation of science stop and reevaluate the place of humankind in nature and our relationship to those other species with whom we share nature. Today, 26 years after that first flame was lit, scientist and nonscientist alike are familiar with those discoveries: tool use, tool making, eating of meat, hunting, warfare, cannibalism, maternal behavior, mourning of a mother's death, auntie behavior, compassion between siblings, the developmental stages thought to be so unique to humans, the unique personalities of individual chimpanzees and their contributions to their community, and so on. If, for a moment, we look beyond the brilliance of the discoveries, we realize that the manner in which Jane approached her task is also extremely important. Consider the following passage on chimpanzee/human relations and compare it to the attitude toward the subjects of experiments which, too often, one finds in research. It occurs in a section of Jane's latest book where she writes about the relationship between ". . . chimpanzees and observers":

> We should also ask how disturbing it is for a chimpanzee to be followed
> through the forest by one or two humans, sometimes for days on end.
> Some chimpanzees show what appears to be a total lack of concern, of
> interest even, in close proximity of one or more humans. Others are far

more anxious. In part it depends on the behavior of the human observers: those who are insensitive to the behavior of the animal they are following, who move noisily when he is trying to listen, who make sudden movements when he is resting, who approach too closely when he is traveling, and so on, are likely to affect the behavior of their subject. . . . If a particular subject appears nervous or irritable during a follow, the observers are instructed to fall back and watch from farther away, *even if this means losing their target animal* [emphasis added]. (Goodall, 1986, p. 57)

At the heart of this methodology is a sensitivity and receptiveness to the chimpanzee as chimpanzee. It speaks of an ethological approach that has the humility in the presence of another creature to allow it to tell its story on its own terms, at its own pace, in its own time (Fig. 1). This is of great scientific importance.

Other researchers have shown this quality of receptivity in that they were willing to let their subjects teach them about themselves rather than vice versa. In her book *The Chimpanzees of Gombe* (Goodall, 1986), Jane cites contemporary colleagues such as Itani, Nishida, and Kortlandt, as well as those who went before. So Jane's findings do not seem to be earthshakingly new discoveries, as she presents them, but merely the next logical step in the progress of science. One of those who came before Jane, and who is well recognized in her book, is Wolfgang Köhler. What is interesting about Köhler is that he explicitly makes the point that Jane implicitly makes, namely that:

Lack of ambiguity in the experimental setup in the sense of an either-or has, to be sure, unfavorable as well as favorable consequences. The decisive explanations for the understanding of apes frequently arise from unforeseen kinds of behavior, for example, the use of tools by the animals in ways very different from human beings. If we arrange all conditions in such a way that, so far as possible, the ape can only show the kinds of behavior in which we are interested in advance, or else nothing essential at all, then it will be less likely that the animal does the unexpected and thus teaches the observer something. (Köhler, 1971, p. 215)

Often it is the case that when scientists ignore Köhler's warning in their approach to the study of ape behavior, we end up finding out more about the mental capacity of the scientist than about the ape he or she claims to be studying. Jane's approach to the study of ape behavior is, of course, the antithesis of the type Köhler warns us against. In Jane's own words:

Only by working with immense patience, understanding, insight, and skill with individual chimpanzees in captivity is it possible to investigate questions that relate to the upper limits of chimpanzee intellectual ability. (Goodall, 1986, p. 16)

Photo: H. van Lawick

Figure 1. The infant Flint, curious about his human observer.

Another important emphasis of the Gombe research is on the presence of behavior rather than the absence of behavior. Those influenced by Descartes have sought to find discontinuities in nature, ways to separate beast from beast, and man from nature and beast alike; to place man, and occasionally some of his personally favorite beasts, outside of nature; and to create supernatural beings out of human beings by noting the absence of those uniquely human behaviors in the beasts. Many scientists today who are still influenced by this philosophy seem to have forgotten that the absence of evidence is not the evidence of absence.

Jane discovered the nature of chimpanzee behavior rather than what chimpanzees did not have relative to humans. Her discoveries were remarkable because they demonstrated continuity in a world that had assumed discontinuity. The chimpanzees that Jane described in the Gombe forest were not Cartesian automata. With Jane's discoveries, unique human traits became interesting primate traits.

Another pair of scientists who have made discoveries about apes by being receptive and sensitive to the capabilities of the apes, R. A. and Beatrix T. Gardner, state the following in regard to the notion of discontinuity:

> Truly discontinuous, all or none phenomena are rare in nature. Histori-
> cally, the great discontinuities have turned out to be conceptual barriers
> rather than natural phenomena. They have been passed by and abandoned
> rather than broken through in the course of scientific progress. (Gardner
> and Gardner, 1983, p. 1)

When one examines Jane's historical trail, one finds it is littered with theories of
human uniqueness that have been abandoned because of her discoveries. Her discoveries have forever changed the way humans will perceive themselves in relation to
nature. We are no longer a solo violin playing alone on this planet; we are instead part
of the great orchestra of nature playing the music of the spheres.

Have I completely evaluated the influence of the Gombe research? Perhaps not,
judging by S. A. Barnett's standards. In his text entitled *Modern Ethology*, he lists
three criteria that scientific effort must satisfy:

> . . . (a) it must be logical; (b) it must conform with existing knowledge;
> and (c) it must be morally acceptable to the people of the society in
> which it is used. (Barnett, 1981, p. 503)

We have dealt with the first two criteria, that research and the scientific approach must
be (a) rational in their logical assumptions, and (b) they must be empirically sound.
The third criteria of being moral is too often left out of experimental consideration in
favor of empirical possibility. This moral quality is very much in evidence in Jane's
research.

Indeed, Jane's technique of unbiased observation epitomizes a willingness ever to
be the student ready to learn from any aspect of nature. And when one reads Jane's
research, it is as if one is in Gombe – she is a superb writer of natural history. This
ability is founded in an unbounded compassion and sensitivity for chimpanzees and
other living beings. In her own words:

> I readily admit to a high level of emotional involvement with individual
> chimpanzees – without which, I suspect, the research would have come
> to an end many years ago. (Goodall, 1986, p. 59)

Thus from Gombe comes a deeply moral message: Jane's work teaches the world
of human beings that our most important characteristic is one we share with chimpanzee beings – namely that of our "beingness" – and that "human" is merely an adjective
for that "beingness."

REFERENCES

Barnett, S. A. 1981. *Modern Ethology: The Science of Animal Behavior.* New York: Oxford University Press.

Gardner, R. A., and B. T. Gardner. 1983. Early signs of reference in children and chimpanzees. Unpublished manuscript, University of Nevada, Reno, Department of Psychology.

Goodall, J. 1986. *The Chimpanzees of Gombe: Patterns of Behavior.* Cambridge: Harvard University Press.

Köhler, W. 1971. Methods of psychological research with apes. In M. Henle, ed., *The Selected Papers of Wolfgang Köhler*, pp. 197–223. New York: Liveright.

RESEARCH AT MAHALE

Toshisada Nishida

Japanese study of African great apes in the natural environment began in 1958, when Kinji Imanishi and Junichiro Itani ("the gorilla expedition" sponsored by the Japan Monkey Center) traveled around equatorial Africa to look for a good field site for long-term study of wild gorillas. Imanishi had long been interested in the origins of animal societies. The successful study of Japanese monkeys turned his interest more toward the origin of the human family. He believed that the study of great ape societies would provide clues to the question of the origin of human societies. The gorilla study, however, could not be continued because of the political conditions in Zaire. As a result, Imanishi had to change the study target to chimpanzees.

In 1961, Imanishi organized the Kyoto University Africa Primatological Expedition (KUAPE) with the financial support of the Ministry of Education, Science and Culture (Monbusho), and he began a study of wild chimpanzees at Kabogo Point, in western Tanzania. However, chimpanzees were not seen often, and Itani moved the base camp to the Kasakati Basin in 1963. There, good socio-ecological observations could be made, and some male chimps were individually identified. However, opportunities to observe chimpanzees were not frequent enough to habituate most of the chimpanzees to humans.

Itani, who succeeded to the leadership of the expedition in 1965, decided to establish two more camps, at Filabanga and Mahale. Three methods of habituating chimpanzees were attempted: "habituation" by T. Kano at Filabanga, "impressing" by K. Izawa at Kasakati, and "provisioning" by T. Nishida at Mahale. Kano openly observed wild chimpanzees without using any incentives. Izawa obtained a male infant chimpanzee, "Brucy," in Nairobi and attempted to "show off" to wild chimps how well the human and the chimpanzee were interacting. Nishida, with the assistance of village people of Kasoje, cut a tract of elephant grass where chimpanzee groups regularly passed through, and opened a large sugarcane plantation in order to attract chimpanzees.

From March 1966, the chimpanzees of Kasoje began to visit the sugarcane plantation regularly. Consequently, three other camps (Kabogo, Kasakati, and Filabanga) were closed permanently in 1967. KUAPE continued to send researchers to Mahale until 1974, though only intermittently.

In 1975, the Japan International Cooperation Agency (JICA) began to send prima- tology experts to Mahale as an international cooperative nature conservation project. This project was based on the Tanzanian government's keen interest in proposals by Itani and Nishida that the Mahale Mountains be conserved as a reserve of fauna and flora, and chimpanzees in particular. At the same time, the Tanzanian government (Game Division) established the Kasoje Chimpanzee Research Station (KCRS), which has cooperated closely with Japanese experts. This project enabled Japanese experts to monitor two chimpanzee unit-groups continuously until early 1988. In 1979, KCRS was reorganized as one of the centers of the Serengeti Wildlife Research Institute and was newly christened "Mahale Mountains Wildlife Research Centre." In 1985, the Mahale Mountains National Park was finally established as the first "foot- walking" national park of Tanzania.

In addition, the University of Tokyo Ape Expedition to Africa led by Nishida, and funded by Monbusho, has been sending students to Mahale every two years since 1979. Researchers from the Universities of Stirling, St. Andrews, Michigan, and Dar-es-Salaam have also contributed to the development of chimpanzee studies at Mahale. Although the research work at Mahale was at first heavily "sociological," many researchers are now studying, or planning to study, various aspects of chim- panzee behavior, including postures and locomotion, vocal communication, and food ecology.

Provisioning at Mahale has never been as extensive as at Gombe, except during the earliest period. Provisioning has been drastically reduced since 1981 (and completely abandoned since 1987) for fear of poaching and the transmission of human disease to chimpanzees.

REFERENCES

Imanishi, K. 1961. The origin of human family: a primatological approach. *Japn. J. Ethnol.* 25:119–130 (in Japanese with English abstract).

Itani, J., ed. 1974. *The Chimpanzees,* 744 p. Kodansha, Tokyo (in Japanese).

SOCIAL INTERACTIONS BETWEEN RESIDENT AND IMMIGRANT FEMALE CHIMPANZEES

Toshisada Nishida

INTRODUCTION

Compared with male-male relationships, much less attention has been paid to the social relationships of adult female chimpanzees. There are several reasons for this: First, females, particularly those with offspring, rarely interact with each other, and tend to spend most of their time within their own core areas (Hasegawa, 1987; Wrangham and Smuts, 1980). Second, sociobiology has emphasized the cooperative relationships of animals, and researchers studying wild chimpanzees have shown more interest in the study of cooperative behavior among male chimpanzees (e.g., Busse, 1977; Bygott, 1974; Nishida, 1983; Riss and Busse, 1977; Riss and Goodall, 1977; Wrangham, 1975).

Authors have described some characteristics of social relationships among female chimpanzees, which appear to differ subtly, and these very nuances illustrate the difficulties involved in defining female relationships.

First, as to dominance relationships, Bygott (1974) stated that it may not be meaningful to describe female-female agonistic relationships in terms of dominance, because although females rarely interact, they do have a high frequency of two-sided fights. De Waal (1982) stated from a study of the Arnhem colony that the female hierarchy seems to be based on respect from below rather than intimidation from above and that acceptance of dominance is probably more important than imposing dominance. Goodall (1986) stated that females cannot be ranked in a clear-cut dominance hierarchy, but that there are females who are clearly very high ranking (one of whom often emerges as alpha) and others who rank very low. Nishida (1979) reported that a linear dominance hierarchy was likely to exist among frequently observed females, with newly immigrated females being in the lowest ranks, although this conclusion was based only on observations of K-group at the feeding place. Most authors admit that age is an important factor in female rankings (Bygott, 1974; Nishida, 1979; de Waal, 1982). Goodall (1986) stated that for a female's rank, the status of her relatives and offspring, as well as her health and age, is a very significant factor.

Second, as to female associations and coalitions, Nishida (1979) reported that female association is characterized as "passive aggregation," by which he meant that females meet each other rather inadvertently when they seek to accompany adult males, or when they come to food patches. A baby-sitting relationship is also a kind of passive aggregation because the baby-sitting female seeks contact with an infant rather than with its mother. Goodall (1986) emphasized the strong supportive strategy of a high-ranking aggressive female for her daughter during agonistic interactions with other females. However, de Waal (1982) documented cases of strong coalitions between unrelated females in the Arnhem colony.

Third, newly immigrated females have been seen to be attacked by resident adult females (Goodall, 1986; Nishida and Hiraiwa-Hasegawa, 1985; Pusey, 1980). Pusey (1978) suggested that females may compete for food resources, hence the establishment of core areas.

Few systematic studies have been published on social relationships among adult female chimpanzees, and this paper aims to fill an important gap regarding the social structure of chimpanzees. Special emphasis will be laid on the interactions between resident and newly immigrated females, since these interactions appear to influence the reproductive success of female chimpanzees.

METHODS

Chimpanzees of the Mahale Mountains, Tanzania, have been studied since 1965, and long-term demographic data are available. Formerly the main study group was K-group. However, adult males of K-group disappeared one by one until, in 1978, the number of adult males in the group had dropped to only two. Therefore, many cycling females began to associate frequently with M-group, a larger neighboring unit-group. By the end of 1979, all the fertile cycling females had transferred to M group, radically changing their annual range. This erosion resulted in the near extinction of K-group (Nishida et al., 1985). Since then, M-group has been our major study group.

M-group consisted of about 100 individuals in 1981 (Hiraiwa-Hasegawa et al., 1984). From 1981 to 1985, ten young adult or older adolescent females immigrated into M-group, one from K-group, and nine from other unknown groups. In addition, two females from K-group, one middle-aged and one old, accompanied by their juvenile offspring, began to interact with M-group chimpanzees. Also, at least three older adolescent females disappeared from M-group and probably transferred to other unit-groups, since they were last seen in good health.

I made three four-month studies: in August-December 1981; September 1983–January 1984; and October 1985–January 1986. In 1981, I followed only cycling or pregnant females, of which there were five recent immigrants and two residents. One of the immigrants was an older adolescent female. In both 1983–84 and 1985–86, I followed only lactating mothers, including three immigrated females. Over the three periods, a total of 844.2 hours of focal sampling was collected on 14 females (Table 1). Data on agonistic interactions between females were also collected on an ad libitum basis. By "immigrated females" I mean those females who entered

Table 1. List of focal females and hours of observation.

Name	Presumed year of birth	Year of immigration	Year of study			Total focal hours
			1981	1983	1985	
PL	1971	1981	0	0	31.9	31.9
NK	1970	1981	26.5	0	0	26.5
ND	1963	1972	0	13.8	31.1	44.9
FT	1963	—	0	32.4	29.4	61.8
GW	1962	1979	42.3	0	0	42.3
WD	1961	1981	17.2	14.1	0	31.3
SA	1960	1972	0	9.0	4.9	13.9
WO	1958	—	32.0	31.0	30.6	93.6
CH	1958	1979	61.8	52.3	30.9	145.0
WE	1956	—	0	31.1	—	31.1
SL	1955	—	0	7.6	31.6	39.2
WA	1954	1978	49.5	14.8	30.4	94.7
SO	<1953	—	53.0	39.8	32.0	124.8
BO	<1952	—	0	31.7	31.5	63.2
Totals			282.3	277.6	284.3	844.2

M-group after 1978; by "newly immigrated," those who had been in M-group for less than three years; and by "resident females," those who had been members of M-group since at least 1977.

Provisioning with sugar cane was done at the camp or in the bush in 1981; however, the amounts offered were very small, with only one or two sticks of sugar cane given to 20 to 40 chimpanzees a day (Hasegawa and Hiraiwa-Hasegawa, 1983). Almost no provisioning was done in 1983–84 and 1985–86. Observational data for the period when chimpanzees were being given sugar cane are *not* included in this paper.

RESULTS
Ranging, Association, and Grooming Patterns of Adult Females

First, the ranging, association, and grooming patterns of adult females will be briefly described. In 1981 all prime females, both targeted (CH, GW, WA, WD) and non-targeted (WS, WP, GM), who immigrated from K-group in 1979–81 had individual core areas in the northern part of M-group's range, which originally was the overlapping area of K-group's and M-group's ranges. By contrast, among targeted resident females, one (WO) occupied central, and another (SO) south-central areas of M-group's range. Although the data have yet to be analyzed, it appears that the core areas of these prime females (both ex-K-group and resident) have changed little throughout the whole study period. Only the adolescent immigrant from K-group (NK) showed no strong tendency to stay within the northern part of M-group's range. This might be due, at least partly, to her continuous cycling.

During focal sampling I recorded all the other chimpanzees within a 10-meter radius of the target female. The ten most frequent associates of each target in 1981 are shown in Figure 1. Data are given as the proportion of observation time spent in association. For female associates, ex-K-group prime females (CH, GW, WA) associated almost exclusively with each other, while resident females (WO, SO) associated almost exclusively with other residents.

Figure 1. The ten most frequent associates of targeted females in 1981. WO and SO are resident females; WA, CH, GW, and NK are newly immigrated females. Numbers under females indicate years since immigration into M-group.

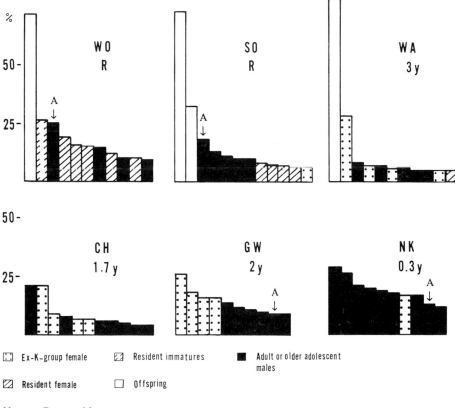

Notes: R = resident.
A = alpha male.
3y, etc. = lapse of years since immigration to M-group.

Figure 2. Grooming profiles of targeted females as percentage of each category in total time spent grooming. WO and SO are resident females; WA, CH, GW, and NK are newly immigrated females. Number above profile indicates years since immigration. Number under profile indicates year of observation.

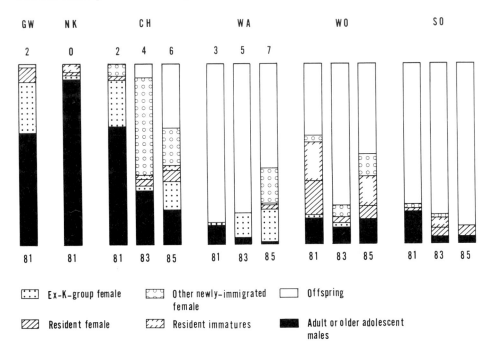

| :::: | Ex-K-group female | | Other newly-immigrated female | | Offspring |
| | Resident female | | Resident immatures | | Adult or older adolescent males |

NK displayed a rather different pattern. Her favorite companions were almost exclusively adult or adolescent males. The data also suggest that resident females preferred to associate with the alpha male (NT) rather than with other adult or adolescent males, with the exception of their own offspring.

In 1981, ex-K-group females of prime age had grooming relationships primarily with other ex-K-group females, while two resident females groomed primarily with other resident females, although two targeted females (SO, WA) were rarely seen to groom with any other females. NK rarely groomed with females, and apparently did not discriminate between ex-K-group and resident females. Thus, the data on grooming are consistent with those for ranging and association (Fig. 2).

For two ex-K-group (CH, WA) and two resident (WO, SO) females, grooming profiles in all three periods of observation are available. During these periods, females experienced childbirth, death of an offspring, or both. Overall grooming profiles differed from year to year, indicating that the number and age of offspring were important factors in determining the time spent grooming other individuals. However, in almost every year, ex-K-group females tended to groom more with each other than with the resident females, and resident females more with each other than with the ex-K-group females (Fig. 2). Thus, even six or more years after their immigration, some parous ex-K-group females still maintained friendships with each other and remained almost completely estranged from resident females.

Dominance Relationships Among Adult Females

Pant-grunt. Pant-grunting has been recognized as a useful criterion for discriminating status relations in both adult males and adult females (Bygott, 1974; de Waal, 1982; Goodall, 1986). In my samples, there were seven pairs of females in which pant-grunting and aggressive episodes were observed at separate times within the same study-period. In all seven pairs, the female who pant-grunted also displayed subordinate responses such as *scream, escape, grimace,* or *appeasement gesture* towards the other individual at other times. Although females sometimes simultaneously pant-grunt to each other (Bygott, 1974; Nishida, 1979; Goodall, 1986), I never saw it in this study. Finally, pant-grunting was always uni-directional, from subordinate to dominant.

However, pant-grunting is rare among adult females. During the 844 hours of focal observation, focal females pant-grunted to other females only 18 times and received pant-grunting from other females only 21 times (or 2.1 and 2.5 times per 100 observation hours). On the other hand, females pant-grunted to adult or older adolescent males 404 times, or 48 times per 100 hours. (No adult or older adolescent male ever pant-grunted to an adult female.) This difference is even more impressive, considering that there were more adult or older adolescent females than adult or older adolescent males (roughly 40 versus 15).

Older females scarcely ever pant-grunted to other females. Although all of the six young females pant-grunted at least once, four of the seven older females never did so. On the other hand, five of the six younger females never received pant-grunting, but six out of seven older females did so at least once.

The analysis of 49 pant-grunting events between adult females (39 focal and ten ad-lib sampling) shows that age, parity, and tenure of females are factors influencing the direction of pant-grunting between females.

Nulliparous or primiparous females tended to pant-grunt to multiparous females more often, and multiparous females tended to pant-grunt to nulliparous or primiparous females less often than expected from the numbers of each class (Table 2).

Females with shorter tenure pant-grunted 39 times to those with longer tenure, compared with only four times in the other direction. Six pant-grunts occurred between females of similar tenure.

Younger females pant-grunted to older females 40 times, compared with four times in the other direction. Five pant-grunts occurred between females of like age.

When females with longer tenure pant-grunted to those with shorter tenure, the latter were older in two of the four cases. In the other two cases, the recipients apparently were more aggressive, showing male-like displays, although they were of similar age or younger. When older females pant-grunted to younger ones, the former were newly immigrated. Moreover, when older females pant-grunted to younger females with similar tenure, the latter were found to show more male-like displays.

Toshisada Nishida

Table 2. Pant-grunting or greeting and parity.

		Greeted		
		Nulliparous or primiparous	Multiparous	Total
Greeter	Nulliparous or primiparous	4 (5.2)	25 (13.1)	29
	Multiparous	3 (13.1)	17 (17.5)	20
	Totals	7	42	49

Note: Figures in parentheses are expected values of pant-grunting by each parity class calculated from the ratio of habituated females of each parity class (3 nulliparous or primiparous: 5 multiparous), based on the assumption that each parity class should display an equal amount of pant-grunting to each other. Since the sampling of data was not always systematic, the results should be regarded as indicating only a general tendency.

Table 3. Frequency of agonistic episodes.

	Overall frequency	Frequency per 100 hrs
1981	17	6.0
1983	9	3.2
1985	13	4.6
Total	39	4.6

Thus, younger, nulliparous, and short-tenured females tended to pant-grunt to older, multiparous, and longer-tenured females. Since parity and tenure usually parallel the age of females, age is likely to be the single most important determinant of female dominance, but personality (such as aggressive competitiveness or self-assurance) and longer tenure can, in a few cases, modify the pattern of age-dependent dominance. Bygott (1974) stated that in relationships which showed evidence of dominance, it was generally the older individual who was dominant. De Waal (1982)

74

Table 4. Aggression among parity classes.

		Aggressed		
		Nulliparous or primiparous	Multiparous	Total
Aggressor	Nulliparous or primiparous	10 (5.5)	2 (13.7)	12
	Multiparous	31 (13.7)	8 (18.2)	39
	Totals	41	10	51

Note: Figures in parentheses are expected values of aggression by each parity class calculated from the ratio of habituated females of each parity class (3 nulliparous or primiparous: 5 multiparous), based on the assumption that each parity class should display an equal amount of aggression to each other.

reported that personality and age seem to be the determining factors of female dominance in his captive study population.

Conflict among adult females. During the three periods, focal females were observed to attack or threaten or to be attacked or threatened by other adult females only 39 times, or 4.6 times per 100 hours. Agonistic episodes were observed most frequently in 1981, when five of seven focal individuals were newly immigrated, cycling females (Table 3). In addition, non-focal females attacked other non-focal females 17 times. Five of 56 conflicts were communal attacks, in which two or three females chased one female simultaneously. Thus, 51 conflicts began as dyadic interactions.

Multiparous females tended to be aggressive towards nulliparous or primiparous females more often than expected relative to the numbers in each class (Table 4). This is not surprising given that multiparous females usually are dominant to nulliparous females, as was shown above.

However, it should also be noted that nulliparous or primiparous females tended to be aggressive towards each other more often than expected, and that multiparous females were aggressive towards each other less often than expected. This suggests that females tend to compete with those of like age frequently when they are young, but that they avoid aggression with each other when they become older. The adaptive significance of this may be that young females are in the process of establishing their own core areas, and, therefore, that their interests are more likely to be competitive.

Females with longer tenure were aggressive towards those with shorter tenure 39 times, but only six times was the opposite observed. Moreover, aggression occurred between females of like tenure on six occasions.

Table 5. Frequency and distribution of dominance interactions.

Year	Number of agonistic interactions observed	Number of focal females*	Number of possible dyads	Number of dyads with agonistic interactions observed (%)
1981	17	7	21	5 (23.8)
1983	9	9	36	6 (16.7)
1985	13	9	36	7 (19.4)

* With observation of over 10 hours.

Table 6. Dominance matrix among 13 focal females.

						Agonism received							
Agonism by	WO	SL	SO	WE	BO	FT	WA	CH	GW	SA	ND	PL	NK
WO		O			O				Δ	O	O	(Δ)	Δ
SL			Δ	Δ		Δ		Δ		O		Δ	Δ
SO				O								O	Δ
WE						O	O	O		O	O	O	
BO								O					
FT							O				O	O	(Δ)
WA										O		O	Δ
CH									O	Δ			
GW							O				(Δ)	Δ	(Δ)
SA											Δ		
ND												(Δ)	Δ
PL													Δ
NK													

O = Column females pant-grunt to row females.
Δ = Column females are attacked or threatened by row females.
(Δ) = Column females pant-grunt to, and are attacked or threatened by, row females.

Older females were aggressive towards the younger females 42 times, but only seven times was the opposite observed. Aggression occurred twice between females of like age.

"Anomalous" cases mentioned above (shorter-tenured or younger females aggressive towards longer-tenured or older ones) are classified into three categories: aggressors were older but shorter-tenured females on four occasions; younger but longer-tenured on four occasions (three times, adolescent resident females were aggressive towards newly immigrated females); and shorter-tenured and younger on two occasions. In one of the two most anomalous cases, both of the participants were newly immigrated females: PL aggressed to GW, but ran away because of GW's retaliation. Another case involved two resident females (SA and SO), which will be described below (see "Intervention in conflicts among adult or adolescent females by adult or adolescent males").

Thus, older, multiparous females with longer tenure tended to be aggressive towards younger, nulliparous or primiparous females with shorter tenure. This is consistent with the results of pant-grunting.

Dominance hierarchy. There were seven to nine targeted females in each study period, which means 21 to 36 possible combinations (dyads) of two females. However, dominance interactions were observed in only one-fifth of the combinations in each year (Table 5). As will be seen below, there was only one possible case for dominance reversal between periods, and so data from the three periods were pooled. Among 13 focal females, agonistic interactions including pant-grunting were seen at least once in 40 dyads (namely 51% of the 78 possible combinations), including data collected by T. Hasegawa and M. Huffman (Table 6).

According to this overall dominance matrix combining all measures, there was no evidence contradictory to the assumption of a linear dominance hierarchy among focal females, except the possible case of GW and CH. Second, a prime female (WO) was the only female who neither pant-grunted to, nor was attacked by, any other female, so she appeared to be the alpha female. She was dominant even to another prime female (SO) who had adult and adolescent sons. Third, the older generation dominated the younger: all females that were born in the 1950s were dominant to those born in the 1960s or later.

There were about 40 adult or older adolescent females in M-group. This means that there were 780 possible combinations of two females. However, dominance interactions were observed for only 83 dyads (11%), even if all the available data both from focal and ad-lib sampling are incorporated (Table 7). This low number of cases so far makes it impossible to provide an overall picture of dominance relationships among females. For example, WO (who seems to have been "the alpha female" in my sample) was never observed to have agonistic interactions with the six oldest females of M-group. So it cannot now be said who, if anyone, was the alpha female of the entire group.

Table 7a. Dominance matrix in 1981.

Name	Year of immi- gration	Presumed birth year	NG	GP	WS	AD	SO	WA	SL	WE	WO	CH	SA	WD	GK	GW	WP	FT	ND	IK	GM	TY	MR	NK	PL	MM	PA
NG	r	1930 s												*													
GP	r																							+			
WS	1979	Late 1940 s																									
AD	r																										
SO	r																							+			
WA	1978	Early			+										**			#		#				**	**		
SL	r	1950 s																					+	+	*		
WE	r						*											*									
WO	r	Late										+	**		+					*			*		*+	*	
CH	1979	1950 s													*										+		
SA	1972																										
WD	1981																										
GK	r																										
GW	1979	Early																			+			*+#		*	
WP	1979	1960 s																			+						
FT	r																										
ND	1972																		+						*++			
IK	r	Late																									
GM	1979	1960 s																							+			
TY	n																										
MR	n																								+			
NK	1981	Early																										
PL	1981	1970 s													#										+			
MM	n																										
PA	n	1975																							+			

Notes:
- r = resident at least since 1973–74.
- n = probably natal in M-group.
- * = column females pant-grunt to row females (nine of the pant-grunts were recorded by T. Hasegawa).
- + = column females display submissive responses to row females.
- # = two-sided fights.

Table 7b. Dominance matrix in 1983.

Name	Year of immigration	Presumed birth year	NG	BO	SO	WA	SL	WE	WO	CH	SA	GW	WP	FT	ND	IK	NK	PL	TM	PU	OP	VL	PK	NP
NG	r	1930 s								*														
BO	r	Late 1940 s								*														
SO	r							*																+
WA	1978	Early																						
SL	r	1950 s			+						*					*								
WE	r	Late				*				**	*				*				***					+
WO	r																+	+	*	*			+	
CH	1979	1950 s										*								*++			+	
SA	1972																							
GW	1979	Early								*				*						*				
WP	1979	1960 s																						
FT	r														*		*+	*						
ND	1972																				*			
IK	r	Late 1960 s																						
NK	1981	1970																						
PL	1981																							+
TM	1982																							
PU	1982																							
OP	1983	Early																						
VL	1983	1970 s																						
PK	1983																							
NP	1981																							

Notes: r = resident at least since 1973–74.

* = column females pant-grunt to row females.

\+ = column females display submissive responses to row females.

Table 7c. Dominance matrix in 1985.

Name	Year of immi-gration	Presumed birth year	NG	BO	SO	WA	SL	WO	CH	SA	GW	FT	ND	NK	PL	TM	PU	OP	VL	PK	PT	LS
NG	r	1930 s							*				+							*+		
BO	r	Late 1940 s						*														
SO	r																					
WA	1978	Early 1950 s																				*
SL	r							*+							*		*					+
WO	r	Late 1950 s	*									+			+		*+			++		+
CH	1979									*	*#				*++			*	+	*		
SA	1972				+										+	*				+		
GW	1979	Early 1960 s											*+			+				*		
FT	r													**	*++							*+
ND	1972																				+	
NK	1981	1970																				
PL	1981																					*
TM	1982																					
PU	1982																					
OP	1983	Early 1970 s																				+
VL	1983																					
PK	1983																					
PT	1984																					
LS	n	1975																				

Notes:
r = resident at least since 1973–74.
n = probably natal in M-group.
* = column females pant-grunt to row females (five pant-grunts were recorded by M. Huffman).
+ = column females display submissive responses to row females.
# = two-sided fights (five agonisms were recorded by H. Hayaki).

However, the following conclusions could be drawn from Table 7:

1. Agonism among females was concentrated in the youngest age class, while old females rarely were aggressive towards each other.

2. Young, newly immigrated females occupied the lowest ranks in the female hierarchy.

3. Relative dominance in particular dyads was quite stable for several years; in nine out of ten dyads whose dominance interactions were observed in separate study periods, the relative dominance did not change. In the only exception, a sterile ex-K-group female (GW) received pant-grunting from another ex-K-group female (CH) in 1983; CH had been dominant to GW in 1981, and then CH gave birth in 1983. However, in 1985 the ranks again reversed, and the original dominance rank order was restored.

4. There is no indication that females fall to very low rank in advanced age: very old females have not been seen to be attacked or threatened by young or prime females.

Intervention in Conflicts by a Third Party

Process of conflict among females. Protection of offspring and competition over food were two main sources of female conflicts. Similar conclusions were drawn by Goodall (1986). Other causes of aggression included competition over mates or over infants for handling, reunion, re-direction of aggression, responses to pant-grunting, etc. (Table 8). However, the causes of female aggression were often not apparent to human observers.

In 32 out of 51 (63%) of the episodes observed among females, the female who was aggressed against ran away screaming, and the conflict soon ended without intervention by a third party (Fig. 3).

Table 8. Causes of aggression among females.

Protection of offspring	14
Competition for food	14
Competition for mate	3
Reunion	2
Join other's aggression	1
Response to pant-grunt	1
Competition for play-mothering	1
Response to pant-hoot	1
Redirection	1
Not reciprocating grooming	1
Response to tickling	1
Reason unknown	16
Total	56

Toshisada Nishida

Figure 3. Process of female conflict.

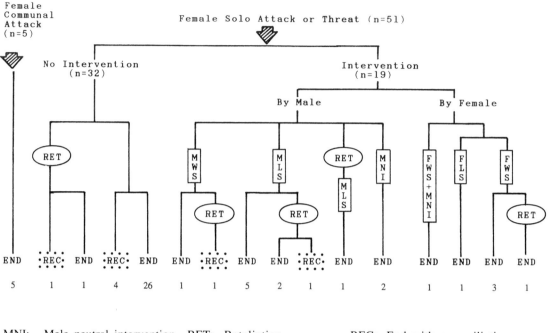

MNI: Male neutral intervention   RET: Retaliation   REC: End with reconciliation
MWS: Male winner support   FWS: Female winner support END: End without reconciliation
MLS: Male loser support   FLS: Female loser support

Retaliation. Retaliation by the subordinate female occurred once before, and five times after, intervention by a third party. Moreover, retaliation occurred twice without intervention by a third party (Fig. 3). Thus, the rate of two-sided fights was 15.7% (8/51). In addition, prolonged vocal protest (such as exaggerated screaming) frequently occurred, and this attracted distant adult or older adolescent males to the scene, causing them to intervene between the females. In the following episode, abbreviations are used to denote ex-K-group females and the year of their transfer to M-group: for example, (ex-K, 79) means that the female immigrated from K-group in 1979.

Episode 1: Dec. 31, 1981. 11:55, GM (ex-K, 79) chased and threatened NK (ex-K, 81) by stamping and shaking a large bough. CH (ex-K, 79) and an adult male (MU) came to the scene. Seeing this, NK retaliated against GM, screaming. MU charged at GM, who ran away. After this event, NK began to follow MU.

Reconciliation. After the conflict one of the parties concerned approached and touched the other, who usually reciprocated the action. Sometimes they then engaged in prolonged grooming. Reconciliation occurred after seven (or 13.7%) of 51 episodes, five times without the intervention of a third party, and twice after intervention by an adult male.

Intervention in conflicts among adult females by a third female. On six occasions, an adult female (twice with another adult female, once with an adult male) intervened in conflicts between two adult females. Five times the third party supported the winner and chased after the loser (i.e., winner-support). In all these cases losers were newly immigrated females, and the aggressor and supporter(s) were resident females or immigrants with longer tenure. Since special relationships were not apparent between the aggressor and supporter(s), the cooperation is likely to have been mostly opportunistic.

Episode 2: Dec. 20, 1983. 15:01, NK (ex-K, 81) pant-grunted to a resident female (FT), who immediately responded by attacking NK. ND (ex-K, 72) came and combined forces in chasing NK.

On the other hand, ex-K-group females sided with other immigrant females, who were being aggressed by resident females.

Episode 3: Sept. 12, 1981. 13:50, a nulliparous resident female (IK) chased a nulliparous immigrant female WP (ex-K, 79) in competition for anting sites in a tree, near which there were adult males. WP ran away screaming, whereupon GW (ex-K, 79), who was under the tree, barked at the resident aggressor, apparently supporting her long-term associate.

Communal attack by females. Communal attacks on a resident female by ex-K-group females were observed twice. First, two ex-K-group females (WS and GW) violently chased a middle-aged resident female communally. Second, when three ex-K-group females (CH, GW, and WP) were feeding on the fruits of *Pseudospondias*, a resident mother accompanied by her infant and juvenile offspring began to climb into the tree. Seeing this, the three females rushed to the scene almost simultaneously and chased the mother and her offspring, who climbed to the ground and fled.

On three occasions two or three resident females communally chased immigrated females: a newly immigrated female (WD) was subjected to a gang attack by adult males and at least three adult females (Nishida and Hiraiwa-Hasegawa, 1985). In the same year CH was chased by two resident females (WE and FT). In 1985 PL was chased by two resident females (SA and WO).

Intervention in conflicts between newly immigrated females and resident adolescents by resident females. On two occasions, resident mothers (GP and WO) intervened in conflicts between their juvenile daughters (WC, PA) and newly immigrated females (NK, PL) and supported their offspring. A resident female (SO) supported an adolescent male (TW) who was attacking a newly immigrated female (NK).

In addition, after a resident mother (WO) attacked a newly immigrated female (PK), her eight-year-old daughter (PA) violently chased the latter, supporting her mother.

Thus, what little evidence is available suggests that newly immigrated females are vulnerable to concerted attacks by resident females or immigrated females with longer tenure, and that immigrated females who had associated for a long time in their original group sometimes combine forces in attacking resident females.

As illustrated above, mother and adolescent daughters occasionally were observed to combine forces against other females. However, effects of mother-adult daughter association or female sibling association on female conflicts appear limited, at least in M-group. There are only two known cases where mature daughters did not transfer but remained to give birth in the natal M-group (Hiraiwa-Hasegawa et al., 1984). One is TY, the daughter of BO, one of the focal females. No communal attack by them was observed. There is only one known case of co-residence of two adult sisters in the same unit-group: NK and SA were sisters born to WN of K-group. SA transferred to M-group in 1972 and NK in 1981. When NK was born in 1970, SA was still in K-group and had opportunities to interact with NK. However, SA was never observed to help NK, who was attacked frequently by other females. On the contrary, SA was one of the females who had the least association with NK throughout the study period. It was likely that SA did not remember her.

Intervention in conflicts among adult or adolescent females by adult or adolescent males. On 14 occasions, adult or adolescent males intervened in conflicts between adult females. On nine occasions, adult males clearly supported aggressed or losing females (i.e., loser-support). In all of these cases, the supported females were newly immigrated females, and the aggressors were resident or immigrated females with longer tenure.

Episode 4: Dec. 15, 1981. 12:39, ND (a resident female) bit NK (ex-K, 81). NK screamed. DE (an adult male) ran to and sat between them. ND fled and climbed a tree. 12:45, ND climbed to the ground and pant-grunted to DE.

Episode 5: Dec. 13, 1983. 12:44, NK, in estrus, approached BB (an infant male). Seeing this, WO (BB's mother) chased NK to the end of a branch and shoved her to the ground. Hearing NK's screaming, DE (adult male) appeared on the scene. 12:45, NK approached WO, emitting "protest screams." WO shoved her off again. At that moment, DE climbed the tree and bit WO, who screamed. Owing to WO's resistance, DE fell to the branch below. 12:55, WO pant-grunted to DE. 12:57, WO approached DE and reached towards DE. 13:00, WO groomed NK one-sidedly for 4.2 minutes. 13:08, WO and DE engaged in mutual grooming for over six minutes and then DE groomed WO for two minutes.

Adult or adolescent males supported the aggressors twice. Once an adolescent male (LL) supported his mother (WA); and once an adult male (DE) supported a resident female (WO) against a newly immigrated female (PK). This was the only case in which an adult male supported a resident female against a newly immigrated female.

On three occasions it was not clear which side an adult male supported ("male neutral intervention" in Fig. 3). In the first case, no sooner had a one-year-old male (CP) left his mother CH (ex-K, 79) to recover a half-eaten fruit on the ground than GW (ex-K, 79) picked him up with one hand and headed into the bushes with him on her belly. Seeing this, CH followed GW, screaming. GW also screamed. An adult male (BA) ran to the scene, and both females screamed. BA stood and rocked bipedally, with hair erect. CH eventually recovered her son. In the second case, a weaning infant (SV) fell into a violent temper tantrum in response to her mother's (SL) rejection of her attempt to suckle, and SV, while screaming, lunged into an innocent female (FT) who happened to be nearby in the same tree. Perhaps misunderstanding her infant's situation, SL charged towards FT, who ran and climbed to the ground screaming. The alpha male (NT) then rushed to the scene, with hair erect, and stood still in front of them without attacking either of the females. The quarrel ended.

In addition, a subordinate female (SA), who was distressed by an attack on her adolescent son (NS) by the dominant female (SO) and her family, availed herself of a high-ranking adult male (KI) who inadvertently approached while giving a charging display nearby. Then SA chased after SO, while screaming.

In summary, adult males show a strong tendency for loser-support or "reactor-alliance" (de Waal, 1977), thus supporting weaker, newly immigrated females against females with a stronger footing in the group. This sharply contrasts with the supportive strategy of adult females who favored resident females against weaker, newly immigrated females.

Male Aggression Towards Immigrated Females

This paper does not deal with male-female relationships systematically. However, some remarks are useful in considering the possible factors having influences on female relationships.

Adult males tend to attack strange females who are anestrous or not fully estrous, although other males sometimes intervene in and prohibit such attacks (Nishida and Hiraiwa-Hasegawa, 1985). Moreover, once during my focal observation and four times outside my study period between 1979 and 1985, five mothers with newborn babies were attacked by adult males, and the babies were killed and four of them cannibalized. Also, injuries were observed on two mothers on the first day after they lost newborn babies (Hiraiwa-Hasegawa et al., 1984; Kawanaka, 1981; Masui, 1986; Nishida and Kawanaka, 1985; Norikoshi, 1982; Takahata, 1985). In all of the cases, the mothers were immigrants from other unit-groups, and they were ranging mostly in the peripheral parts of the M-group range for at least a few months before and after delivery. The ranging and association patterns of mothers during the critical period may be a key factor in determining the survival of infants born to immigrated females (Nishida and Kawanaka, 1985).

DISCUSSION

Although social hierarchy among females has not been elucidated in the study group as a whole, it has been shown that clear-cut social rank appears to exist among at least some habituated females. Undoubtedly female rank often is more difficult to define than male rank, as many authors assert, but this is not to say that female rank does not exist. There appears to be a propensity for females of a unit-group to be organized into a linear dominance hierarchy mainly based on age. However, the female hierarchy may not always be clearly established, especially when there are many females, as in M-group. One reason for this is the fact that females, particularly young and/or newly immigrated females, tend to solicit interventions by adult males when they come into conflict with other females. Moreover, the dominance relationships among older, multiparous females are very stable (see below) and they do not display aggression to each other often enough for human observers to detect their relative rank.

Perhaps, unlike males, females are less predisposed to risk severe physical battle, which might expose their young offspring to danger. For resident females who are well settled in their core areas, it would be useless and harmful to quarrel with each other. However, newly immigrated females may threaten to deplete the resources of the resident females. Therefore, resident females may attack new immigrants when their aggression does not incur great cost (e.g., when an aggressor is very high-ranking, when an immigrant is young and small-sized and not protected by adult males). Aid in combat may help lower the risk of aggression, but since aggression itself is not frequent, the female coalition is often only opportunistic. Long-term female coalitions documented in the Arnhem colony may be explained by the constant availability of allies in the captive conditions.

What characterizes female relationships is the stability of social rankings over a long period. De Waal (1982) stated that, compared with the male hierarchy, the female hierarchy has been stable for years. Unlike males, whose reproductive success depends upon social status (Hasegawa and Hiraiwa-Hasegawa, 1983; Nishida, 1983; Tutin, 1979; de Waal, 1982), female reproductive success may depend primarily upon acquiring a core area near the center of the unit-group's territory. Therefore, females who have acquired their own core areas have no pressing reason to strive for higher rank. Thus, a female's rank will be more or less fixed sometime after her immigration. Thereafter, her promotion in rank will be caused mainly by the death of senior high-ranking females and by the addition of younger low-ranking females to the hierarchy. Hence "gerontocracy" is a general rule in the female rank.

Does the aggression of resident females towards newly immigrated females have any influence on the reproductive success of the latter? Female conflicts were observed on average less than once every 20 hours. This was partly due to observational bias: if more young immigrant females had been chosen as focal subjects, I would have seen aggressive episodes more often. When I followed an adolescent immigrant (NK) in 1981, she was subjected to female aggression as many as eight times per 20 hours.

Although more data are needed, I speculate that aggression by resident females towards immigrant females has at least indirect effects upon the reproductive success of the latter: aggression by resident females will make it difficult for new immigrants to settle into the established core areas of resident females, and the immigrants will be forced to retreat to more peripheral areas of the unit-group's range. Thus, they will

Table 9. Reproductive profiles of immigrated females as of 1986.

Reproductive status at immigration	Total tenure in years	Total no. of births	Total no. of surviving offspring	Infanticide observed	Infanticide strongly suspected
Adolescent (n = 10)	61 (3-14)	14 (0-5)	7 (0-3)	2 (0-1)	0
*Adult Nulliparous (n = 3)	20 (6-7)	2 (0-2)	0	0	0
*Prime Multiparous (n = 4)	25 (3-8)	12 (1-5)	2 (0-1)	3 (0-1)	2 (0-1)
*Old Multiparous (n = 2)	10 (3-7)	0	–	–	–

Notes: * All of the fully mature immigrant females came from K-group.
Figures in parentheses indicate the range.

incur a greater risk of infanticide by resident adult males who suspect the paternity of the infants born to these females, as well as a greater risk of inter-group aggression.

At any rate, the reproductive success of immigrated females appears to be extremely low, and abuse by resident males of infants born to immigrant females is a prominent cause of infantile death (Table 9).

Since adult males intervene in conflicts between females, and almost always support weaker, newly immigrated females who gradually become higher-ranking as they grow older, these immigrants will eventually manage to establish themselves around the center of the group's range. However, this process may require many years of social contact with the adult males. Perhaps prolonged years of adolescent sterility and the longer maximal tumescence phase characteristic of adolescent females may be an adaptation to this stressful period in the life cycle of female chimpanzees.

ACKNOWLEDGMENTS

I should like to express my deepest gratitude to the Director-General of the Tanzania National Scientific Research Council and the Coordinator of the Serengeti Wildlife Research Institute for permission to do research; the Acting Director of the Mahale Mountains Wildlife Research Centre and his staff for their cooperation in the field work; M. Bunengwa, M. Hamishi, R. Kijanga, and R. Nyundo for their assistance in the field work; T. Hasegawa, H. Hayaki, and M. Huffman for permission to use their unpublished data; M. Hasegawa, T. Hasegawa, H. Hayaki, M. Huffman, K. Kawanaka, Y. Takahata, H. Takasaki, and S. Uehara for their cooperation in the field work; D. Hill and M. Huffman for comments and English revision of earlier drafts; and Y. Endo for preparing figures. I am extremely grateful to W. C. McGrew for his critical comments and English revision of the final draft. The original paper was read at the symposium "Understanding Chimpanzees," held at the Chicago Academy of Sciences on November 7–9, 1986. I thank J. Goodall and P. Heltne for their invitation to the symposium. The study is one of the results of the University of Tokyo Ape Expedition to Africa, financed by the Grants-in-Aid for the Overseas Scientific Research of the Ministry of Education, Science and Culture, Japan (#56041018, 57043014, 58041025, 59043022, 60041020, and 61043017 to T. Nishida).

REFERENCES

Busse, C. D. 1977. Do chimpanzees hunt cooperatively? *Amer. Natur.* 112:767–770.

Bygott, D. 1974. Agonistic behaviour and dominance in wild chimpanzees. Ph.D. diss., Cambridge University.

Goodall, J. 1986. *The Chimpanzees of Gombe.* Cambridge, Massachusetts: Harvard University Press.

Hasegawa, T. 1987. Sexual behavior of wild chimpanzees in the Mahale Mountains, Tanzania. Ph.D. diss., The University of Tokyo.

Hasegawa, T., and M. Hiraiwa-Hasegawa. 1983. Opportunistic and restrictive matings among wild chimpanzees in the Mahale Mountains, Tanzania. *J. Ethol.* 1:75–85.

Hiraiwa-Hasegawa, M., T. Hasegawa, and T. Nishida. 1984. Demographic study of a large-sized unit-group of chimpanzees in the Mahale Mountains, Tanzania: a preliminary report. *Primates* 25:401–413.

Kawanaka, K. 1981. Infanticide and cannibalism in chimpanzees with special reference to the newly observed case in the Mahale Mountains. *Afr. Stud. Monogr.* 1:69–99.

Masui, K. 1986. An infant chimpanzee was captured and lethally injured by the first-ranking male: the seventh case of infant-killing at Mahale Mountains National Park, west Tanzania. Mahale Mountains Chimpanzee Research Project. *Ecological Report* No. 46.

Nishida, T. 1979. The social structure of chimpanzees of the Mahale Mountains. In D. A. Hamburg and E. R. McCown, eds., *The Great Apes,* pp. 73–122. Menlo Park, California: Benjamin/Cummings.

———. 1983. Alpha status and agonistic alliance in wild chimpanzees *(Pan troglodytes schweinfurthii). Primates* 24:318–336.

Nishida, T., and M. Hiraiwa-Hasegawa. 1985. Responses to a stranger mother-son pair in the wild chimpanzee: a case report. *Primates* 26:1–13.

Nishida, T., and K. Kawanaka. 1985. Within-group cannibalism by adult male chimpanzees. *Primates* 26:274–284.

Nishida, T., M. Hiraiwa-Hasegawa, T. Hasegawa, and Y. Takahata. 1985. Group extinction and female transfer in wild chimpanzees in the Mahale National Park, Tanzania. *Z. Tierpsychol.* 67:284–301.

Norikoshi, K. 1982. One observed case of cannibalism among wild chimpanzees of the Mahale Mountains. *Primates* 23:66–74.

Pusey, A. 1978. The physical and social development of wild adolescent chimpanzee. Ph.D. diss., Stanford University.

———. 1980. Inbreeding avoidance in chimpanzees. *Anim. Behav.* 28:543–552.

Riss, D. C., and C. Busse. 1977. Fifty day observation of a free-ranging adult male chimpanzee. *Folia Primatol.* 28:283–297.

Riss, D. C., and J. Goodall. 1977. The recent rise to the alpha rank in a population of free-living chimpanzees. *Folia Primatol.* 27:134–151.

Takahata, Y. 1985. Adult male chimpanzees kill and eat a male newborn infant: newly observed intragroup infanticide and cannibalism in Mahale National Park, Tanzania. *Folia Primatol.* 44:121–228.

Tutin, C. E. G. 1979. Mating patterns and reproductive strategies in a community of wild chimpanzees (*Pan troglodytes schweinfurthii*). *Behav. Ecol. Sociobiol.* 6:29–38.

de Waal, F. B. M. 1977. The organization of agonistic relations within two captive groups of java-monkeys *(Macaca fascicularis)*. *Z. Tierpsychol.* 44:225–282.

———. 1982. *Chimpanzee Politics.* New York: Harper & Row.

Wrangham, R. W. 1975. The behavioural ecology of chimpanzees in Gombe National Park, Tanzania. Ph.D. diss., Cambridge University.

Wrangham, R. W., and B. B. Smuts. 1980. Sex differences in behavioural ecology of chimpanzees in Gombe National Park, Tanzania. *J. Reprod. Fert. (Suppl.)* 28:13–31.

SEXUAL BEHAVIOR
OF IMMIGRANT AND RESIDENT
FEMALE CHIMPANZEES AT MAHALE

Toshikazu Hasegawa

INTRODUCTION

Probably one of the most important findings in recent research on wild chimpanzees in Mahale is the "within-group infanticide" (Takahata, 1985; Nishida and Kawanaka, 1985; Hiraiwa-Hasegawa, 1987); that is, the killing of conspecific infants who were born in the same unit-group in which the killing occurs. Infanticide represents the dark side of chimpanzee society, but it is also a key to understanding the social structure of chimpanzees.

In Mahale, five cases of "within-group infanticide" have been observed and two other cases have been suspected on the basis of strong circumstantial evidence since 1977 (Kawanaka, 1981; Takahata, 1985; Nishida and Kawanaka, 1985; Norikoshi, Masui, pers. comm.). In her recent review, Hiraiwa-Hasegawa (1987, 1988) pointed out three major characteristics of "within-group infanticide" in Mahale:

1. The killers were adult males, which is different from cases in Gombe where all the "within-group infanticides" were committed by females (Goodall, 1977).

2. The killing is biased according to sex of victim. In all cases, only male infants were killed.

3. The mother of each victim was an immigrant female.

The functional explanation of the "within-group infanticide" in Mahale is thought to be related to severe antagonisms between males of different unit-groups (Nishida and Kawanaka, 1985; Hiraiwa-Hasegawa, 1987).

So the question is: How do the immigrant females, whose coming infants will, very likely, be killed, have sexual relations with males of the new group? Are there any differences in sexual interactions between the immigrant and the resident females? So far, little has been mentioned about detailed features of sexual behavior of the immigrant females in Mahale (however, see Hasegawa and Hiraiwa-Hasegawa, 1983; Takasaki, 1985). In this paper, I would like to compare some aspects of sexual behavior between the resident and the immigrant females.

STUDY GROUP AND METHODS

Since 1978, many cases of female transfer have been observed in Mahale. An especially notable series of transfers occurred during the periods from 1980 to 1982 while I was in Mahale, when all the receptive females of a smaller group (K-group) immigrated one by one into a bigger neighboring group (M-group). In our earlier paper, we called this phenomenon "a large-scale female transfer" (Nishida et al., 1985). This rather unusual transfer was thought to be the result of a sharp decrease in the number of adult males in K-group which might have been caused by intergroup fighting among the males. As I had been conducting research on chimpanzees of M-group, I was able to observe the sexual behavior of nine immigrant females from K-group.

M-group consisted of about 100 chimpanzees during periods from 1980 to 1982 (Hiraiwa-Hasegawa et al., 1984), including 27 receptive females who were observed to copulate at least once. In this paper, I have classified these receptive females into two life history categories: (1) Resident females (n = 17) who have been observed with M-group males up to and including 1977; (2) Immigrant females (n = 10) who have been observed with M-group males only since 1978. Except for one adolescent female (PL), all these immigrant females were from K-group. As PL was observed to copulate only once and did not become pregnant during the study periods, her data were excluded from this paper. Females were also categorized by age classes: a younger female refers to a female younger than or equal to 15 years old, and an older female refers to a female equal to or older than 16 years of age.

In the M-group there were 26 males who were observed to copulate at least once. Among them, 12 were adult males and 14 were immature males. Adult males were divided into three age groups: young (ca. 14 to 19 years old, n = 4), middle (ca. 20 to 30 years old, n = 5), and old (over 30 years old, n = 3); and they were divided into four rank classes: alpha (n = 1), high (n = 3), middle (n − 4), and low (n = 4). Immature males included infants (from birth to weaning age, five years old), juveniles (weaning to the beginning of puberty, five to eight years old), and adolescents (puberty until almost full-size body, nine to 13 years old).

Sequences of females' swelling were checked by the staff of the Mahale Mountains Wildlife Research Centre, and were recorded on daily census sheets with three categories: full, one-half, and no swelling. My definition of terms describing mating pattern and behavioral pattern followed Tutin (1979). A copulation was defined as an interaction between a female and a male which included at least one intromission. Some copulations included multiple intromissions, and, to be separate copulations, intromissions had to be separated by at least ten minutes. I employed the ad-libitum sampling method in the course of general survey. However, I tried to record all the copulations which I witnessed.

PROCESS OF TRANSFER

Before stating results on sexual behavior, I will briefly describe some features of female transfer. Figure 1 shows the names of subject immigrant females and their

Figure 1. Process of female transfer from K-group to M-group between 1978 and 1982. Data were collected primarily by Uehara, Kawanaka, Nishida, Takahata, Hiraiwa-Hasegawa, and the author.

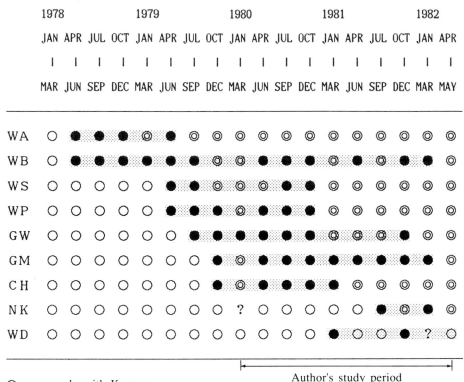

○ : seen only with K-group
◎ : seen only with M-group
● : seen with both K- and M-group
⋯⋯ : transitional periods

Author's study period

process of transfer from K-group to M-group from 1978 to 1982. An open circle indicates the period when each female was found only with K-group chimpanzees. A double circle shows the period when she was only with M-group chimps. A closed circle represents the period when she was seen with both K-group and M-group chimpanzees. As seen straight away, it takes some time for each female to immigrate into another group, although the length of time differs to some extent from individual to individual. Transitional periods last from six months to more than two years. This is in striking contrast to the cases of gorillas in which female immigration is completed within a few minutes (Harcourt, 1978). During my study period, all except WA (who transferred earliest) were in the stage of transition. In Figure 2, chimpanzee associations during 1980–81 are expressed in a maximum spanning tree. Clearly, the immigrant females are located between K-group and M-group. In general, immigrating females were more often observed in the new group (M-group) when they were in

Figure 2. Maximum spanning tree of chimpanzee association in M-group (1980–81).

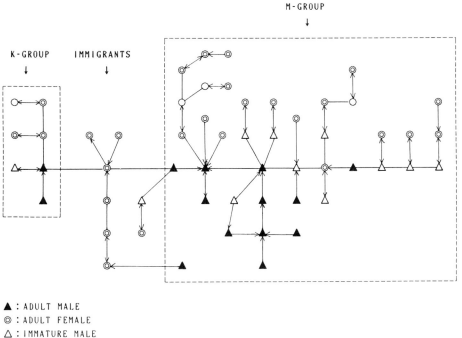

▲ : ADULT MALE
◎ : ADULT FEMALE
△ : IMMATURE MALE
○ : IMMATURE FEMALE

estrus than when they were in a non-receptive period (Nishida et al., 1985). Also, the immigrant females had a tendency to return to the K group during the period of pregnancy or in seasons when fruits abounded in the K-group range. Social conflict between immigrants and residents is discussed in detail by Nishida (this volume).

COMPARISON OF SEXUAL BEHAVIOR
BETWEEN IMMIGRANT AND RESIDENT FEMALES
The Sexual Cycle

The most accurate way to assess the length of sexual cycle is to measure an interval of successive menstruations. In field study, however, menses is not always detectable. Therefore, female cycles were estimated by the intervals between swellings. In this study, I chose the first day of maximal tumescence as the point to measure the length of a cycle. Median length of 23 sexual cycles for eight older females was 31.5 days. There was no significant difference between immigrant (median: 31.5 days, n = 17) and resident females (median: 33.3 days, n = 6) (U-test; U = 43, ns). Median length of maximal tumescence during which almost all the copulations were observed was 12.5 days (27 cases for eight older females). There was also no difference between immigrants and residents (U-test; z = .91, N_1 = 22, N_2 = 5, ns).

Figure 3. Relative copulation rate (RCR) and female age.

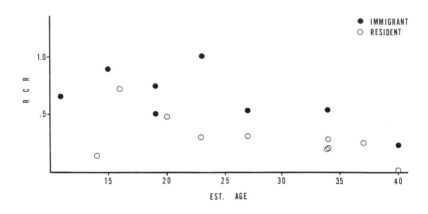

Copulation Rate

As I employed the ad-libitum method for data sampling, I could not calculate an individual's hourly copulation rate. Instead of this, however, I estimated a relative copulation rate which takes into consideration the number of copulations observed for a female and her observability (Table 1). I specifically defined the relative copulation rate (RCR) as the number of copulations observed for a certain female divided by the sum of observation hours of subgroups which included that female during her period of maximal tumescence. Membership of a subgroup was checked at least every 30 minutes. Though I recorded all the copulations that I witnessed, I could not see all the copulations within the subgroup which I followed. Consequently, the RCR is thought to be lower than the actual hourly copulation rate.

Figure 3 represents the RCR and the estimated age for 18 females who were observed in a subgroup for over 1,000 minutes when their swellings were full-size. Eight immigrant females were plotted with closed circles respectively, and ten resident females with open circles respectively. A tendency can be seen for the RCR to become lower as the female gets older. When the data of the immigrants and the residents were combined, female age negatively correlated with the RCR (Kendall's tau = –0.490, p < .01, two-tailed test). Although there was not significant correlation between age and RCR when the data of the immigrants and the residents were calculated separately, it may be due partly to the small sample size.

When testing differences of RCR between the immigrants and the residents, the immigrants' RCR was significantly higher than that of the residents (U-test; U = 10, N_1 = 10, N_2 = 8, p < .01, two-tailed test). When immigrants were compared with residents within age classes divided every five years of age (e.g., 13–17, 18–22, 23–27, 28–32, etc.), the immigrants' RCR was higher than the residents' in any age class.

Then, taking into consideration the age of the male partner, I compared the RCRs of immigrant middle-aged females (ca. 16–37 years old) with resident middle-aged females. When the RCRs for middle-aged females with younger males (from infant to young adult) were compared, the immigrants' RCR was significantly higher than that of residents (U-test; U = 5, N_1 = 6, N_2 = 8, p < .01). However, when the data for copulations with middle-aged and old adult males were compared, the difference was not statistically significant (U = 12, ns).

Five out of the eight immigrant females were also observed to copulate with K-group males (Table 1). All but one, the oldest female (WB), copulated more frequently in M-group than in K-group. Although the K-group was not observed for a sufficient time for statistical tests, the immigrant females tended to copulate more actively in their new group than in their original group.

Table 1. Relative copulation rate (RCR) among M-group females.

	Female's name (est. age in 1982)	Total time of observation in estrus in M-group (number of 10 min. periods)	RCR in M-group (freq/hr)	RCR with young males	RCR with middle and old aged males	RCR in K-group
	WB (40+)	148	0.243	0.162	0.081	0.600
	WS (34)	251	0.550	0.335	0.215	–
	WA (27)	169	0.533	0.213	0.319	–
	CH (23)	493	1.083	0.742	0.340	–
Immigrant[1]	GW (19)	833	0.756	0.526	0.231	0.392
	WP (19)	725	0.505	0.364	0.141	0.286
	GM (15)	430	0.907	0.726	0.181	0.571
	NK (11)	199	0.663	0.603	0.060	0.638
	NG (40+)	177	0.034	0.034	–	–
	UM (37)	233	0.283	0.103	0.180	–
	SO (34)	194	0.309	0.124	0.185	–
Resident	LN (34)	174	0.207	0.138	0.069	–
	BH (34)	145	0.207	0.041	0.165	–
	TI (27)	193	0.311	0.124	0.186	–
	WO (23)	200	0.300	0.150	0.150	–
	GK (20)	307	0.489	0.215	0.274	–
	IK (16)	480	0.725	0.563	0.163	–
	TY (14)	229	0.157	0.157	–	–

Note: Time of observation means the sum of observation minutes of subgroup which included the female in maximal tumescence. Females observed over 1,000 minutes are listed. Dashes indicate "not observed to copulate" in this category.
1. WD was not considered an immigrant (see Figure 1).

Table 2a. Differences in the age of mating partner between the immigrant and resident females: observed and expected frequencies of copulations.

		♂ Old and middle (n = 263)	♂ Young adult (n = 164)	♂ Adolescent (n = 159)	♂ Juvenile and infant (n = 230)
♀ Younger	Immigrant (n = 95)	7 (30.6)	10 (19.1)	19 (18.5)	59 (26.8)
	Resident (n = 30)	0 (9.7)	8 (6.0)	4 (5.8)	18 (8.5)
♀ Older	Immigrant (n = 456)	140 (147.0)	112 (91.6)	104 (88.9)	100 (128.5)
	Resident (n = 235)	116 (75.7)	34 (47.2)	32 (45.8)	53 (66.2)

Notes: Upper figures indicate the number of observed copulations. Figures in parentheses are expected frequencies.

Mating Patterns and Mating Partners

I classified mating patterns into two categories, following Tutin (1979): opportunistic mating (non-competitive and temporary mating) and restrictive mating (a continuous sexual relationship, which includes possessiveness and consortship, between a particular pair).

Out of 826 copulations in M-group observed by Hasegawa, Hiraiwa-Hasegawa, and Nishida during the periods from January 1980 to May 1982, 778 (94%) occurred during opportunistic matings and 42 (5%) occurred during restrictive matings. Mating patterns were unknown for the other six (1%) cases. The results showed that opportunistic mating was a dominant pattern. This mostly agreed with the results found in Gombe (Tutin, 1979).

I compared mating patterns between the immigrant and the resident females, separating data on individuals into two categories. It was found that the rate of opportunistic mating was higher among the immigrants, and, conversely, that restrictive mating was observed at a higher rate among the residents. Out of the 551 immigrant female copulations, 548 (99%) represented opportunistic mating and only 3 (1%) were restrictive matings. On the other hand, out of the 264 resident female copulations for which mating patterns were known, 225 (85%) were opportunistic and 39 (15%) were restrictive. The difference was highly significant (p < .001, Fisher's exact probability test). A similar result was obtained from analysis of individual data.

Table 2b. Differences in the age of mating partner between the immigrant and resident females: results of statistical tests of age differences relative to mating partners.

		♂ Old and middle	♂ Young adult	♂ Adolescent	♂ Juvenile and infant
♀ Younger	Immigrant	- - (‡‡)	– (‡)	n s	+ + (‡‡)
	Resident	- - (‡‡)	n s	n s	+ + (‡‡)
♀ Older	Immigrant	n s	+ (‡)	n s	– (‡)
	Resident	+ + (‡‡)	n s	– (‡)	n s

Notes: ‡ = p < 0.05; ‡‡ = p < 0.01; + or + + indicates significant excess of observed over expected; – or - - indicates significant deficiency of observed relative to expected copulations; ns = no significant difference between observed and expected.

There were 8 resident and 12 immigrant females who were observed to copulate at least 5 times. Among the immigrants, only 1 female was observed to copulate in a restrictive mating. Eight out of 12 resident females showed restrictive matings at least once. Furthermore, all 7 females who copulated with the alpha male, and with whom he showed possessiveness, were residents.

There were differences in the age of mating partners between the immigrant and the resident females. Table 2a shows a contingency table of the number of observed copulations, arranged by female age (younger or older) and life history (immigrant or resident) and by male age. Upper figures indicate the number of observed copulations, and figures in parentheses are expected frequencies. Results of the statistical test using the cell-by-cell method are shown in Table 2b. Cells marked with ++ or -- represent highly significant differences between observed and expected frequencies. It can be seen that while the older residents tend to copulate with older males relatively frequently, the younger females, both immigrant and resident, copulate relatively often with infants and juveniles and rarely mate with males over middle-age. The copulation frequency of the older immigrants mostly accords with what was expected. Their mating tendency seems to be close to random mating.

In an earlier paper, Hasegawa and Hiraiwa-Hasegawa (1983) pointed out that the rate of copulations with high-ranking males increases as ovulation approaches. This, however, holds true for the resident females and not for the immigrant females.

IMMIGRANT FEMALES

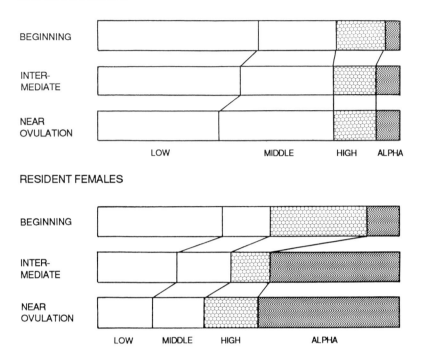

RESIDENT FEMALES

Figure 4. Changes in rank of mating partner during stages of females' estrous. Low, Middle, High, and Alpha represent rank of adult males.

Figure 4 shows the rank of mating partners relative to a female's estrous phase. The female estrous phase was divided into three stages: beginning, intermediate, and near ovulation. The upper figures are for the immigrants and the lower for the residents. Although the ratio of copulation with higher-ranking males, especially with the alpha male, increases among the residents as estrus progresses, the ratio was largely stable through all stages among the immigrants.

Female Courtship Behavior

The immigrant females "exhibited" more actively in order to solicit males for sexual interactions than did the resident females. During the study period, ten types of positive courtship display by females were observed: leaf clip (n = 19), branching (n = 3), stamp (n = 2), rocking, cushion make, self-hit, slap, pull, genital touch (n = 1, respectively). Most of these elements were common to male courtship, and females displayed them in the same way as males. The element that was most often observed was "leaf clip" which was a courtship display unique to Mahale and also most frequently observed among juvenile males (Nishida, 1980).

Six females were seen to court males a total of 28 times using the above mentioned types of display. All six were older females, four immigrants and two residents. Immigrants carried out 26 of the 28 acts of courting. Out of the 26 cases of courtship by the immigrants, 15 cases were successful, that is, led to copulations. Target males were adult in 13 cases, adolescent in one case, and juvenile and infant in 12 cases. It should be noticed that the immigrant females often solicited even immature males. The resident females (n = 2) courted only adult and adolescent males. The rate of successful courtship by the immigrant females toward adult males (31%, 4/13) was lower than that toward pre-adolescent males (85%, 11/13).

These results imply that the immigrant females were more willing than the resident females to have sexual relations with males.

Copulatory Vocalization

During a male intromission, females often utter a characteristic high-pitched scream. Although I do not discuss the function of this vocalization in detail here, it is true that the scream is easily detected by other members in a group. Therefore, this vocalization seems to be a form of female sexual expression. It was found that the ratio of the copulatory vocalization differed between the immigrant and resident females.

Frequency of screaming was studied in eight immigrant and 11 resident females, each of whom had been observed in at least five copulations. The copulatory vocalization rate of the immigrant females was significantly higher than that of the resident females (U-test; U = 16, N_1 = 8, N_2 = 11, p < .05). As individual data consisted of relatively small samples, I grouped the data into categories for the following analysis. Table 3a shows the differences in the vocalization rate among females. First, the younger females uttered screams at a higher rate than the older females, both within the immigrant group (χ^2 = 19.52, df = 1, p < .01) and within the resident group (χ^2 = 7.44, df = 1, p < .01). Second, within the younger females there was not a significant difference between the immigrant and the resident group (χ^2 = .098, df = 1, ns). On the other hand, within the older females, the immigrants vocalized more than the residents (χ^2 = 6.28, df = 1, p < .05). Although the difference was not significant between the younger residents and the older immigrants (χ^2 = 2.68, df = 1, ns), it may be due partly to the small samples of the former.

Table 3b shows the relationship between vocalization rate and female estrous stages for older females. As regards the immigrant females, the rate did not differ across estrous stages (χ^2 = 1.54, df = 2, ns). But with the resident females, the rate became higher as ovulation approached (χ^2 = 8.23, df = 2, p < .05). A significant difference in the rate between the resident and immigrant was found only in the beginning stage (beginning: χ^2 = 6.38, df = 1, p < .01; intermediate: χ^2 = 0.08, df = 1, ns; near ovulation: χ^2 = 0.06, df = 1, ns). These results suggest that the older immigrant females vocally signal successful intromission throughout their receptive periods.

Table 3a. Rate of female copulatory scream: results of analysis by age and life history.

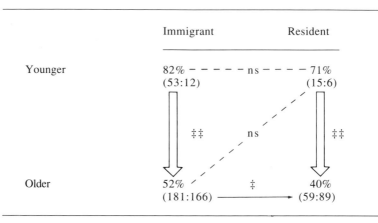

	Immigrant	Resident
Younger	82% – – – – ns – – – 71%	
	(53:12)	(15:6)
	‡‡ ns ‡‡	
Older	52%	40%
	(181:166) ———— ‡ ————➤ (59:89)	

Notes: ‡ = p < .05, ‡‡ = p < .01, ns = no significant difference between observed and expected.

Table 3b. Frequency of screams during copulation analyzed by estrous stage and life history for older females.

	Immigrant	Resident
	‡	
Beginning	45% ————————➤ 23%	
	(43:52)	(11:36)
Intermediate ns	52% – – – ns – – – – 50% ‡	
	(69:63)	(26:26)
Near ovulation	53% – – – ns – – – – 46%	
	(56:49)	(19:22)

Notes: ‡ = p < .05, ns = no significant difference between observed and expected.

DISCUSSION

The above results are summarized in Table 4. Although female age affects several differences in sexual behavior, my main conclusions concern older females' characteristics.

Broadly speaking, although physiological conditions did not differ between the immigrants and the residents, the immigrants showed a much more active sexuality in many behavioral aspects.

Does active sexuality relate to "infanticide"? If immigrant females copulate with many males, the number of probable fathers will increase. Could this tend to reduce "infanticide" to some degree?

Among those females who gave birth during my research, the mating patterns during the periods of conception were known for only three. Two resident females are thought to have conceived during the restrictive mating, one with the alpha male and the other with the second ranking male. The third female was an immigrant who copulated with many males in an opportunistic manner during the period of conception. There is no evidence that the infant of the immigrant female was killed, but the infant was lost before it was three months old.

In this study, I could not determine probable fathers of the victims of infanticide. However, if we consider the characteristics in mating behavior of the immigrant females that I mentioned above, the immigrant females whose infants were killed likely copulated with many males promiscuously. If that was the case, then some males may have killed their own infants. In fact, Kawanaka (1981) observed that a male, who was thought to have killed and then eaten an infant, copulated with the victim's mother at the period of her conception. Also, in this study an immigrant female (CII) had repeatedly copulated (although not at the time of conception) with the male (KZ) who later killed her infant. It seems to suggest that experiences of copulations cannot restrain the male from killing his partner's infant. As females usually mate several hundreds of times in order to conceive an infant, males cannot be certain of their paternity with opportunistic mating.

Table 4. Summary of results.

	Immigrant	Resident
Sexual cycle	no difference	
Copulation rate	higher	lower
Mating pattern	more opportunistic	more restrictive
Mating partners	indiscriminate	mainly older males
Female courtship	often	rarely
Copulatory scream	higher rate	lower rate

On the other hand, as the immigrant females tend to live alone during pregnancy in the area near the range of their former group (Takahata, 1985; Nishida and Kawanaka, 1985), males might suspect that females are copulating with males of the enemy group. Moreover, if an infant appears to be a male, males become more suspicious about the paternity, because a male infant sired in another group would break the important kin-relationship which ties the males together in each group. Promiscuity exhibited by the immigrants may, at worst, raise more doubts about cuckoldry.

From the facts and situations described above, the more active sexuality of the immigrant females does not seem to be explained as a counter strategy against infanticide. So, what is the reason for the heightened sexual activity? It may be due partly to "preferences of unfamiliar mates" (Harcourt, 1978) or the "strange female effect" (Allen, 1981). At the same time, however, I believe that, through active sexual interactions, the immigrant females will gradually form social relationships with males and enhance their social status in the new community. Nishida (this volume) related that immigrant females are often attacked by resident females (see also Goodall, 1986; Pusey, 1980) and that adult males show a strong tendency to support losers. When resident and immigrant females are in competition, the active sexuality shown by immigrant females is thought to serve to gain support from adult males. Usually, it will take two to three years for a new immigrant to become an established member of the group. Two years after this research, Takasaki conducted his research on mating relationships in Mahale and observed that the same immigrant females of this research had begun to copulate with males more restrictively (Takasaki, 1985). Also, the probability of killing the second and later infants of the immigrant females is much less than the probability of killing the firstborn (Hiraiwa-Hasegawa, 1987). The reproductive status of older females observed in Mahale seems to be very vulnerable. Because females normally leave their natal group during the period of adolescence, I was unable to fully analyze the behavior of young females in this study. Further investigations will undoubtedly reveal the characteristics of young immigrant females in more detail.

ACKNOWLEDGMENTS

I would like to express my gratitude to the Government of Tanzania, especially the Serengeti Wildlife Research Institute and the Wildlife Division, for allowing me to conduct research in Mahale, and to the Mahale Mountains Wildlife Research Centre (MMWRC) for supporting the field work. Special thanks are due to Dr. T. Nishida, Dr. S. Uehara, and Dr. M. Hiraiwa-Hasegawa for their continuous advice and for permission to use their unpublished data, and to Drs. H. Hayaki, Y. Takahata, and H. Takasaki for their cooperation in the field work. I deeply acknowledge the continuing help of the staff of MMWRC during my stay in Mahale. I also thank Dr. J. Goodall, Dr. P. Heltne, and L. Marquardt for their invitation to the symposium "Understanding Chimpanzees." This study was financed by the Japan International Cooperation Agency.

REFERENCES

Allen, M. 1981. Individual copulatory preference and the "strange female effect" in a captive group-living male chimpanzee *(Pan troglodytes)*. *Primates* 22:221–236.

Goodall, J. 1977. Infant killing and cannibalism in free-ranging chimpanzees. *Folia Primatol.* 28:259–282.

_____. 1986. *The Chimpanzees of Gombe: Patterns of Behavior.* Cambridge, MA: Harvard University Press.

Harcourt, A. H. 1978. Strategies of emigration and transfer by primates, with particular reference to gorillas. *Z. Tierpsychol.* 48:401–420.

Hasegawa, T., and M. Hiraiwa-Hasegawa. 1983. Opportunistic and restrictive matings among wild chimpanzees in the Mahale Mountains, Tanzania. *J. Ethol.* 1:75–85.

Hiraiwa-Hasegawa, M. 1987. Infanticide in primates and a possible case of male-biased infanticide in chimpanzees. In Y. Ito, J. Brown, and J. Kikkawa, eds., *Animal Societies,* pp. 125–139. Tokyo, Japan: Scientific Societies Press.

_____. 1988. Adaptive significance of infanticide in primates. *Trends in Evolution and Ecology* 3:102–105.

Hiraiwa-Hasegawa, M., T. Hasegawa, and T. Nishida. 1984. Demographic study of a large-sized unit-group of chimpanzees in the Mahale Mountains, Tanzania: a preliminary report. *Primates* 25:401–413.

Kawanaka, K. 1981. Infanticide and cannibalism in chimpanzees, with special reference to the newly observed case in the Mahale Mountains. *African Study Monogr.* 1:69–99.

Nishida, T. 1980. The leaf-clipping display: a newly discovered expressive gesture in wild chimpanzees. *J. Human Evol.* 9:117–128.

Nishida, T., and K. Kawanaka. 1985. Within-group cannibalism by adult male chimpanzees. *Primates* 26:274–284.

Nishida, T., M. Hiraiwa-Hasegawa, T. Hasegawa, and Y. Takahata. 1985. Group extinction and female transfer in wild chimpanzees in the Mahale National Park. *Z. Tierpsychol.* 67:284–301.

Pusey, A. 1980. Inbreeding avoidance in chimpanzees. *Anim. Behav.* 28:543–552.

Takahata, Y. 1985. Adult male chimpanzees kill and eat a male new-born infant: newly observed intragroup infanticide and cannibalism in the Mahale National Park, Tanzania. *Folia Primatol.* 44:161–170.

Takasaki, Y. 1985. Female life history and mating patterns among the M-group chimpanzees of the Mahale National Park, Tanzania. *Primates* 26:121–129.

Tutin, C. E. G. 1979. Mating patterns and reproductive strategies in a community of wild chimpanzees *(Pan troglodytes schweinfurthii)*. *Behav. Ecol. Sociobiol.* 6:29–38.

SEX DIFFERENCES
IN THE BEHAVIORAL DEVELOPMENT
OF CHIMPANZEES AT MAHALE

Mariko Hiraiwa-Hasegawa

INTRODUCTION

Differences between sexes have been reported in several aspects of adult chimpanzee behavior. In ranging patterns and day-range lengths, females are confined to a more limited ranging area than males, and the mean distance traveled per day is shorter for females than males (Wrangham and Smuts, 1980). In feeding behavior, males specialize in hunting mammals and females specialize in gathering ants (McGrew, 1979). In western Africa, females, using tools, crack open hard nuts more frequently than males (Boesch and Boesch, 1981). However, the development of these sex differences has not yet been fully described. If there are sex differences in certain behaviors of adults, when do they begin? Do they exist from the period of infancy?

Few analyses have been made of sex differences in the behavioral development of the chimpanzee, either in captivity or in the wild. This is due not only to the difficulty of collecting sufficient data on a large number of male and female infants of comparable ages, but also due, probably, to the fact that the researchers feel chimpanzee infants are too unique individually for simple generalizations to be made under sex categories.

This report attempts to describe the development of sex differences in three areas of behaviors in which the differences in adults are clear: ant-eating behavior, daytime bed-making behavior, and the features of greeting behavior. The analysis of greeting behavior is still preliminary, based on the ad-lib sampling method (Altmann, 1974) with a small number of subjects.

METHODS

The field study, totaling two years and six months (from July 1979 to January 1980, and from June 1980 to June 1982), was carried out on the chimpanzees of the M-group and K-group of Mahale Mountains National Park, Tanzania. The total hours of observations on both groups amounted to about 1200 hours.

When a group of chimpanzees was encountered, it was followed as long as possible. The group members and their activities – traveling, resting/socializing, and feeding (with the name of the food species and the part eaten) – were recorded every 15 minutes. In addition, various social behaviors, such as grooming, aggression, and sexual behavior were recorded by the ad-lib sampling method.

As the main theme of my study was the mother-infant relationship, special attention was paid to the infants under weaning age. Sixteen infants (10 females and 6 males) were observed by using the focal animal sampling method. However, sufficient observational hours were obtained on only 5 females and 4 males; thus the analyses were made only on these individuals. The focal animal sampling method was not applied to the older individuals. But for each older individual, the amount of observational hours was calculated for the subgroup in which the individual was included. Therefore, the results for the infants and for the older individuals cannot be compared directly.

RESULTS

General Pattern of Development

In Gombe and Mahale, the patterns of infant development are similar. In general, the chimpanzee infant weans at about age 5, matures at about age 15, and lives as long as 35–40 years or more. "Infant" refers to the period birth to 5 years, "juvenile" refers to 5 to 8 years, "adolescent" refers to 8 to 15 years, and "adult" to 15 years or older (Goodall, 1983; Hiraiwa-Hasegawa et al., 1984). In this report, however, the 6-year-old individuals without a younger sibling were included in the infant category.

Development of Sex Differences in Ant-Eating Behavior

In Mahale, chimpanzees feed on various species of ants, but the majority of ants taken by them are small ants of *Crematogaster* spp. and wood-boring ants of *Camponotus* spp. (Nishida, 1973).

Camponotus ants are sometimes taken by using tools. Sex differences have been reported in the frequency and duration of this "ant-fishing behavior" (Goodall, 1968a; Nishida, 1973; McGrew, 1979). Adult females ant-fish more frequently and in longer duration than adult males do. In addition to the greater frequency and duration, it is possible that they are also superior to adult males in their technique, at least in the M-group. For example, the alpha male of the M-group, NT, and an adult female, IK, were once observed to fish ants at the same time in the same tree. IK began to fish ants at one site and fished out 15 ants in 5 minutes. Then NT pushed her out and took her site. He continued to fish for 1 minute without getting any ants and stopped. Then IK returned and fished 13 ants in 3 minutes. NT pushed her again and recovered the site, but he could fish only 5 ants in 9 successive minutes. NT's fishing rod was limp, and compared to IK, he was very clumsy inserting the rod in the hole.

In this report, however, analysis was made only on the feeding on *Crematogaster* ants and *Camponotus* ants by direct licking from the broken branches or stems, not on the ant-fishing behavior. The amount of intake of other ant species was negligible during my observation period.

From the age of about 4 months, infants were observed to show interest in the solid food which their mothers were eating. And they were observed to actually eat the solid food from the age of about 5 months. Infants usually begin with some soft, small fruits or leaves. The ant is not the type of food they eat from the very first stage of their solid food intake. The youngest infant who was observed to eat *Crematogaster* ants was 11 months old.

Since the difference in the amount of ant intake between the adult members of the K-group and those of the M-group, as well as between the two sexes, has been reported by Uehara (1986), the analysis in this study centers on the individuals of the M-group only.

For the mothers and infants who were observed by the focal animal sampling method, all the foods they ate during the observation were recorded. When they took one food item (a certain part of a certain food species, i.e., fruits of *Saba florida),* it was counted as one feeding bout. And if they fed on the same food item continuously, one count was added every successive 15 minutes. For individuals other than the

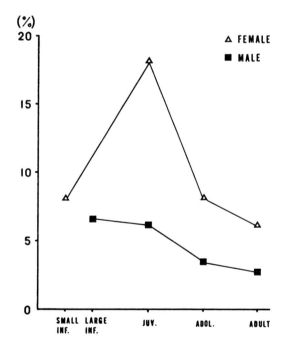

Figure 1. Ant-eating bouts as a percentage of all the feeding bouts for four age/sex classes of chimpanzees.

target mother-infant pairs, the food items were recorded at the time of routine 15-minute checks of membership and activities of the subgroup.

The percentage of ant-eating bouts in the total number of feeding bouts was calculated for each individual. Figure 1 shows the average percentages of ant-feeding bouts for each age/sex class. Among adults whose observational hours exceeded 100 hours, the average percentage of ant-feeding bouts was 2.9% (24/817) for males (n = 6) and 6.5% (39/598) for females (n = 8). The difference between adult males and females is statistically significant (χ^2 = 10.43, df = 1, p < 0.01). Among adolescents, the percentage for males was 3.5% (8/231) (n = 2), and for females was 8.3% (16/192) (n = 3) (χ^2 = 4.65, df = 1, p < 0.05). Among juveniles, the percentage was 6.2% (35/569) for males (n = 3), and 18.7% (56/300) for females (n = 2) (χ^2 = 32.82, df = 1, p < 0.01). In these two age classes, the sex differences are statistically significant. Among infants the difference is not significant: 6.6% (13/197) for males (n = 2) and 8.1% (20/247) for females (n = 4) (χ^2 = 0.36, df = 1, ns).

In summary, it can be concluded that, first, the frequency of ant-feeding appears higher in both sexes when they are young and decreases with age. Second, in all age classes, females eat more ants than males do, and the difference becomes clear when they are juveniles.

Development and Sex Differences in Daytime Bed-Making Behavior

A chimpanzee makes a bed on a branch and sleeps in it at night. The infant, before weaning, sleeps in its mother's bed with the mother. Two female infants, PA and LS, and a male infant, TB, were all observed to sleep with their mothers at the age of 4 years 11 months, and another male infant, KB, up to age 5. PA was observed to make her own bed and sleep separately from her mother at the age of 5 years 3 months. Thus, infants are thought to begin sleeping independently at night several months before their younger sibling is born. In one case, a child, BU, was observed to sleep in her mother's bed even after the younger sibling was born. In this case, the mother's interbirth interval was assumed to be less than 5 years, though the precise date of birth of the elder sibling was not known. BU was still sleeping in her mother's bed with her mother and her younger brother when the younger one was 6 months old. It is not confirmed when she began to sleep in her own bed.

During the daytime, chimpanzees sometimes make a bed or a cushion which is usually simpler than the bed made at night. Infants are observed to make a day bed fairly frequently. This may serve as practice to make their own bed. For the three infants (PN, AM, TL) whom I continuously observed from birth, daytime bed-making attempts were first observed at the age of 12 months, 13 months, and 14 months respectively (Fig. 2). At first, the infants were not able to bend branches, and they just gathered some vines and twigs with leaves, crumpled them into a cushion-like mass, and sat on it. However, the infants were usually very clumsy, and the cushions easily fell apart. Also, the infants were not good at selecting a suitable place to set a cushion. In one observation, PN, when she was 1 year 10 months old, collected some vines with leaves and made a bed on a small branch. After she finished crumpling the material,

Figure 2. Thirteen-month-old PN tries to make a day bed.

she sat on it. But, as it was on a small branch which was not stable, she almost fell down to the ground.

Among adults there seems to be a sex difference in the frequency of bed-making behavior during daytime. For the 9 adult males of the M-group, the sum of the observational hours on the subgroups in which each of them was included amounted to 2235 hours. These males were observed to make day beds only 3 times during that period. This represents a frequency of 0.01 times per 10 hours. The 8 adolescent and juvenile males were observed to make day beds 8 times in 1429 hours of observation. The frequency was 0.06 per 10 hours. However, among the 9 adult females, other than the target mothers, daytime bed-making was observed 11 times in 1442 hours of observation. The frequency was 0.08 per 10 hours. Although I could not apply a statistical test due to the small sample size of data on males, sex differences may exist in the frequency of daytime bed-making.

On the other hand, the target infants were observed to make day beds a total of 97 times. Figure 3 shows the frequencies of daytime bed-making behavior per 10 hours for the males and females according to their ages. As the observational hours on the 4-, 5-, and 6-year-olds were small, data on these individuals were combined.

Considering the two sexes together, the frequencies of bed-making behavior per 10 hours are 1.0 for 1-year-olds, 1.8 for 2-year-olds, 4.4 for 3-year-olds, 1.9 for 4-year-olds through 6-year-olds. The frequency for all infants is 2.0. The frequency of bed-making of the target mothers during the same period was 0.15 per 10 hours (19/1277 hours, n = 8). Therefore, the infants made beds 10 times more frequently than the mothers did. Considering the data on the two sexes separately, in all age groups female infants make beds more frequently than male infants do. In conclusion, the frequency of bed-making during daytime seems to be high in young of both sexes and to decrease with age. From the onset of this behavior, females appear to make day beds more frequently than males do.

Development and Sex Differences in the Greeting Behavior

The chimpanzee greets a dominant individual by pant-grunting and pant-barking with various gestures: extending hand, soft-biting, kissing, embracing, touching, presenting, bobbing, etc. These greeting behaviors, usually of a nonaggressive nature, can be observed mainly when two individuals meet again after some duration of separation (Goodall, 1968b).

From as early as 4–5 months of age, an infant begins to show some kinds of gestures when its mother greets others: extending hand to them or trying to touch them, sometimes with some vocalizations. These behaviors, however, are difficult to classify as "greeting." The youngest infant who was observed to make an apparent pant-grunt in the greeting context was PN, when she was 14 months old. Extending hand and soft-biting were also first performed by PN when she was 14 months old.

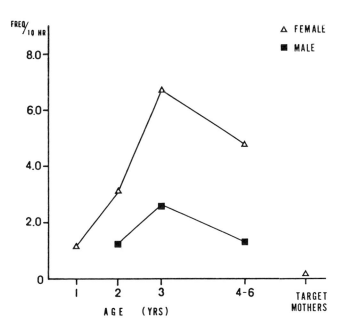

Figure 3. Frequencies of daytime bed-making behavior of infants and mothers.

Figure 4. An adult female, GW, greets an adult male, JG, by extending hand.

Embracing was first performed by MA when he was 2 years 10 months old, and bobbing was first observed in TB at 3 years 10 months of age. Immature females were observed to direct the greeting behavior not only to adult males but also to adult females, while immature males were never observed to greet the adult females.

The greeting behavior was divided into seven categories: (1) pant-grunting with no gesture, (2) embracing and arm-around, (3) biting and kissing, (4) touching and extending hand (Fig. 4), (5) bobbing,[1] (6) presenting, and (7) brief grooming and other behaviors.

Greetings to adults were recorded a total of 673 times in the M-group. In Table 1 and Figure 5, the percentage of each type of greeting to all the greetings with gestures was calculated for four age/sex classes. Here, "immature" means infants, juveniles, and adolescents altogether. In Figure 5, the immature females' greetings which were addressed to adult males were combined with those addressed to adult females because of the small sample size. Differences between the sexes were found in the features of greeting behavior.

Figure 5. Six types of greeting gestures as a percentage of all greetings with gesture in the four age/sex classes.[1]

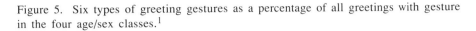
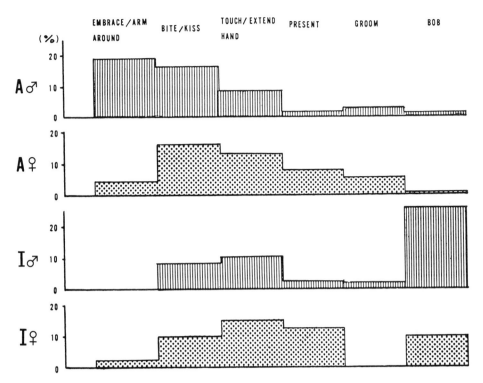

Notes: A = adult, I = immature (infant, juvenile, and adolescent grouped together).
1. The greetings given by adult males, adult females, and immature males include only those addressed to adult males. The greetings given by immature females include those addressed to both adult males and adult females.

In all the age/sex classes, about half of the greetings were accompanied by some kind of gestures. However, the most frequent component of greeting gestures differed according to the age/sex classes. In adult males, embracing/arm around and biting/kissing were performed most frequently, and bobbing, presenting, and brief grooming were seldom performed. On the other hand, adult females performed biting/kissing and touching/extending hand most frequently; presenting and grooming were less frequent. Embracing/arm around and bobbing were seldom observed.

Among immatures, the picture is different. In more than half of their greetings with gesture, immature males bobbed. They also performed touching/extending hand, and biting/kissing, but seldom did embracing/arm around, presenting, and grooming. Among immature females, except for embracing/arm around and grooming, the remaining categories of gestures were observed rather evenly. Although the sample size

Table 1. Percentages of the six types of greeting gestures used by the four age/sex classes of chimpanzees of Mahale.

Age/Sex Classes	N_1	N_2	% of N_2 to N_1	Embrace/ Arm Around	Bite/ Kiss	Touch/ Extend Hand	Bob	Present	Groom etc.
				% to N_2					
AM to AM	134	70	50.7	38.6	32.9	17.7	2.9	2.9	5.7
IM to AM	214	100	46.7	1.0	17.0	21.0	52.0	5.0	4.0
AF to AM	285	129	45.3	9.3	32.6	27.1	1.6	16.3	11.6
IF to AM	29	12	41.4	8.3	16.7	16.7	25.0	33.3	0.0
IF to AF	11	8	72.7	0.0	25.0	50.0	12.5	12.5	0.0

Notes: N_1 = Total number of greetings observed. AM = Adult Male
  N_2 = Number of greetings with gestures. IM = Immature Male
  AF = Adult Female
  IF = Immature Female

is small compared to other age/sex classes, gestural components of immature females' greetings toward adult males were different from those toward adult females (Table 1). Toward adult males, presenting and bobbing were the most frequently recorded gestures. Toward adult females, however, they adopted touching/extending hand and biting/kissing most frequently but not presenting or bobbing.

Therefore, bobbing seemed to be a characteristic of the immatures' greetings toward adult males, and the tendency to bob was especially strong in the immature males. The same tendency was reported in Gombe (Goodall, 1968b).

WHY THE SEX DIFFERENCES?

The sex differences in the ant-feeding behavior among adult chimpanzees are interpreted in relation to the sex differences in the meat-eating behavior among adults: the nutritional counterbalance of meat and ants (McGrew, 1979; Uehara, 1986).

Immature animals have virtually no opportunity to hunt mammals by themselves. However, they can sometimes eat meat by begging and recovery. So far, there are no systematic data comparing the amount of meat intake between immature males and females. It seems unlikely, however, that there is much difference between them, though it is possible that the immature males' tendency to follow the adult males might give the immature males more chance than immature females to take meat by recovery.

Another explanation attributes the differences in ant-feeding behavior to the differences in the sociability of the two sexes (Boesch and Boesch, 1984). Adult males

Figure 6. A two-year-old female attempting to ant-fish at dusk.

spend more time with one another than the adult females do; they participate in the various social behaviors more frequently than the adult females do; they are more alert to one another's activities than the adult females are. All these factors are thought to reduce the adult males' time for feeding on insects, which is time-consuming compared to the amount of nutritional gain.

From this point, Uehara's (1986) results of the fecal analysis is suggestive: the percentage of the feces containing ants was smaller in the females in estrus than in anestrous females. As it seems unlikely that the nutritional requirement changes according to the females' estrous phases, the results suggest that the occurrence of ant-feeding behavior is affected by the social condition of the individual. Females in estrus might not have enough time for gathering ants, or might not be in the mood to concentrate on gathering ants because they range long distances and wide areas with males, constantly seeking the opportunity to mate.

Sex differences in the ant-eating frequency among immatures might be interpreted on the same basis. After weaning, male infants tend to spend their time more and more with adult males than with their mothers. As a general behavioral pattern, males are active and socially alert, and females are sedentary and solitary. And the differences in the ant-eating frequency among immatures might reflect these differences in overall behavioral tendencies between the two sexes (Fig. 6).

The sex differences in the bed-making behavior during the daytime may also be explained as a reflection of the sex differences in the overall behavioral pattern. Adult males taking a rest during daytime usually lay down on the ground. Perhaps it enables them to react more quickly to others' movements than lying on a bed in a tree. Or perhaps they do not indulge in bed-making because it is rather superfluous during the daytime.

Finally, what is the meaning of the age/sex differences in the features of greeting behavior? The particular gestures employed during a greeting depend on the length of separation between the animals concerned, the rank gap between the two, individuality, and the animal's mood at that time, as well as the age/sex classes of the animals. These varied gestures may represent different degrees of tension on the part of the subordinates.

The bobbing gesture may reflect the actor's ambivalence between its fear of the adult male and its strong desire to make a greeting gesture to him. This may explain why the immature females performed bobbing more frequently toward adult males than toward adult females.

It is reasonable to suppose that the immature males have stronger incentive than do immature females to join the adult male band. Thus the ambivalence may be stronger in immature males than in immature females, and this may be reflected in the differences in the features of their greeting behavior, especially in the frequencies of bobbing behavior.

ENDNOTE

1. There might be a "cultural" difference in the style of bobbing between Gombe and Mahale. In Gombe, the behavior is described as follows: "The individual drops back onto four limbs and as a continuation of the movement flexes its elbows until its chest is close to the ground. It then jerks itself back to the bipedal position prior to repeating the entire sequence several times" (unpublished manuscript of the Behavioural Glossary of Chimpanzees of Gombe). In Mahale, the gesture is almost the same but the chimpanzee flexes its knees rather than its elbows.

ACKNOWLEDGMENTS

I would like to express my gratitude to the Government of Tanzania, especially the Tanzania National Scientific Research Council, the Serengeti Wildlife Research Institute, and the Wildlife Division, for permission to conduct the research, and to the staff of the Mahale Mountains Wildlife Research Centre for supporting my field work. Special thanks are due to Drs. T. Nishida, S. Uehara, and T. Hasegawa for their valuable comments and discussions. The research was financially supported partly by the Japan International Cooperation Agency and partly by the Ministry of Education, Science and Culture, Japan (Grant-in-Aid for Overseas Scientific Research No. 404130 of 1979).

REFERENCES

Altmann, J. 1974. Observational study of behavior: sampling methods. *Behaviour* 49: 227–267.

Boesch, C., and H. Boesch. 1981. Sex differences in the use of natural hammers by wild chimpanzees: a preliminary report. *J. Human Evol.* 10:585–593.

_____. 1984. Possible causes of sex differences in the use of natural hammers by wild chimpanzees. *J. Human Evol.* 13:415–440.

Goodall, J. van Lawick. 1968a. The behavior of free-living chimpanzees in the Gombe Stream Reserve. *Animal Behav. Monogr.* 1:161–311.

_____. 1968b. A preliminary report on expressive movements and communication in the Gombe Stream Chimpanzees. In P. Jay, ed., *Primates: Studies in Adaptation and Variability,* pp. 313–374. New York: Holt, Rinehart, and Winston.

Goodall, J. 1983. Population dynamics during a 15 year period in one community of free-living chimpanzees in the Gombe National Park, Tanzania. *Z. Tierpsychol.* 61:1–60.

Hiraiwa-Hasegawa, M., T. Hasegawa, and T. Nishida. 1984. Demographic study of a large-sized unit-group of chimpanzees in the Mahale Mountains, Tanzania: a preliminary report. *Primates* 25:401–413.

McGrew, W. C. 1979. Evolutionary implications of sex differences in chimpanzee predation and tool use. In D. A. Hamburg and E. R. McCown, eds., *The Great Apes,* pp. 441–464. Menlo Park, Calif.: Benjamin/Cummings.

Nishida, T. 1973. The ant-gathering behavior by the use of tools among wild chimpanzees of the Mahale Mountains. *J. Human Evol.* 2:357–370.

Uehara, S. 1986. Sex and group differences in feeding on animals by wild chimpanzees in the Mahale Mountains National Park, Tanzania. *Primates* 27:1–13.

Wrangham, R. W., and B. B. Smuts. 1980. Sex differences in the behavioural ecology of chimpanzees in Gombe National Park, Tanzania. *J. Reprod. Fert. Suppl.* 28:13–31.

FEEDING ECOLOGY OF CHIMPANZEES IN THE KIBALE FOREST, UGANDA

G. Isabirye-Basuta

INTRODUCTION

Optimal foraging theory has so far been applied to few ecological studies of primates (Hamilton et al., 1978; Milton, 1980; Glander, 1981; Harrison, 1984). The reasons for this are not clear since models of optimal foraging theory can provide testable predictions about foraging strategies that have been observed in the field (Harrison, 1984). Optimal foraging theory is based on the premise that, as a result of natural selection, an animal maximizes feeding efficiency. Feeding efficiency can be measured in several currencies (Post, 1984), the most commonly used being energy (Schoener, 1971). Using this currency the animal maximizes energy intake while foraging. Post (1984) has discussed the various problems associated with optimality models as applied to the study of feeding strategies of primates.

One of the earliest optimality models concerned choice of diet in a patchy environment (MacArthur and Pianka, 1966). This model, based on the analysis of cost-benefit ratio of food items, appears to be suitable for studying some aspects of chimpanzee foraging strategies. The model predicted that (a) as the availability of high-ranking food items increases, the optimal diet should include fewer food items; (b) the animal should become more selective when profitable food items are common; and (c) the animal should ignore unprofitable food items regardless of their abundance.

Other models concerning food choice in a patchy environment have also been proposed (e.g., Krebs, 1978). While it is most likely that none of the foraging strategies of any primate can fulfill all the predictions of the optimality models (Harrison, 1984; Post, 1984), it is nonetheless possible that some of the predictions are fulfilled by some primates. In this paper, two aspects of chimpanzee feeding ecology are studied using optimal foraging theory: (1) the relationship between fruit abundance and diet composition and diversity, and (2) diet composition in relation to the abundance of plant species eaten.

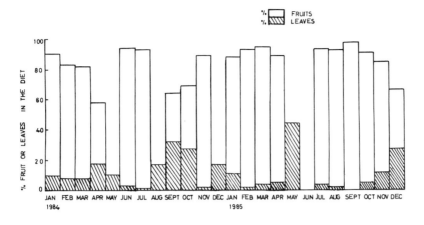

Figure 1. The proportion of leaves and fruit (based on feeding scores) in the monthly diet of chimpanzees at Kanyawara, Kibale Forest.

FRUIT ABUNDANCE
IN RELATION TO DIET COMPOSITION AND DIVERSITY

Wherever they have been studied, chimpanzees have been found to be predominantly frugivorous (Goodall, 1968, 1971; Reynolds and Reynolds, 1965; Nishida, 1968; Sugiyama, 1973; Wrangham, 1975; Ghiglieri, 1984; and others). They are highly specialized frugivores, exploiting food patches that occur at low densities (Ghiglieri, 1984). For example, in the Kanyawara area of Kibale Forest, only two important food-tree species, *Uvariopsis congensis* (Robyns and Ghesquiere) and *Celtis durandii* (Engl.), are among the most common tree species in the forest. In the Kanyawara area, fruit accounts for at least 60% of the monthly diet in most cases (Fig. 1). The availability and abundance of fruit is therefore likely to have a major influence on diet composition and diversity. In terms of the MacArthur and Pianka model outlined above, fruit is a high-ranking food item in the diet of chimpanzees. In the Kanyawara area, fruit of several plant species eaten by chimpanzees is usually present at any given time (Fig. 2). However, abundant fruit at any given time in a localized area belongs to one or two species. Fruit of different types present at the same time may also be ranked according to its abundance. From Figure 2 it is also clear there are periods of fruit scarcity. It should be at such times that low-ranking food (mainly non-fruit) items should form a major portion of the chimpanzee diet.

From the above observations, it follows that:

1. the number of food items should be inversely correlated with fruit abundance;

2. fewer plant species should be utilized for fruit when abundant fruit is available;

3. overall, fewer plant species should be utilized for food when fruit is abundant than when it is scarce (this is because few plant species utilized for non-fruit food items will be included in the diet when fruit is abundant); and

4. the proportion of the diet made up of fruit should be positively correlated with fruit abundance.

Figure 2. The availability and abundance of fruit (based on phenological scores) at Kanyawara, Kibale Forest. Absence of entry indicates no fruit.

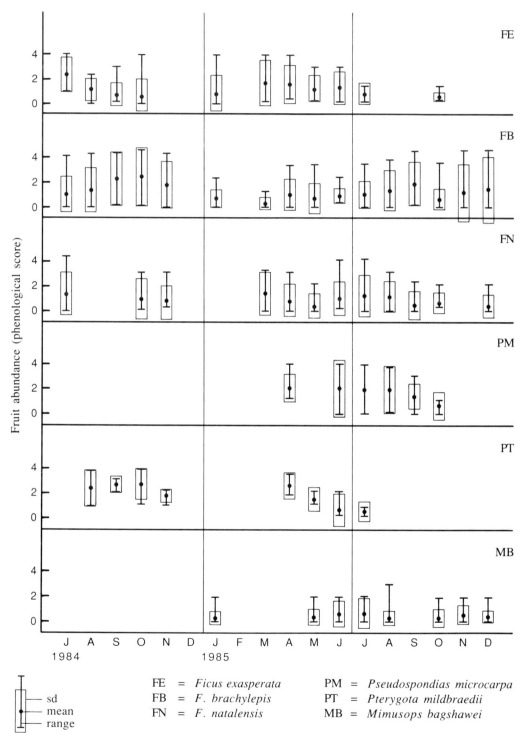

Fruit abundance in this paper only refers to fruit belonging to plant species eaten by chimpanzees. Fruit abundance was assessed on individual trees using a simple phenological score (PS), ranging from 0 for no fruit to 4 for maximum fruit possible. This was done starting July 1984 with a sample of 34 trees, mainly figs. This sample was subsequently expanded to 55 trees, through addition of trees belonging to rare species, such as *Pseudospondias microcarpa* (A. Rich.) (Engl.) and *Ficus dawei* (Hutch.). I also had access to phenological data, collected by the staff of Kibale Forest Project, on 88 *Mimusops bagshawei* (S. Moore) trees. However, no phenological data on fruit abundance were available on some important tree species, such as *Celtis durandii*, *Uvariopsis congensis*, *Linonciera johnsonii* (Baker), and *Pancovia turbinata* (Radlk.). Furthermore, the sample size used for most tree species, especially figs, was too small to give an adequate picture of fruit abundance. For example, in some months fruit was available on non-sample trees, but not on sample trees of the same species. Consequently, fruit abundance was underestimated for most tree species sampled.

For each month, an index of fruit abundance was calculated. This was the weighted mean for all tree species that had fruit on them that month. Using 10 months of data for 1985, the number of food items in the monthly diet was inversely (but not significantly) correlated with fruit abundance ($r = -0.48$, $p > 0.05$, $df = 8$, Fig. 3). This lack of significance may be related to the fact that phenological data on some important tree species were not included. But this could also be due to other factors, such as those related to fruit quality. For example, some *Ficus exasperata* (Vahl) trees had abundant fruit during March and April 1984, and yet this fruit was not eaten by chimpanzees despite the fact that there were hardly any alternative sources of fruit. The five months of data available for 1984 also showed a different trend; the number of food items eaten each month was positively (but not significantly) correlated with fruit abundance ($r = +0.20$, $p > 0.05$, $df = 3$).

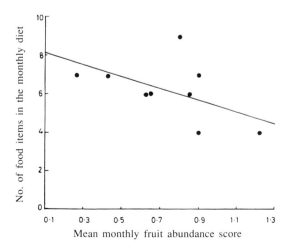

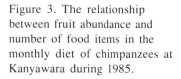

Figure 3. The relationship between fruit abundance and number of food items in the monthly diet of chimpanzees at Kanyawara during 1985.

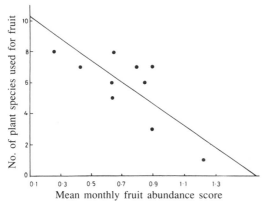

Figure 4. The relationship between fruit abundance and the number of plant species utilized for fruit each month by chimpanzees at Kanyawara during 1985.

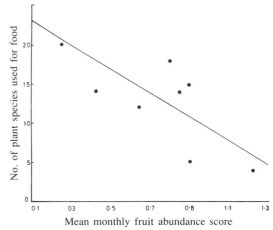

Figure 5. The relationship between fruit abundance and the number of plant species utilized for food each month by chimpanzees at Kanyawara during 1985.

Significantly fewer plant species were utilized for fruit each month when fruit was abundant than when it was scarce ($r = -0.74$, $p < 0.05$, df = 8, Fig. 4) during 1985. Similarly, significantly fewer plant species were utilized for food each month when fruit was abundant than when it was scarce ($r = -0.72$, $p < 0.05$, df = 8, Fig. 5). In spite of this, chimpanzees did not appear to have foraged optimally. For instance, chimpanzees did not exclude food items occurring at low densities (and apparently low ranking) when fruit was abundant. This was particularly so with regard to the short forest fig *(Ficus urceolaris* [Welw. ex Hiern]).

The proportion of fruit in the monthly diet was not significantly correlated with fruit abundance ($r = +0.19$, $p > 0.05$, df = 8). This lack of significance may be associated with the difficulties of measuring fruit abundance discussed above. But it is also confounded with problems associated with food preference. For example, in some seasons, one fruit type (e.g., *Conopharyngia holstii* [K. Schum Slapf]) was heavily utilized when no alternative fruit source was available. But this was not the case in other seasons when alternative sources of fruit were available. Furthermore, it is also possible that chimpanzees, being specialist frugivores, spent more time eating fruit than other food items regardless of fruit abundance. If this were the case, then chimpanzees were utilizing fruit in a non-optimal manner.

However, neither the number of plant species nor the number of food items utilized each month are good measures of diet composition. This is because neither indicates the proportions of each plant species (or food items) utilized. It was therefore necessary to calculate monthly indices of diet diversity. Several methods of measuring biological diversity based on H´ indices are now available (Pielou, 1969). Though some workers (e.g., Hill, 1973; and Swindel et al., 1984) have argued that there is no

Figure 6. The indices of diet diversity calculated for plant species and number of food items eaten by chimpanzees at Kanyawara during 1984 and 1985.

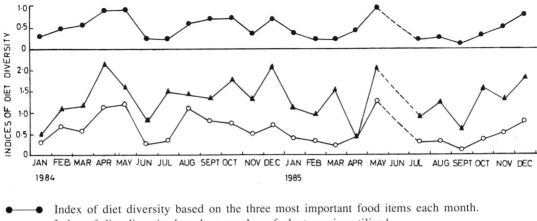

●——● Index of diet diversity based on the three most important food items each month.
▲——▲ Index of diet diversity based on number of plant species utilized.
○——○ Index of diet diversity based on number of food items.

biological basis for using one index as opposed to another, the Shannon-Wiener index is the most commonly used. Here:

$$H' = \sum_{i=1}^{n} (p_i) \ln (p_i)$$

where p_i is the proportion of the total sample represented by plant species or food item i. Still other workers have argued against using H' indices in determining diversity on the grounds that such indices lump disparate phenomena and confound biological interpretation (e.g., Hurlbert, 1971). However, the diversity index can be improved upon by calculation of an equitability index. For example, Hill (1973) used the following equitability index:

$$E = \frac{\text{Exp} (H')}{S}$$

where Exp (H') is the number of equally common plant species that would produce the observed H' value, and S is the number of plant species actually observed.

Two diet diversity indices were calculated using the Shannon-Wiener equation, one based on the number of food items and the other on the number of plant species utilized each month. Indices of diet diversity based on plant species were in the majority of cases higher than those based on food items (Fig. 6). Both types of indices fluctuated greatly over the two years. In the case of those based on food items, some of the fluctuation can be eliminated if the indices are calculated using the three most important food items each month (Fig. 6, upper graph).

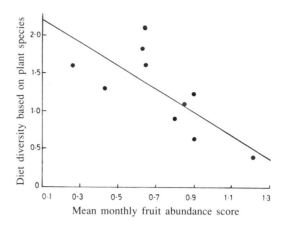

Figure 7. The relationship between fruit abundance and diet diversity based on plant species used for food each month at Kanyawara during 1985.

Table 1. Summary of correlations among different variables.

Variables	Correlation coefficient (r)	df	Level of significance
Fruit abundance and number of food items in monthly diet	−0.48	8	n s
Fruit abundance and number of plant species utilized for fruit	−0.74	8	*
Fruit abundance and number of plant species utilized for food each month	−0.72	8	*
Fruit abundance and proportion of fruit in monthly diet	0.19	8	n s
Diet diversity (plant species utilized each month) and fruit abundance	−0.70	8	*
Diet diversity (food items eaten each month) and fruit abundance	−0.23	8	n s
E. I. (based on food items) and fruit abundance	0.18	8	n s
E. I. (based on number of plant species eaten each month) and fruit abundance	0.16	8	n s
Diet diversity (based on food items) and proportion of diet made up of fruit	−0.96	21	***
Diet diversity (based on plant species) and proportion of diet made up of fruit	−0.62	21	**
E. I. (of food items) and proportion of diet made up of fruit	−0.69	21	**
E. I. (number of plant species) and proportion of diet made up of fruit	−0.54	21	**
Diet diversity (food items) and proportion of diet made up of leaves	0.74	21	**
Diet diversity (number of plant species) and proportion of diet made up of leaves	0.58	21	**
Selection ratio and food plant species relative abundance	−0.45	20	*

Notes: * = p < 0.05, ** = p < 0.01, *** = p < 0.001, ns = not significant.
     E. I. = equitability index, df = degrees of freedom.

If chimpanzees were foraging optimally with regard to fruit, then (1) diet diversity should be negatively correlated with fruit abundance (this should be the case because fewer plant species and food items will be utilized when fruit is abundant than when it is scarce), and (2) the proportion of the diet made up of fruit should be negatively correlated with diet diversity, while that of leaves, the second most important food item, should be positively correlated. This will be so because the proportion of the monthly diet made up of fruit was found to be inversely correlated with that made up of leaves ($r = -0.84$, $p < 0.001$, df = 21).

Using data for 1985, diet diversity based on the number of plant species utilized each month was significantly inversely correlated with fruit abundance ($r = -0.70$, $p < 0.05$, df = 8, Fig. 7). However, this was not the case when diet diversity was based on food items eaten each month ($r = -0.23$, $p > 0.05$, df = 8). Equitability indices were not significantly correlated to fruit abundance ($r = 0.18$, $p > 0.05$, df = 8, for those based on food items; $r = 0.16$, $p > 0.05$, df = 8, for those based on plant species eaten).

Diet diversity based on food items was highly inversely correlated with the proportion of the diet made up of fruit ($r = -0.96$, $p < 0.001$, df = 21). This was also the case for those based on plant species eaten ($r = -0.62$, $p < 0.01$, df = 21). In addition, equitability indices were also strongly inversely correlated with the proportion of the diet made up of fruit ($r = -0.69$, $p < 0.01$, df = 21, for those based on food items; and $r = -0.54$, $p < 0.01$, df = 21, for those based on plant species). The proportion of leaves in the monthly diet was positively correlated with diet diversity based on food items ($r = 0.74$, $p < 0.01$, df = 21), and that based on plant species ($r = 0.58$, $p < 0.01$, df = 21). The results of correlation coefficient of various variables are shown in Table 1.

Though some of the data presented in this section support some of the predictions, overall, chimpanzees did not appear to be following a simple optimal foraging model.

DIET COMPOSITION IN RELATION TO PLANT SPECIES ABUNDANCE

Though the chimpanzees did not appear to utilize fruit sources optimally, it is still possible that they foraged optimally with regard to the few plant species in which they specialized. If this were the case, individual plant species would have been utilized in proportion to their abundance. One way of investigating this possibility is through calculation of plant selection ratios (SR) for each important plant species on a monthly basis. This can be done as follows:

$$\text{SR for species A} = \frac{\% \text{ of feeding scores on species A} \times 100}{\text{basal area (m}^2\text{/ha) of species A} \times \text{PS for sp A} \times 1000}$$

The division by 1000 is only for purposes of reducing the calculated SR to reasonable proportions. This ratio follows Harrison (1984) but differs from his in several important respects.

Table 2. The densities (stems/ha) and basal areas (m²/ha) of some of the important food tree species in the diet of chimpanzees at Kanyawara, Kibale Forest, in compartments K30, K14, and K15.

Tree species	Mean basal area (m²)	Density (stems/ha) K30	K14	K15	Mean density (no/ha)	Mean basal area (m²/ha)
Mimusops bagshawei	0.78 [11][a]	2.1[e]	3.0[c]	2.2[e]	2.4	1.90
Ficus exasperata	1.27 [3]	0.8[d] 1.0[c]	2.4[d] 0.6[c]	0.2[d]	0.8	1.04
Ficus dawei	3.20 [4]	1.4[b]	–	0.2[e]	0.53	1.71
Ficus brachylepis	3.95 [4]	1.6[f]	0.8[d] 1.0[c]	0	0.85	3.36
Pseudospondias microcarpa	1.14 [70]	2.0[f] 1.4[b]	0	0	0.85	0.97
Ficus natalensis	5.0 [9]	0.8[d]	0	0	0.27	1.33
Celtis durandii	0.13 [93]	35.6[c] 34.4[b]	38.8[f] 29.5[d]	31.0[c] 73.9[e]	77.7[e]	54.9	7.14
Celtis africana	0.23 [5]	2.0[c] 16.20[e]	8.4[c] 15.2[e]	13.2[e]	11.4	2.61
Uvariopsis congensis	0.02 [189]	25.2[b] 4.5[c]	74.7[f] 10.9[e]	13.5[c] 10.9[e]	0	13.7	0.27

a. Numbers in brackets indicate the number of specimens measured.
b. Struhsaker, T.T. (1975)   e. Kasenene, J. M. (pers. comm.)
c. Oates, J. F. (1974)   f. My own tree enumeration results.
d. Skorupa (1988)

In his calculation of SR, Harrison (1984) used total available canopy cover as the denominator, arguing that this was an absolute measure, since it incorporated canopy sizes, phenological scores, and tree densities. I did not use total canopy cover for several reasons. First, the community range of the Kanyawara chimpanzees is about 17 km². This makes it impossible to determine total canopy cover even for a few important food-plant species. I therefore used basal area (m²/ha) as an index of canopy cover. Second, while total canopy cover for any given species may not vary much from one season to another, the amount of food (particularly fruit) produced by the plant species may vary tremendously from one season to another. It was for this reason that I multiplied basal area for each species by the mean phenological score for the same species. This has, however, meant that tree species for which no phenological data on fruit abundance are available have been excluded from the analysis below.

The data used for the calculation of stem density and mean basal areas for various plant species were obtained from several sources (Table 2). For some tree species no data on stem density are available for some compartments. A density of 0 stems/ha has been recorded in such cases. Mean stem density for the whole area was based on the weighted mean for the three compartments. This practice probably led in some

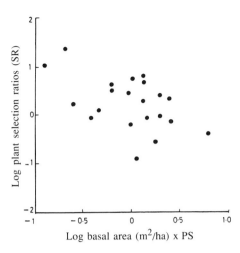

Figure 8. The relationship between relative abundance of plant species used for fruit by chimpanzees at Kanyawara and plant selection ratios (SR).

cases (e.g., *Uvariopsis*) to obtaining a lower, and in other cases (e.g., *F. dawei*) higher, density than the actual stem density.

Plant species which form a high proportion of the monthly diet but have a low relative abundance will have high SR. The plant selection ratios obtained varied tremendously, even for the same plant species. This implies that chimpanzees had a wide range of selectivity for plant species they utilized for fruit. The highest SR were obtained for *Mimusops* during February 1985; *Ficus natalensis* for September 1985; *F. exasperata* for August 1984 and April 1985; and *Pseudospondias* for September 1985.

However, plant selection ratios may be misleading as each value is determined largely by availability of the plant species and very little by its proportion in the diet (Harrison, 1984). This bias can be eliminated by plotting the SR calculated against abundance of trees that bore fruit (i.e., basal area [m²/ha] x PS). Plant selection ratios were found to be significantly and negatively correlated with plant-species abundance (r = –0.45, df = 20, p < 0.05, Fig. 8). In other words, chimpanzees did not utilize the various plant species for fruit in proportion to their abundance.

DISCUSSION

In this paper some aspects of chimpanzee feeding ecology were investigated in terms of the MacArthur and Pianka (1966) model. Chimpanzees appeared to have foraged according to some of the predictions of this model. For example, fewer plant species were utilized for fruit when fruit was abundant than when it was scarce. Furthermore, chimpanzees utilized fewer plant species for food during periods of fruit abundance than in those of fruit scarcity. However, chimpanzees, on the whole, did not forage optimally. For example, chimpanzees continued to exploit food patches occurring at low densities (e.g., those of *F. ureolaris*) for fruit at times when other species bore abundant fruit. Food patches occurring at low densities were usually exploited when chimpanzees were travelling from one rich fruit patch to another. Chimpanzees also ate substantial amounts of leaves in some months when abundant fruit was available. This was particularly so with regard to the young leaves of *Chaetacme* and *Celtis africana*. These were probably eaten to balance the diet. For example, young leaves of *C. africana* are very rich in protein (T. Struhsaker, pers. comm.). Chimpanzees also appeared to have eaten large quantities of fruit regardless of fruit abundance. However, this may be connected to sampling methods used. For example, fruit formed a high proportion of the diet in April 1985, but this was because the few times chimpanzees were seen that month, they were found eating fruit. However, the fruit of some plant species was definitely eaten out of proportion with its abundance. This was particularly so for *Mimusops* and *Pseudospondias* in some months.

Several interesting problems which require further investigation emerged from this study. For example, why did chimpanzees eat substantial amounts of leaves when one type of fruit was abundant but not when another type of fruit was abundant? This was probably because some fruit types provide more nutrients than others. It will also be interesting to find out why the same type of fruit is utilized more in some seasons but not in others, as was the case with *Ficus exasperata*. However, before these questions can be answered, it will be necessary to collect more data on fruit abundance on all important food-tree species and on non-fruit food items. It also appears that feeding scores on food items eaten by chimpanzees from terrestrial herbaceous vegetation were under-represented. This led to fruit (a food item eaten from trees, where visibility was much better) being over-represented in most samples. There is also a need to develop more reliable methods of assessing fruit abundance.

SUMMARY OF RESULTS

Two aspects of chimpanzee feeding ecology were investigated using optimal foraging theory. These were the relationship between fruit abundance and diet composition and diversity, and diet composition in relation to the relative abundance of plant species utilized for fruit.

1. The number of food items eaten by chimpanzees each month was inversely (but not significantly) correlated with fruit abundance.

2. Chimpanzees utilized significantly fewer plant species for fruit when fruit was abundant than when it was scarce. Similarly, significantly fewer plant species were utilized for food when fruit was abundant than when it was scarce.

3. Indices of diet diversity based on the number of food items eaten each month were lower, in most cases, than those based on the number of plant species utilized each month.

4. Diet diversity based on the number of plant species utilized each month was significantly inversely correlated with fruit abundance. Furthermore, indices of diet diversity were inversely correlated with the proportion of fruit in the monthly diet.

5. Chimpanzees did not utilize the plant species they exploited for fruit in relation to the plants' relative abundance.

REFERENCES

Ghiglieri, M. P. 1984. Feeding ecology and sociality of chimpanzees in Kibale Forest, Uganda. In P. S. Rodman and J. G. H. Cant, eds., *Adaptations for Foraging in Non-Human Primates.* New York: Columbia University Press.

Glander, K. E. 1981. Feeding patterns in mantled howling monkeys. In A. G. Kamil and T. D. Sargent, eds., *Foraging Behaviour: Ecological, Ethological and Psychological Approaches*, pp. 231–259. New York and London: Garland STPM Press.

Goodall, J. 1968. Behaviour of free-living chimpanzees of the Gombe Stream area. *Anim. Behav. Monogr.* 1:163–311.

———. 1971. *In the Shadow of Man.* Boston: Houghton Mifflin.

Hamilton, W. J., R. E. Buskirk, and W. H. Buskirk. 1978. Omnivory and utilization of food resources by chacma baboons, *Papio ursinus. Am. Nat.* 112:911–924.

Harrison, J. H. S. 1984. Optimal foraging strategies in the diet of the green monkey, *Cercopithecus sabaeus,* at Mt. Assirik, Senegal. *Int. J. Primatol.* 5:35–470.

Hill, M. O. 1973. Diversity and evenness: a unifying notation and its consequences. *Ecology* 54:427–432.

Hurlbert, S. H. 1971. The nonconcept of species diversity: a critique and alternative parameters. *Ecology* 52:577–586.

Krebs, J. R. 1978. Optimal foraging rules for predators. In J. R. Krebs and N. B. Davies, eds., *Behavioural Ecology,* pp. 23–63. Sunderland, MA.: Sinauer Associates.

MacArthur, R. H., and E. R. Pianka. 1966. On optimal use of a patchy environment. *Am. Nat.* 102:381–383.

Milton, K. 1980. *The Foraging Strategy of Howler Monkeys: A Study in Primate Economics.* New York: Columbia University Press.

Nishida, T. 1968. The social group of wild chimpanzees in the Mahale Mountains. *Primates* 9:167–224.

Oates, J. F. 1974. The ecology and behaviour of the black-and-white colobus monkey (*Colobus guereza* Ruppel) in East Africa. Ph.D. thesis, University of London, London.

Pielou, E. C. 1969. *An Introduction to Mathematical Ecology.* New York: Wiley-Interscience.

Post, D. G. 1984. Is optimization the optimal approach to primate foraging? In P. S. Rodman and J. G. H. Cant, eds., *Adaptations for Foraging in Non-Human Primates,* pp. 280–304. New York: Columbia University Press.

Reynolds, V., and F. Reynolds. 1965. Chimpanzees in the Budongo Forest. In DeVore, ed., *Primate Behaviour: Field Studies of Monkeys and Apes,* pp. 368–424. New York: Holt, Rinehart and Winston.

Schoener, T. W. 1971. Theory of feeding strategies. *Ann. Rev. Ecol. Syst.* 2:369–404.

Skorupa, J. P. 1988. Effects of selective timber harvesting on a community of primates in Kibale Forest. Ph.D. dissertation. University of California-Davis.

Struhsaker, T. T. 1975. *The Red Colobus Monkey.* Chicago: University of Chicago Press.

Sugiyama, Y. 1973. The social organization of wild chimpanzees: a review of field studies. In R. P. Michael and J. H. Crook, eds., *Comparative Ecology and Behaviour of Primates,* pp. 375–410. London: Academic Press.

Swindel, B. F., L. F. Conde, and J. E. Smith. 1984. Species diversity: concept, measurement, and response to clear cutting and site-preparation. *For. Ecol. and Manage.* 8:11–22.

Wrangham, R. W. 1975. Behavioural ecology of chimpanzees in the Gombe National Park, Tanzania. Ph.D. thesis, Cambridge University.

RECENT RESEARCH ON CHIMPANZEES IN WEST AFRICA

William C. McGrew

INTRODUCTION

This paper has two aims: (1) to give an overview of recent field work on wild chimpanzees in West Africa; (2) to describe in more detail the studies in Gabon by my associates, C. Tutin and M. Fernandez.

The first is a daunting task: by rough reckoning, there are at least 29 countries in Africa which, at least in principle, could have populations of wild chimpanzees (see Table 1). Twenty-one of these are in what is bio-geographically West Africa. They contain three of the four recognized forms of the chimpanzee: *Pan troglodytes troglodytes*, *P. t. verus*, and *P. paniscus,* thus leaving only *P. t. schweinfurthii*.

Before going on, some guidelines are needed: by "West Africa" is meant basically Africa west of the Great Rift. This can be further subdivided into central-western Africa, from the Rift to the Dahomey Gap, and far-western Africa, from the Dahomey Gap westwards. For "recent," the 1980s have been arbitrarily chosen. "Could have" means only that at least some part of the country has suitable habitat in terms of vegetation and rainfall. Finally, it bears remembering that political boundaries need not map meaningfully onto biotic ones, but research opportunities depend more on the former than on the latter, and so are used here.

First, it should be noted that of the 21 countries in West Africa, not all have seen equal amounts of research, and some have seen little or none. Thus, I will refer not only to accomplishments but also to gaps in knowledge. Second, the countries can be classed conveniently into thirds: seven in which long-term field studies have been done ("long-term" means studies lasting at least one annual cycle); seven in which chimpanzees are known to occur now and have been studied in the short-term; and seven in which chimpanzees are known to be absent or have not been recently confirmed to be present. I will concentrate on the first two classes. Third, researchers who have worked in some of these countries are represented in this volume, and so I will defer to their firsthand reports (Kano, Kuroda, Malenky et al., and White on Zaire; Sugiyama on Guinea; Teleki on Sierra Leone).

Long-term studies of apes are to be preferred on many grounds, not least of which are the variety, complexity, and longevity of the subjects. Minimally, at least one

128

Table 1. African countries that have or may have chimpanzee populations in the wild.[1]

West *(P. t. verus)*	Central *(P. paniscus)* *(P. t. troglodytes)*	East *(P. t. schweinfurthii)*
* Ivory Coast	* Equatorial Guinea	* Tanzania
* Senegal	* Zaire	* Uganda
* Guinea	* Gabon	Zaire
* Sierra Leone	Cameroon	Sudan
Liberia	Central African Republic	(Rwanda)
Mali	Congo	(Burundi)
Ghana	(Nigeria)	(Zambia)
Guinea–Bissau	(Angola–Cabinda)	(Malawi)
(Gambia)		(Kenya)
(Burkina Faso)		
(Togo)		
(Benin)		
(Niger)		

Notes: * = Site of a long-term study, i.e., more than 12 months. ( ) = Absent, or current presence unconfirmed.
1. Zaire appears in the table twice.

annual cycle is needed to study subjects living in seasonal environments which may present yearly "bottlenecks," such as dry-season shortage of water. Even better are studies of two or more such cycles, so that year-to-year variation can be assessed. Finally, systematic data collection over time reduces the risk of superficial or even unrepresentative impressions.

FIELD STUDIES IN WEST AFRICA

The most notable long-term study of the 1980s in West Africa is that of C. and H. Boesch in the Tai Forest of Ivory Coast (e.g., Boesch and Boesch, 1983). Their findings on the use of hammers of wood and of stone to crack open several species of nuts have been truly revelationary. To cite but one example, prehistorians of Africa will never again be able to look in the same way at the archaeological record. Here are non-perishable tools of a lithic industry found in a limited area of apparent cultural diffusion. However, the Boeschs' results are not just on tool use; their recently presented data on frequent social hunting of red colobus monkeys are equally important. Further, they have been able to habituate wild chimpanzees to human observation without provisioning and in the most demanding of habitats for this, the evergreen forest. This is a great accomplishment.

Elsewhere in Ivory Coast further work remains to be done, perhaps most interestingly in the Comoe National Park in the north. Chimpanzees occur in this large savanna park, but no details are known. A preliminary survey focusing on chimpanzees is scheduled.

For Senegal, on the far northwestern edge of the species' distribution, there are only a few chimpanzees in the southeastern part of the country, in Senegal Oriental. Studies have so far concentrated on Mt. Asserik in the Niokolo-Koba National Park (McGrew et al., 1981). Since the Stirling African Primate Project finished there in 1980, I know of only one follow-up study, a short-term one by Magda Bermejo, a student of Sabater Pi's at Barcelona. This remains unpublished. Given the vulnerability and small numbers of apes in Senegal, more work needs to be done there. We found chimpanzees outside the park, between its southern boundary and the border with Guinea, in Pays Bassari. Chimpanzees likely exist to the east of the park too, along upstream tributaries of the Gambia River. Both areas need surveying. Finally, the extensive development project involving several dams on the Gambia could, for example, markedly affect riverine forest in the area, and this needs to be monitored.

Guinea has the longest record of chimpanzee field study of any African country, from Nissen's pioneering efforts (1931), to Albrecht and Dunnett's later work (1971), to Sugiyama's current research (this volume). Much has centered on Bossou in the Mt. Nimba region in the southeast of the country. However, other parts of Guinea need attention too, especially the Fouta Djallon region to the north. De Bournonville's excellent survey (1967), made 20 years ago, needs replication and extension.

Sierra Leone has been more newsworthy, recently, with regard to the conservation of chimpanzees than to field research on them. The news has been both good and bad; and, in a sense, Sierra Leone pointedly exemplifies the conflicting forces which exist in many countries where wild chimpanzees live. One hopes that scientific research on chimpanzees, such as that at Kilimi in the northwest (Harding, 1984), will continue to progress.

In west-central Africa, much research was done by Sabater Pi in Equatorial Guinea (then Rio Muni), but this is now dated. It remains to be seen if viable populations of chimpanzees survive at such sites as Okorobiko, so that more intensive follow-up research can be done. This is especially important in Equatorial Guinea, where the first attempts at comparative studies of sympatric chimpanzees and gorillas were made (Jones and Sabater Pi, 1971).

In Zaire studies of bonobos are underway at two productive sites, Lomako and Wamba. Both sites are well represented here by their workers, so further details will come best from them. In a nutshell, Lomako builds on the founding work of A. and N. Badrian, eschews provisioning, and emphasizes the research viewpoint of American physical anthropology. Wamba makes use of provisioning, and exemplifies the breadth of behavioral focus of Japanese field primatology. In spite of differences in method and outlook, one hopes for collaboration close enough for truly comparative research to emerge. Lessons learnt from the obvious parallel of Gombe and Kasoje (Goodall, Nishida, this volume) in Tanzania could well be applied.

Further, it needs pointing out that the common chimpanzee appears to occur widely in Zaire too. Sadly, Kortlandt's (1962) early field work in eastern Zaire has not been followed up, even in places where gorillas have been studied, and where comparative studies of the two great apes could, at least in principle, be done, such as in the Kahuzi-Biega National Park. Apparently totally neglected is the southeastern part of Zaire (Shaba-Katanga) where wild chimpanzees may reach their southernmost point on the continent. One hopes for the day when political stability will allow research to begin.

Field study in Gabon is covered in the second half of this paper, as indicated above.

Of the four far western countries in which at least some chimpanzees are still left in the wild, Ghana and Guinea-Bissau are little known to me. Clarification is needed. In Mali, J. Moore's short survey of 1984 (Moore, 1985) has confirmed the presence of chimpanzees in the Bafing area in the southwest of that country. The hot, dry, and open habitat looks to be an intriguing one for further study of savanna-dwelling chimpanzees.

In Liberia, the only work so far on wild chimpanzees was a three-month survey in the Sapo National Park in 1982 (Anderson et al., 1983). Though short, it revealed meat eating and stone-tool use much like that found at Tai. Further study for point-by-point comparison with the Boeschs' work would be most useful.

Of the three central-western countries, Cameroon has received the most attention from field students of primates, in the past as well as in the present. However, this research has focused on other species, and I know only of Sugiyama's (1985) recent short study there on chimpanzees. Given Cameroon's variety of habitats in roughly a north-south gradient and the considerable conservational and scientific investment already made, including study of lowland gorillas at Campo in the southwest, it would seem a high priority to push on with further fieldwork on chimpanzees in Cameroon.

Of Congo (Brazzaville) and Central African Republic, next to nothing was known of the status of chimpanzees until very recently. So far as I know, nothing is yet in print, though initial work is under way in both. This is good news, as both countries would be expected to have large populations of both chimpanzees and gorillas.

Finally, of the seven countries where chimpanzees are known to be absent, or where their present status is unconfirmed, the data vary enormously. Some clearly have no chimpanzees (e.g., Gambia); some had chimpanzees until recently but up-to-date data are lacking (e.g., Nigeria); some apparently had chimpanzees at the time of European contact (e.g., Angola); and some may never have had chimpanzees (e.g., Niger). Needless to say, further information should be sought.

FIELD STUDY IN GABON

Gabon is an extraordinary country in which to study wildlife. About 85% of its surface area of 267,667 km^2 is covered in tropical rain forest, most of it undisturbed. The human population numbers 1.2 million at most, and large areas away from roads and rivers remain unused, except in minimal, traditional ways. A wealth of other natural resources means that pressure from logging and agriculture is relatively low.

Field study in Gabon of wild chimpanzees and gorillas was minimal until recently. Only Hladik's (1973) study of semifree-ranging chimpanzees released onto an island provided data on naturalistic life. Then, in 1980, Tutin and Fernandez began an ongoing study in two stages: (1) nationwide survey and census of habitats and ape populations (December, 1980 to February, 1983), and (2) focused comparative ecology and ethology of sympatric chimpanzees and gorillas in the Lope-Okanda Reserve (September, 1983 to the present [1986]). Both stages have been financed by the Centre International de Recherches Medicales de Franceville (Gabon), in collaboration with the University of Stirling (Scotland). In the second stage, further finance has come from the L.S.B. Leakey Foundation (U.S.A.) and collaboration from the University of Edinburgh (Scotland).

The first part of the survey and census was done in an initial survey sector of just over 6,000 km^2 in northeastern Gabon. This area was divided into a grid of 83 squares (or partial squares) with sides 10-km long. Within each of these, up to 10 km of transects were made to record vegetation types and sleeping nests. In all, 783 km of transects in 15 types of vegetation yielded 1,606 chimpanzees' nests. From these data, mean densities of apes per km^2 were calculated for various types of habitats. The rest of the country was then divided into 23 sectors, and further transecting was done there. These data, plus existing knowledge of vegetation distribution, and adjustments to take account of such factors as hunting pressure, allowed for the eventual computation of numbers of wild chimpanzees in Gabon. The result was a figure of 64,000 ± 13,000 (Tutin and Fernandez, 1984). This was simply the most thorough large-scale survey and census ever done on nonhuman primates.

The results of the survey were used to choose an optimal site for long-term socioecological study. This led to the founding of the Station d'Etudes des Gorilles et Chimpanzees at Lope. Research began in September, 1983, and the construction of the camp was completed in early 1984. It now comprises permanent housing for five researchers, and its facilities include a library and laboratory with electricity, refrigeration, and running water. The camp is fully equipped, with two 4-wheel drive vehicles and both long- and short-range ("walkie-talkie") radios.

Much time was spent in the beginning in exploring and mapping the study area and in setting up long-term data collection systems for climatology and plant phenology. From the beginning, indirect evidence, such as nests and fecal samples, as well as encounters with the apes, has been systematically recorded. No provisioning is being used, so habituation depends on cumulative familiarity through meetings in the forest.

Early results on the chimpanzees at Lope suggest that they differ in unexpected ways from forest-dwelling chimpanzees elsewhere. Apparently the Lope apes range more widely and live at lower densities than do, for example, chimpanzees in the Budongo and Kibale Forests of Uganda. Some notable aspects of chimpanzee life recorded elsewhere in West Africa are present, such as eating of colobus monkeys; but others appear to be absent, for example, use of hammers to break open *Panda* nuts. Interestingly, there seem to be differences between chimpanzee populations within Gabon: at Belinga in the northeast, fishing probes apparently are used to obtain *Macrotermes* termites (McGrew and Rogers, 1983; Tutin and Fernandez, 1985); at Lope in central Gabon this has yet to be found. Finally, it appears that chimpanzees and gorillas at Lope compete for certain food items; for instance, individuals of each species of ape have been seen eating fruits in the same tree on consecutive days.

Overall, my conclusions are not surprising: Studies of chimpanzees in West Africa have finally "taken off," though they have a long way to go before matching those on the eastern shore of Lake Tanganyika. Many opportunities still exist for further studies in several of the countries noted above. It is up to field workers to pursue them.

ACKNOWLEDGMENTS

I thank C. Tutin and M. Fernandez for permission to cite their unpublished data; and the University of Stirling Research Fund, Boise Trust, and Carnegie Trust for the Universities of Scotland for support of my work in Gabon in 1981 and 1985.

REFERENCES

Albrecht, H., and S. C. Dunnett. 1971. *Chimpanzees in Western Africa.* Munchen: Piper.

Anderson, J. R., E. A. Williamson, and J. Carter. 1983. Chimpanzees of the Sapo Forest, Liberia: density, nests, tools, and meat-eating. *Primates* 24:594–601.

Boesch, C., and H. Boesch. 1983. Optimisation of nut-cracking with natural hammers by wild chimpanzees. *Behaviour* 83:265–286.

de Bournonville, D. 1967. Contribution à l'étude du Chimpanzé en République de Guinée. *Bulletin de l'Institut Fondamental d'Afrique Noire* 39A:1188–1269.

Harding, R. S. O. 1984. Primates of the Kilimi area, northwest Sierra Leone. *Folia Primatologica* 42:96–114.

Hladik, C. M. 1973. Alimentation et activite d'un group de chimpanzes reintroduits en foret gabonaise. *Terre et Vie* 27:343–413.

Jones, C., and J. Sabater Pi. 1971. *Comparative ecology of* Gorilla gorilla *(Savage and Wyman) and* Pan troglodytes *(Blumenbach) in Rio Muni, West Africa.* Basel: S. Karger.

Kortlandt, A. 1962. Chimpanzees in the wild. *Scientific American* 206:128–138.

McGrew, W. C., P. J. Baldwin, and C. E. G. Tutin. 1981. Chimpanzees in a hot, dry, and open habitat: Mt. Asserik, Senegal, West Africa. *J. Hum. Evol.* 10:227–244.

McGrew, W. C., and M. E. Rogers. 1983. Chimpanzees, tools, and termites: new record from Gabon. *Am. J. Prim.* 5:171–174.

Moore, J. 1985. Chimpanzee survey in Mali, West Africa. *Prim. Conservation* 6:59-63.

Nissen, H. W. 1931. A field study of the chimpanzee. *Comparative Psychology Monographs* 8:1–122.

Sugiyama, Y. 1985. The brush-stick of chimpanzees found in south-west Cameroon and their cultural characteristics. *Primates* 26:361–374.

Tutin, C. E. G., and M. Fernandez. 1984. Nationwide census of gorilla (*Gorilla g. gorilla*) and chimpanzee (*Pan t. troglodytes*) populations in Gabon. *American Journal of Primatology* 6:313–336.

_____. 1985. Foods consumed by sympatric populations of *Gorilla g. gorilla* and *Pan t. troglodytes* in Gabon: some preliminary data. *Int. J. Prim.* 6:27–43.

POPULATION DYNAMICS OF CHIMPANZEES AT BOSSOU, GUINEA

Yukimaru Sugiyama

ENVIRONMENT AND ADJACENT POPULATIONS

A major part of the core area where the Bossou group of chimpanzees *(Pan troglodytes)* lives consists of different stages of secondary forest, an area of 5–6 km² (Fig. 1). A lesser part is primary forest, and a small part is cultivated field. Within that range and its periphery, people neither kill nor capture the Bossou chimpanzees (except a rare case in 1976), and the author was able to obtain much information from local people (Sugiyama and Koman, 1979a). The core area of the Bossou group does not overlap that of any neighboring group of chimpanzees (Fig. 2). Between the core areas of the chimpanzee groups are belts 5–6 km wide, on the average, consisting of savannah and patched secondary forests. These forests are not useful for the chimpanzees, but the barrier is not too great for individual migrants to pass through. At an adjacent chimpanzee habitat to the east of Bossou, the Nimba Mountains area, poachers hunt animals, and the chimpanzee population density there is presently low, especially in the southern area along the frontier of Liberia. About 1984, a very small group of chimpanzees began to reside to the west of Bossou, in Gba. For further information on the environment and behavior of chimpanzees at Bossou, refer to Sugiyama (1981), and Sugiyama and Koman (1979a, 1979b).

BOSSOU GROUP OF CHIMPANZEES

Since 1967, the Bossou group of chimpanzees has averaged about 20 members. From 1976 to 1986, all members of the group were individually identified during four study periods, which included 410 days of field observations (Figs. 3a-g) (Sugiyama and Koman, 1979a; Sugiyama, 1981, 1984). The number of adult females and offspring in the group at any one time – 17 or 18 individuals – has changed little throughout the last ten years (Fig. 4). The members of the adult female population (seven individuals, one of whom was the senior adolescent at the beginning of the study) never changed, so seven kin-groups and their members can be clearly recognized. On the average, each mother had 1.63 offspring in the group at a given time (the

Figure 1. Center of the core area – Bossou group of chimpanzees.

beginning of each study period). Neither emigration nor immigration of adult females occurred, and none of them died. Female visitors have never been found within the core area of the Bossou group.

On the other hand, members of the adult male population varied from one to four individuals during the study periods, and only the male second in dominance at the beginning of the study stayed in the group throughout the ten years (Fig. 5). Altogether, five different adult males were confirmed in the group during the four study periods; two of them were confirmed as to their immigration, and at least one was confirmed as to his emigration. None of the adult males appeared to have relatives in the Bossou group, judging from social behavior. The mean sex ratio of matured females to males during the four study periods was 3.71.

Old sleeping beds assumed to have been used by the group members were occasionally found at the periphery of the Bossou group core area, but no evidence of an unidentified "peripheral resident" was suggested by my observation or by information supplied by local people. During the course of field studies, no chimpanzees other

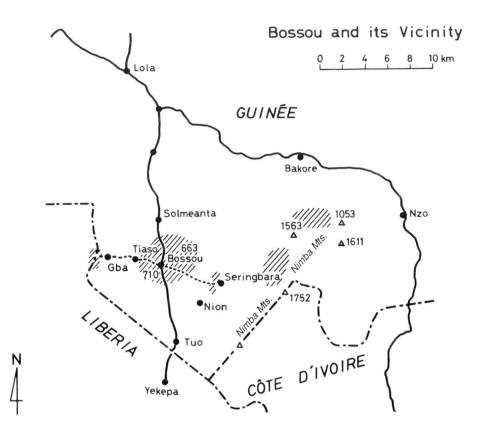

Figure 2. Bossou and its vicinity. Shaded areas indicate confirmed presence of chimpanzees or their sleeping beds. The core area of the Bossou group of chimpanzees is confined to a small range around Bossou. Chimpanzees found at Seringbara were assumed to be migrants from Bossou. The Nimba Mountains extend from Liberia to the Bakore-Nzo road, and the presence of chimpanzees is confirmed only in the northern part.

Figures 3a-g. Some adult members of the Bossou group.

Figure 3a. Bf: dominant male during the 1976-77 study who disappeared before the 1979-80 study.

Figure 3b. Sf: old male who stayed only three weeks in the group.

Figure 3c. Ta: only male who has been in the group throughout the ten years.

Figure 3d. Nn: old female with peculiarly shaped fingers on her left hand.

Figure 3e. Jr: middle-aged female who always has more than one offspring in the group.

Figure 3f. Yo: middle-aged female.

Figure 3g. Vl: middle-aged female who sometimes traveled alone.

Figure 4. Genealogy of chimpanzees in the Bossou group. Each offspring is connected by a broken line to its mother at the estimated year of birth. Adult males are not shown because their kinships are not determined; they are not counted in the total number of chimpanzees indicated at the right of this figure. Confirmed sighting in the Bossou group is indicated by a thick line; the assumed residence of an individual is shown by a broken line. Confirmed or nearly confirmed death is marked as "D," and disappearance from Bossou is shown as "d." Two successive "d" (or D) marks for an individual indicate that disappearance (or death) occurred during that time period; the date of death was placed midway in the interval. "P" shows the first delivery for a given female.

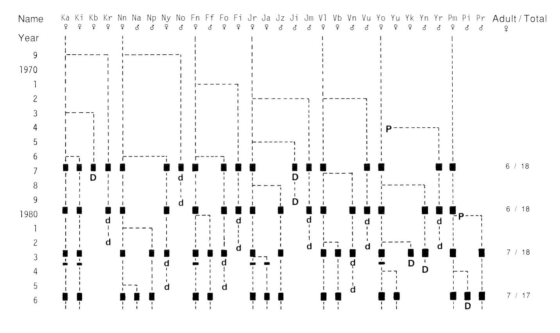

Notes: D = death, d = disappearance, P = primiparity, ■ = confirmed sighting, --- = extrapolated.

than group members listed in Figures 4 and 5 were found by observers or by local people, except an adolescent male visitor in January 1977 (Sugiyama and Koman, 1979a). A carcass of an adult male chimpanzee was discovered by local people within the range of the Bossou group in 1981, but it could not be identified as a group member or a temporary visitor.

Some birth years for new babies and some years of disappearance were confirmed, but other birth years and disappearance years were estimated in each study period. The birth years could be estimated by the body size and behavioral development of the chimpanzee, but the estimated year of disappearance was given as the intermediate year between the date last seen and the first confirmed date of disappearance.

Figure 5. Chimpanzee males at Bossou. This figure indicates all adult males who stayed in the Bossou group more than two weeks without encountering aggression.

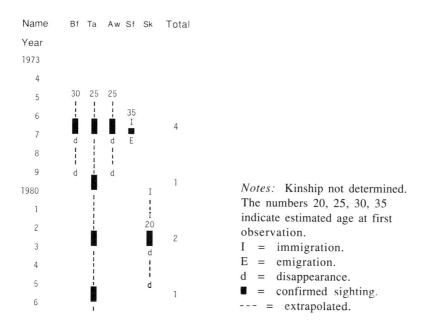

Notes: Kinship not determined. The numbers 20, 25, 30, 35 indicate estimated age at first observation.
I   =   immigration.
E   =   emigration.
d   =   disappearance.
■   =   confirmed sighting.
--- =   extrapolated.

NATALITY RATE AND INTERBIRTH INTERVAL

The natality rate was calculated by comparing the number of adult females living in the Bossou group with the number of infants born between 1972 and 1985. Two younger females were included in the adult class because of their primiparity. Birth years given as 1972 through 1975 were estimated in the year 1976; however, the rate of natality confirmed by the number of infants born between 1976 and 1985 was the same as that of the 1972–1975 period. Therefore, all infants born after 1972 were subsequently used for the calculation of demographic parameters in further studies.

During the 88 adult female years between 1972 and 1985, 20 babies (12 males and 8 females) were born. The natality rate was 0.227 birth/female/yr (Table 1). Thus the

Table 1. Birth record between 1972 and 1985.

Mother		Ka	Nn	Fn	Jr	Vl	Yo	Pm	Total
Offspring	♂		Np, Na	Ff	Jm, Ji, Jz	Vu, Vn	Yr, Yn	Pr, Pi	12
	♀	Kb, Ki	Ny	Fo	Ja	Vb	Yk, Yu		8
Offspring observed 1976-1985		1	3	2	2	2	3	2	15
Adult female reproductive years 1972-1985		14	14	14	14	14	12	6	88
Adult female reproductive years 1976-1985		10	10	10	10	10	10	6	66

Notes: Natality rate = 12/88 + 8/88 = 0.136 ♂ + 0.091 ♀ = 0.227 / ♀ / yr.
Young females are counted as mothers when first observed to be pregnant.

calculated interbirth interval (IBI) is 4.4 years per mother. The observed age gaps between nearest siblings averaged 3.9 years for 13 pairs (range: 2 to 5 yrs). The estimated IBI is 4.2 years for 11 pairs (range: 3 to 5 yrs) where the elder sibling survived at least until the younger one was born. The survival of the elder one was confirmed when the younger sibling was first observed. If it could be determined when a mother became pregnant, the IBI of the latter examples might be shorter than 4.2 years and nearer to 3.9 years.

The natality rate of chimpanzees at Bossou is far higher and the IBI is far shorter than those for chimpanzee groups at Gombe and at Mahale in western Tanzania. The IBI of the Gombe and Mahale groups is about 5.2 to 6 years (Goodall, 1983; Hiraiwa-Hasegawa et al., 1984).

INFANT MORTALITY

The infant mortality rate for the first four years of life was calculated by comparing the number of infants born between 1972 and 1982 and the number of infants who reached four years of age (Table 2). Sixteen babies (ten males and six females) were born, and three (one male and two females) died. The mortality rate was 0.188/4 yrs, and the rate of survival for the first four years was 0.813 (0.950/yr). The number of infants per mother who reached four years of age between 1976 and 1986 was also calculated. We found 0.227 (natality rate) x 0.813 (infant survival rate) = 0.185

Table 2. Mortality of infants (born 1972-1982) and number of offspring reaching four years of age.

Mother		Ka	Nn	Fn	Jr	Vl	Yo	Pm	Survived to 4 yrs	Total born
									Offspring	
Offspring	♂		Np	Ff	Jm, <u>Ji</u>, Jz	Vu, Vn	Yr, Yn	Pr	9	10
	♀	<u>Kb</u>, Ki	Ny	Fo		Vb	<u>Yk</u>		4	6
Adult female reproductive years 1972-1982		11	11	11	11	11	9	3	67	

Notes: Infant mortality for first 4 years = (1 + 2) / (10 + 6) = 0.188 (♂: 0.1, ♀: 0.333).
Infant survival rate for first 4 years = (9 + 4) / (10 + 6) = 0.813 (♂: 0.9, ♀: 0.667).
Infants who died are underlined.

offspring/female/yr reached four years of age. In other words, every 5.41 years one offspring per female reached four years of age.

The infant mortality rate at Bossou is far lower than at Gombe and Mahale (0.3 to 0.5) (Goodall, 1983; Hiraiwa-Hasegawa et al., 1984).

There is a possibility that births and deaths of infants occurred during my absence; if so, however, both the infant mortality rate and the natality rate would go up simultaneously, and the rate of raising offspring to age four per mother would not change.

JUVENILE MORTALITY:
Rate of Raising Offspring to Seven Years of Age

Only one case of juvenile death was recognized: a five- to six-year-old juvenile male carcass was found in 1983. (The death is attributed to the juvenile Yn in Figure 4; however, it is equally probable that the dead individual was Vn.) There were no other disappearances of juveniles aged four to seven. The juvenile mortality rate for the three years from age four to age seven is low (0.083/4 yrs, 0.029/yr) compared to the infant mortality rate (Table 3). The rate of survival from age four to age seven is 0.917/3 yrs (0.972/yr). Using these parameters, the rate of raising offspring to seven years of age per adult female/yr can be calculated as 0.227 (natality) x 0.813 (infant survival rate) x 0.917 (juvenile survival rate) = 0.169.

Table 3. Juvenile mortality and survival rate from age four to age seven.

	Name	Dead / Total	Mortality	Survival rate
♂	No, Jm, Vu, Yr, Vn, Jz, <u>Yn</u>	1 / 7	0.143	0.851
♀	Kr, Fi, Ki, Ny, Fo	0 / 5	0	1
	Total	1 / 12	0.083	0.917

Note: Juvenile who died is underlined.

Table 4. Estimated age at disappearance from Bossou – older juveniles/adolescents.

		Chimps who disappeared						Chimps who did not disappear[a]			
Name	♂	No	Yr	Jm	Vn	Vu	Mean	Jz	Np	Ff	Pr
Estimated age at disappearance		9	7	9	7	9	8.2 yr	8	5	5	5
Estimated age of younger sibling[b]		2	3	3	2	4		2.5	1	—[c]	—[c]
Name	♀	Kr	Ny	Fo	Fi			Ki			
Estimated age at disappearance		12	9	8	10		9.8 yr	10			
Estimated age of younger sibling[b]		5	3	4	5			—[c]			

a. Young individuals (more than four years of age) who remained in Bossou in early 1986; age in 1986.
b. Age of younger sibling when older sibling disappeared.
c. No younger sibling.

DISAPPEARANCE OF OLDER JUVENILES AND ADOLESCENTS

Most older juveniles and adolescents, both males and females, disappeared from the Bossou range (Table 4). Neither the chimpanzees nor carcasses of older juveniles and adolescents have been found since their disappearance. Males disappeared at a younger age (mean: 8.2 yrs old) than females (mean: 9.8 yrs old). When the males and females disappeared from Bossou, their nearest younger siblings were between two and five years of age. In early 1986 there were still two adolescents in Bossou with their mothers. One (an eight-year-old male) had a younger sister 2.5 years old; the other (a ten-year-old female) had no younger siblings as her mother was rather old. Many of the older juveniles and adolescents who disappeared are assumed to have emigrated from Bossou. The factor which drove them away from their natal group must be related primarily to their age and secondarily to the age of their nearest younger sibling.

For this reason, neither the adolescent mortality rate nor the rate of survival can be calculated. If the mortality rate for adolescents between seven and 15 years of age (an eight year span) is the same as that of juveniles (survival rate = 0.972/yr), 0.169 x $0.972^8 = 0.135$ offspring per mother reaches maturity (15 years old) every year and begins to give birth.

REPRODUCTIVE SUCCESS FOR A FEMALE

If all matured females complete their reproductive life (15 to 35 years old as estimated from the records of Gombe and Mahale) and care for their offspring until senility, a female produces 0.135 x 21 = 2.835 offspring, or 1.418 daughters, in her long life span. If only the demographic parameters for daughters born are examined, reproductive success decreases to 0.091 (8/88 = natality) x 0.667 (4/6 = infant survival rate) x 1 (5/5 = juvenile and adolescent survival rate) x 21 (reproductive years) = 1.275 daughters per average female life span. Since the mean reproductive age of a female is estimated to be 25 years old, the population can increase only 1.275 to 1.418 times every 25 years, or 1.0098 to 1.014 times per year.

Although the adolescent mortality rate may be slightly lower than that of juveniles, if the cost of migration and mortality after maturation, as well as predatory pressure, is calculated, the population numbers observed at Bossou must be very near the minimum that provide for stability or very slow increase of the population. If any additional individuals are captured or killed, it will take many years for the population of chimpanzees to recover its size.

It is difficult to believe that the reproductive success of chimpanzees has been so low throughout their evolution of more than several million years. In a certain zoological park where animal treatment is careful, the chimpanzee natality rate is higher than that of Bossou: IBI is 3.17 yrs (range: 2.17 to 5.92 yrs, n = 10 pairs) even when infants were not taken from their mothers; in addition, the age of primiparity for a female is 11.5 years (range: 9.25 to 14.5 yrs, n = 10) (Yoshihara, 1985). Considering the above supplemental facts, it seems most likely that the reproductive success rate of wild chimpanzees in the past, when the habitat supplied much more food, must have been the same or higher than the current reproductive success rate at Bossou.

FACTORS AFFECTING EMIGRATION OF THE TWO SEXES

When resources are restricted, both the males and the females of many species of animals normally leave their mother's range and search for a favorable habitat and group to join. Among primates, many species of Cercopithecidae have female-bonded groups (Wrangham, 1980). In these species, most or all of the males leave their natal group, usually before full maturation, and migrate to another group. By contrast, the females stay in their natal group throughout their lives, bringing up their offspring within the safety of that group.

In the rest of the primate species, including apes (gorillas, orangutans, and gibbons), most females as well as males leave their natal group or range. It can be said that they have a rather generalized type of dispersion.

For a population of chimpanzees like the Bossou group, where the social pressure from neighboring groups does not exist, the carrying capacity of the environment may directly limit the number of females and, as a result, the population size. Then, older juveniles and adolescents must leave their natal group, as seen in many other primate species. This may well be why the population size of chimpanzees at Bossou is little changed, at least in the last 20 years (Sugiyama and Koman, 1979a). If the number of females decreases due to death, adolescent females may continue to stay and mature in the natal group (as seen in Serengeti lions [Bertram, 1975]), or adolescent females may immigrate from neighboring groups. The former is more probable at Bossou, as no female visitor has been observed.[1]

On the other hand, the number of male immigrations into a group is undoubtedly related to the number of resident females. Scarcity of visitors to Bossou, especially females, may also be related to the semi-isolated condition of the habitat. Observations at Bossou indicate that the male population is more likely to be supplemented by occasional immigration.

ACKNOWLEDGMENTS

Field research was financed by the Japanese Ministry of Education, Science and Culture Grant-in-Aid for Overseas Field Research (#60041042), and officially supported by Direction de la Recherche Scientifique et Technique, Republique de Guinee. Jeremy Koman, Aly G. Soumah, and Mouctar Camara joined the field work. Guano Gumi and Tino Camara assisted the research and helped me collect local information. I am grateful to all these organizations and colleagues.

ENDNOTE

1. Ki, the youngest daughter of an old female, Ka, gave birth to her first baby in early 1987, when she was 11 or 12 years old. Grandmother, mother, and baby were still living in their natal group in Bossou as of March 1988.

REFERENCES

Bertram, B. C. 1975. The social system of lions. *Scientific American* May:54–65.

Goodall, J. 1983. Population dynamics during a fifteen-year period in one community of free-living chimpanzees in the Gombe National Park, Tanzania. *Z. Tierpsychol.* 61:1–60.

Hiraiwa-Hasegawa, M., T. Hasegawa, and T. Nishida. 1984. Demographic study of a large-sized unit-group of chimpanzees in Mahale Mountains, Tanzania: a preliminary report. *Primates* 25:401–413.

Sugiyama, Y. 1981. Observations on the population dynamics and behavior of wild chimpanzees at Bossou, Guinea, 1979–1980. *Primates* 22:435–444.

_____. 1984. Population dynamics of wild chimpanzees at Bossou, Guinea, between 1976 and 1983. *Primates* 25:391–400.

Sugiyama, Y., and J. Koman. 1979a. Social structure and dynamics of wild chimpanzees at Bossou, Guinea. *Primates* 20:323–339.

_____. 1979b. Tool-using and -making behavior in wild chimpanzees at Bossou, Guinea. *Primates* 20:513–534.

Wrangham, R. W. 1980. An ecological model of female-bonded primate groups. *Behavior* 75:262–299.

Yoshihara, K. 1985. The data of pregnancy, nursing and growth of chimpanzees, *Pan troglodytes,* at Tama Zoological Park. *Dohsuishi* 27:58–61 (in Japanese).

THE USE OF STONE TOOLS
BY WILD-LIVING CHIMPANZEES

Adriaan Kortlandt

About 20 sites are now known where chimpanzees use (or used in the past) stone hammers and anvils to crack hard nuts and eat the kernel. All these sites are located in West Africa, in a relatively small area ranging from the eastern part of Sierra Leone to southeastern Guinea to the western part of the Ivory Coast. Two types of lithic culture may be distinguished: (1) cracking the nuts of oil palms and (2) cracking the nuts of forest trees (Kortlandt, 1986; Kortlandt and Holzhaus, 1987).

CRACKING THE NUTS OF OIL PALMS

The chimpanzees in the farmland-forest mosaic landscape around Bossou in southeastern Guinea use relatively small hammer stones (weighing 150 g to 1.5 kg) and crack only semi-cultivated oil palm nuts. They do not show the type of tool use described in the next section. Their motor patterns and technology are identical to those of the villagers of Bossou, who also crack oil palm nuts by means of stone tools. Similar behavior has been described for the chimpanzees at Beatty's site in Liberia. No evidence of stone-tool use by chimpanzees has been found around the village of Gba, 6 km west-northwest of Bossou, or in the Nimba Nature Reserve, 6 km east-southeast of Bossou. Elsewhere in Africa, in most of the area where oil palms commonly occur, from Guinea to the Congo of Brazzaville to Tanzania, chimpanzees swallow the fruit without cracking it, digest the outer layer, and defecate the kernels undamaged. The chimpanzee community of Bossou had a more or less sacred status in the past and was fed by the villagers, or at least was allowed to feed even on expensive human crops (e.g., pineapples) until the 1960s, just like Indian temple monkeys. This suggests that these apes have copied the human stone-tool technique of palm-nut cracking as a result of observational learning. The process would have been reinforced by forest cutting for agriculture, which caused a shortage of natural food and the spread of oil palms. These apes and Beatty's population may represent the first identifiable cases of direct cultural transmission of technology from man to animal in the wild.

CRACKING THE NUTS OF FOREST TREES

Chimpanzees within the pentagonal area of Tiwai-Zinta-Toulepleu-Soubré-Tabou (Sierra Leone, Liberia, and Ivory Coast) have been observed to use much heavier hammer stones, up to 15 and even 24 kg, both in the Bossou manner and as pestles, to crack the very hard nuts of certain wild fruit species, but they do not crack the less-hard nuts of the oil palms in their area. At Tai and Tiwai they also use wooden clubs as hammers, and roots and branches of trees as anvils. Effective manipulation of both types of tools, not only on the ground but also high up in the trees, requires an amount of arboreal competence, acrobatic skill, and manual dexterity that far exceeds that observed anywhere else in Africa. A possible explanation might be that this region has been the most important lowland rain forest refuge in West Africa during dry epochs in the Pleistocene. Consequently, for hundreds of thousands, or perhaps for millions of years, these apes may have been genetically selected for, and culturally adapted to, a highly specialized way of life in such habitats.

In chimpanzees, the direction of the blow of the hammer follows the perpendicular to a flat surface of the hammer through its gravity center (with very few exceptions not to be discussed here). No human stone tools showing such a type of wear are known from earliest Paleolithic times. On the contrary, nearly all early hominid stone tools show sharp edges resulting from flaking and were used as cutting, cleaving, and chopping devices over the entire lengths of their edges. It was a meat-eater's technique. *Australopithecus boisei* ("Nutcracker Man") may have cracked all the edible nuts with his teeth so that nothing was left for *Homo habilis* (competitive exclusion). In summary, the use of stone tools by chimpanzees and earliest hominids shows no indications of homology, functional equivalence, similarity of motor patterns, or identity of motivation.

ACKNOWLEDGMENTS

The author wishes to thank Dr. Y. Sugiyama, and Dr. C. and Mrs. H. Boesch, for their help and friendship, which enabled him to develop ideas different from theirs; his assistant Ewald Holzhaus for his participation in the program of 1986; and all who generously offered help and hospitality in Guinea, Liberia, and the Ivory Coast.

REFERENCES

Kortlandt, A. 1986. The use of stone tools by wild-living chimpanzees and earliest hominids. *J. Hum. Evol.* 15:72–132.

Kortlandt, A., and E. Holzhaus. 1987. New data on the use of stone tools by chimpanzees in Guinea and Liberia. *Primates* 28:473-496.

CHIMPANZOO

Jane Goodall

ChimpanZoo is an idea which came to me quite suddenly in the spring of 1984. The way it originated is interesting. I continually get letters from students, from young people whose dream is to come to Gombe to work with me. They offer to wash dishes, do the mending, sweep the floor – *anything*, if they can only come to Gombe. Since the unfortunate kidnapping incident in 1975, it has been deemed unwise to have foreign students and researchers living at Gombe. So I have always had to say, "No, I'm sorry. I can't have you at Gombe. You will have to find something else."

These days, as many of you may know, it is becoming more and more difficult to study primate behavior in the field. Funding is harder to get and study sites are closing down due to political troubles. But with all this enthusiasm, all this good will, all this eagerness which could not but help the cause of conservation and the welfare of chimpanzees, how sad to have to say, "No, I cannot offer you anything."

At the same time, I have been working for many years with some zoos to improve the condition of their captive chimps, to enrich the environment, to raise the money to move the chimps from small, barren, concrete and iron cages out into more suitable exhibits. It suddenly occurred to me as I was going around the country that spring that we have the makings of a brand-new and very exciting research opportunity. What if we study these various captive groups, using the same kind of methods we use at Gombe, and compare the behavior of the chimpanzees in all these different groups across the country? Why shouldn't we do that? This is how the idea came into being.

Several zoos were immediately interested. The first was the Cheyenne Mountain Zoo in Colorado Springs. And then, in rapid succession, the North Carolina Zoo in Asheboro; the Washington Park Zoo in Portland, Oregon; Lion Country Safari in West Palm Beach, Florida; and the San Francisco Zoo. Other zoos which have subsequently joined the program are the Dallas Zoo, the Fort Worth Zoo and, most recently, the Sacramento Zoo. Several other zoos have expressed interest in taking part, including the Lincoln Park Zoo in Chicago.

Most fortunately an ex-Gombe student, Ann Pierce, was free and able to take on the somewhat formidable task of traveling from one zoo site to another, talking to management and keepers, college students and professors, and trying to see that the work at the different zoos followed the same overall design.

Methods of data collection are still not standardized from one site to another, but great progress has been made, particularly during two national workshops, the first at San Jose, the second, much larger gathering, at the Cheyenne Mountain Zoo. The workshops were particularly valuable as they enabled zookeepers to discuss problems of mutual interest, and permitted in-depth discussions between students and professors from the different study sites. Once we have video recordings of all or most of the different behaviors shown by the chimpanzees of the various groups under study, and once these have been passed around to all the different sites, we shall be ready to compile a detailed ethogram. Checksheets for recording data have not yet been finalized. But much information has been collected.

Many scientists who had already tried to do research projects in zoos were doubtful that ChimpanZoo would work. They felt that there would be too many confrontations between zoo management on the one hand, and researchers on the other. Thus, when I originally outlined the guidelines for ChimpanZoo, I took this concern very seriously. Clearly it is necessary that the zoo management be truly committed to the project and, most importantly, be prepared to give the chimpanzee keepers time to collect data. The keepers, who work day in and day out with the animals, know them very well – they understand their behavior, and recognize them as individuals. The keepers are, in almost all cases, dedicated and caring people. It is equally necessary that the research-ers who want to study the chimpanzees realize that they are working in a zoo. The management is responsible for the chimps; they have to pay for the upkeep of the chimps. If it is a management decision to take a chimp from one group and send it to another zoo, the researchers may not like it (I wouldn't like it if a leopard came along and killed and ate one of my chimps; however, there would be nothing I could do about it). I think the researchers should be prepared to accept the management decision. They can discuss it and talk about it and perhaps postpone it, but they should not rant and rave, as so many scientists do, because the zoo does something that interferes with their research program. Also, students coming in should defer to the knowledge of the keepers.

Those were the guidelines originally set up for ChimpanZoo, and ChimpanZoo is working.

We hope to find, as a result of this study, what behaviors in a chimpanzee are so innately, rigidly chimp that they appear in all the different environments – in Gombe, in Mahale, in the Ivory Coast, in the zoo in Chicago, in Lion Country Safari in Florida. Everywhere the same: deep-rooted, basic chimpanzee patterns. And much more fas-cinating, at the opposite end of the spectrum, which behaviors are the most flexible? How adaptable are the chimpanzees? To what extent do we see variation between one group and another, between the wild and between the captive. Just as there are some gestures and postures that are innate in humans, like smiling, there are some that are innate in chimpanzees. If you raise a chimp in isolation (which has been done), these innate postures, gestures, and sounds will appear, although they may appear in inap-propriate contexts and in strange sequences. The chimp has to learn how and when to use them.

Suppose we want to study a gesture like the "extend hand." How many different ways do chimpanzees in different environments use that gesture? How many different things can it mean? These are the kinds of questions we are asking and hoping to find the answers to.

The incredible film of the chimpanzee colony at Arnhem, in the Netherlands, gives some idea of the richness of detail that one can obtain by studying chimps in captivity. We have to realize that chimps in captivity are, in a way, free from environmental pressures. They don't have to search for their food. They have shelter provided for them and their ills are looked after. They have a tremendous amount of free time, and they can and do use that free time in very innovative ways: in sophisticated social interactions and in a great variety of inventive manipulations of the physical environment around them. This makes the study of a chimpanzee group in captivity a very rich and rewarding field in which to work.

We hope that ChimpanZoo will be beneficial in the following ways: First, in cases where chimps are still maintained in cement and iron bar, dungeon-type cages, we hope that the added interest in these apes will lead to an increased understanding on the part of zoo management. We hope that having appreciated how intelligent chimpanzees are, management will realize that the chimpanzee environment should be enriched in many ways and, if possible, that the chimpanzees should be moved into more suitable exhibits. Boredom is a tremendous problem for many captive individuals. Chimpanzees need things to do, and they need a decent social group.

Second, we hope that the keepers will gain an even deeper appreciation of the animals in their care. Not only can they meet with other keepers at ChimpanZoo workshops, but they can get input from the field: I plan to visit each site as often as possible, during my annual tours. Most keepers are really pleased to be able to talk about what they have learned already, to share information with others who realize, as they do, just how intelligent and interesting chimpanzees are.

Third, we hope that the experience of working with chimpanzees, our closest living relatives, will give students a new perspective on life, and help them to understand a little better that many of the attributes once thought to be unique to our own species are, in fact, shown also by the chimpanzees. Thus the students will have a better appreciation of the place of our own species in Nature. They will, hopefully, be more willing to take an active part in helping to conserve our natural wilderness heritage, and in helping to reduce cruelty to captive animals.

Finally, we hope that ChimpanZoo will become a powerful educational tool, that some of the interest and enlightenment that is being engendered in those taking part in the program will spill over into the general public. Already many of the zoos taking part in ChimpanZoo have graphics or pamphlets that point out the different chimpanzee personalities, that explain some aspects of their behavior in the wild and their cognitive abilities. I believe there exists a tremendous potential, in this project, for the education of children.

Thus I see ChimpanZoo as a project which will increase our understanding of chimpanzees, improve the conditions in which they are maintained in captivity, and, at the same time, increase our understanding of ourselves.

2

CURRENT FIELDWORK
Pan paniscus

INTRODUCTION: THE FOURTH APE

Frans B. M. de Waal

In addition to our own species, the superfamily of Hominoidea includes the lesser apes (gibbons and siamangs) and the great apes. Three of the great ape species enjoy wide name recognition – the chimpanzee, gorilla, and orangutan – whereas the fourth species, the bonobo, is much less known both to the general public and to science. Originally, bonobos were classified as chimpanzees, but Schwarz (1929) and Coolidge (1933) recognized them as a separate species. They are now classified as *Pan paniscus*, in the same genus as the chimpanzee, *Pan troglodytes*. Bonobos are found in central Africa, south of the Zaire River, which separates them from one subspecies of chimpanzee living to the east and north of it and from another subspecies to the west (for a distribution map, see van den Audenaerde, 1984).

Bonobos differ from chimpanzees by their relatively long legs, narrow shoulders, small head, reddish lips, and long fine hair on the head, which is neatly parted in the middle. Overall, their appearance is more gracile than that of the chimpanzee, and their movements are strikingly elegant and acrobatic. For detailed anatomical comparisons, including the suggestion of a strong resemblance of bonobos to the common ancestor of humans and apes, see Coolidge (1933) and the volume edited by Susman (1984a).

The name bonobo originates perhaps from a mispronunciation of Bolobo, a town on the Zaire River (Susman, 1984b). The species is also commonly known as the pygmy or dwarf chimpanzee, but this name is increasingly considered problematic. In the first place, bonobos do not differ in size from the smallest subspecies of chimpanzee, the average bonobo male weighing 45 kg and the average female 33 kg (Jungers and Susman, 1984). Secondly, Tuttle (1986) compares calling the bonobo a pygmy chimpanzee to calling the chimpanzee a pygmy gorilla. As the following four papers illustrate, bonobos are really a distinct species, not a diminutive form of chimpanzee. Finally, the name pygmy chimpanzee has forced scientists to refer to its close relative as the "common" chimpanzee, hardly an appropriate name for a primate that is becoming rare in the wild.

REFERENCES

van den Audenaerde, D. 1984. The Tervuren Museum and the pygmy chimpanzee. In R. Susman, ed., *The Pygmy Chimpanzee*, pp. 3–12. New York: Plenum Press.

Coolidge, H. 1933. *Pan paniscus:* Pygmy chimpanzee from south of the Congo River. *Am. J. Phys. Anthropol.* 28:1–57.

Jungers, W., and R. Susman. 1984. Body size and skeletal allometry in African apes. In R. Susman, ed., *The Pygmy Chimpanzee*, pp. 131–178. New York: Plenum Press.

Schwarz, E. 1929. Das Vorkommen des Schimpansen auf den linken Kongo–Ufer. *Rev. Zool. Bot. Afr.* 16:425–426.

Susman, R., ed. 1984a. *The Pygmy Chimpanzee*. New York: Plenum Press.

_____. 1984b. Preface. In R. Susman, ed., *The Pygmy Chimpanzee*, pp. xv-xx. New York: Plenum Press.

Tuttle, R. 1986. *Apes of the World*. Park Ridge, NJ: Noyes.

BEHAVIORAL CONTRASTS BETWEEN BONOBO AND CHIMPANZEE

Frans B. M. de Waal

When I first heard bonobos vocalize, their high-pitched, melodious calls reminded me of gibbons. Gibbon experts would probably disagree, but I was comparing the sound with the deep, vibrating voice of the chimpanzee. Unknowingly, Learned (1925) was the first to conduct a comparative study on the vocal repertoires of the two *Pan* species. As pointed out by Coolidge (1933), one of her two subjects was a bonobo, whereas the other was a "common" chimpanzee. Learned's (1925) phonetic transcriptions of hundreds of vocalizations indicate that the bonobo had a higher voice and used more 'a' and 'ae' vowel sounds than the chimpanzee, which mostly uttered 'oo' sounds.

This is just one of the many contrasts between the two species. Since it is my intention to stress the behavioral differences, it is good to realize first that, basically, bonobos and chimpanzees are similar. They rightfully share the same genus name and were considered the same species until Ernst Schwarz distinguished them as recently as 1929. The situation is very much like Napier's (1975, p. 60) conclusion regarding the relation between humans and apes: "Anatomically, man and the African apes are fundamentally similar although superficially different."

Obviously, if our own species is involved we tend to overestimate the differences in our wish to define human identity. Yet, when comparing two other species this risk exists as well, especially if one species is very familiar while the other is new. Thus, we may presently be going through a phase of exaggerating the bonobo's uniqueness by ignoring the tremendous intraspecies variation in behavior and social organization. Our comparison will become more balanced only when the new species is as well-known as the other. I cannot say that I have reached this point with regard to bonobos, but I did learn to recognize in their vocalizations, for instance, the same fundamental "themes" as in chimpanzee vocalizations.

The observations reported here concern captive apes: the bonobo collection of the San Diego Zoo (USA) and the large chimpanzee colony of Arnhem Zoo (Netherlands). Captive studies have obvious limitations, but one important advantage: *detail*. The individuals can be observed at close range, and social interactions can be recorded without one of the participants disappearing from view. I will focus on the behavior of bonobos, using my more extensive observations of the Arnhem chimpanzees for comparison only. The bonobos were observed for nearly 300 hours in the winter of 1983–84. They lived in three subgroups, two of which were fused in the course of my study. The collection included one adult male, two adult females, two adolescent males, four juveniles, and one infant. For more details on the subjects and the observation methods, see de Waal (1987, 1988).

FACIAL COMMUNICATION

The purpose of the study in the San Diego Zoo was to construct a detailed social ethogram for bonobos. Jordan's (1977) qualitative observations on three European zoo colonies provide the most complete ethogram of the species to date, but this work is only available as an unpublished German dissertation. A second unquantified, but much shorter ethogram of zoo animals was published by Patterson (1979). None of the available descriptions of bonobo behavior, however, match the depth and breadth of the excellent ethograms for the chimpanzee (Goodall, 1968; van Hooff, 1973; Goodall, 1986) and orangutan (Rijksen, 1978). My bonobo ethogram is based on the observation of over 5000 social sequences involving ritualized display behavior. For a complete report with definitions and quantified contextual information on each behavior pattern the reader is referred to de Waal (1988). I will highlight some of the results while ignoring most statistics.

Due to the reduced eyebrow ridges and contrasting lip color (reddish lips in a black face), the bonobo's face is open and remarkably expressive. At close range it appears to communicate many shades of emotion. The amount of face-to-face orientation and eye contact in the species suggests that subtle facial cues are closely monitored by others. Therefore, the following five stereotypical expressions are only the tip of the iceberg of this ape's facial communication.

Silent teeth-baring. Baring of the teeth in a wide grin has the same meaning as in the chimpanzee. It is primarily an expression of fear, nervousness, or hesitation, mostly shown by subordinates who are being threatened or pursued by a dominant (Fig. 1). The expression is not exclusively given by subordinates to dominants, however. More than 20% of the instances occurred in the opposite direction. As suggested by van Hooff (1972) for chimpanzees, the signal may have both an appeasement function, if shown towards a potential aggressor, and a reassurance function, if shown to a frightened subordinate. I observed striking instances in which aggression abruptly stopped after teeth-baring by the recipient.

A second meaning of the bared-teeth face is more typical of bonobos. We may call it the "pleasure grin." Female bonobos often bare their teeth at the climax of copulation (Savage-Rumbaugh and Wilkerson, 1978) or during solitary masturbation (Becker, 1983). Apart from these situations, which were also observed in San Diego, teeth-baring occurred at moments of excitement over food or new objects, and as a fleeting expression during intensive play bouts, in alternation or combination with the play face. While in chimpanzees extensive teeth-baring during play usually signals

Figure 1. A juvenile female bonobo bares her teeth to her older sister who is chasing her. With no escape route, the younger sister is bound to lose her leaves.

pain or fear by the younger partner (e.g., when the play becomes rough), this negative meaning was not obvious in bonobos.

Tense mouth. The eyes fixate the partner, the eyebrows are frowned, while the lips show strong horizontal tension. As in other primates (van Hooff, 1967), the motivation of this display was almost exclusively aggressive. It occurred in bonobos who were performing, or about to perform, an aggressive charge or physical attack. The same expression occurs in chimpanzees, which also show the related "bulging lips face" during intimidation displays (van Hooff, 1973; de Waal, 1982; Goodall, 1986). The latter expression may have a different meaning in bonobos, however. Bulging of lips was not observed during displays or hostilities. It occurred instead at moments of apprehension or tension, e.g., when a huge crane was brought onto the zoo grounds.

Silent pout. The lips are pursed forward and curled outward in front, resulting in a circular opening (for a photograph, see de Waal, 1986a). The lip posture resembles and presumably derives from that of a nursing infant. Early in life the silent pout, often accompanied by soft whimpering sounds, occurs as a request for nipple access, or as a complaint about maternal behavior (e.g., sudden moves). Later in life, the functional context broadens to include begging for food, objects, body contact, and even sex. The predominant motivations remain desire and frustration after rejection. Over 90% of the observations concerned the three youngest individuals in their respective groups. Chimpanzees show the same expression under the same circumstances (Goodall, 1968; van Hooff, 1973).

Pouting often seemed an expression of disappointment. It might appear when a play partner departed to do something else, or when a possessor of food refused to share. The youngest adolescent male, Kalind, could keep a less intense form of the display on his face for ten minutes or more. This idiosyncratic expression, referred to as a "sulk face," usually appeared after the adult male had repeatedly interrupted Kalind's sexual advances to the adult female of their group.

Duck face. This is the only facial expression unknown in chimpanzees. It might be part of the San Diego bonobos' unique grooming culture, described below (Traditions and Games), but it may also be a species-specific expression. The lips are pouted and flattened at the mouth corners over a greater length than in the silent pout, creating a resemblance to a duck bill. At the front, the lips are not curled outward to the extent as in the silent pout, leaving a smaller opening. The duck face occurred exclusively during grooming sessions, both in the grooming individual and in the groomee. Although one juvenile female performed over half of the duck-face displays, it was observed at least once in seven different individuals. It seemed an expression of utmost concentration.

Play face. The mouth is opened with the lips either (a) in a relaxed position, covering the upper teeth completely and the lower teeth partially (Fig. 2), or (b) retracted, without pulling back the mouth corners, resulting in baring of all front teeth (Fig. 3). The two variants alternate and blend so frequently that it is hard to draw a line between them.

While the whole gamut of play faces of the bonobo is observable in the chimpanzee, the chimpanzee bares its teeth less often and more briefly during play than the bonobo. The typical chimpanzee play face with covered upper teeth may not be the most frequent variant of the display in the bonobo. Recently, Chevalier-Skolnikoff (1982) distinguished three facial expressions of apes during play: the "smile" (retracted lips, closed mouth), the "laugh" (silent teeth-baring, open mouth), and the covered-teeth play face (cf. van Hooff's [1973] "relaxed open-mouth display"). According to Chevalier-Skolnikoff (1982), nonvocal smiling and laughing displays

Figure 2. A juvenile male bonobo showing the typical "chimpanzee" play face with covered teeth.

Figure 3. An adolescent male (left) and adult female during a play chase. Both bonobos exhibit laughing, i.e., a play face with opened mouth and bared teeth.

are well-developed in orangutans only. Her comparison did not include bonobos, however, which do regularly show all three play expressions, with or without the panting laugh.

VOCAL COMMUNICATION

Sixteen different vocalizations were distinguished in the bonobo's repertoire. More extensive studies will undoubtedly reveal a greater differentiation. Possibly homologous expressions in the chimpanzee's vocal repertoire (described by van Hooff, 1973; Marler and Tenaza, 1977; Goodall, 1986) are recognizable for almost every bonobo vocalization. There clearly exists morphological and functional continuity. Screaming, for instance, is in both species a high-pitched rasping sound uttered in agonistic situations, especially by victims of attack, with wide baring of the teeth. The main difference is that bonobos scream in a shriller voice, which makes adults sound like infants of the other species (Fig. 4). The greatest similarity is in the softer vocalizations, such as the pout moan and panting laugh, which are hardly distinguishable between the two species. The most striking differences occur in the long-distance hooting calls (Fig. 5).

Rather than treating here the complete vocal repertoire, I will concentrate on the bonobo's three forms of hooting: low hooting, high hooting, and contest hooting.

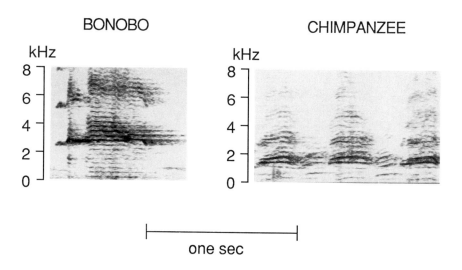

Figure 4. Spectrogram of characteristic scream vocalizations by a bonobo and a chimpanzee (both are 7-year-old males). The greatest concentration of energy occurs at a higher frequency level in the bonobo, resulting in a shriller sounding scream. The chimpanzee vocalization was recorded at the Field Station of the Yerkes Primate Center.

They can be compared with the chimpanzee's pant-hoot displays (van Hooff, 1973; Marler and Hobbett, 1975; Boehm, this volume).

Low hooting. This is the species' lowest-pitched vocalization, uttered in rapid series in which both inspirations and expirations are vocalized. The vocalization may but need not develop into high hooting. Often a couple of high hoots are given interspersed with the low hoots before the display ends. Transitions between low and high hooting are abrupt and may go back and forth; there is not, as in the chimpanzee, a gradual buildup in loudness and pitch towards one or several climax vocalizations. Due, perhaps, to its relative softness, low hooting has never been reported from the wild, and Mori (1983) concluded that the bonobo had lost this part of the display.

High hooting. Two types of high-pitched long-distance whooping calls can be distinguished. Staccato hooting consists of brief, ear-piercingly shrill and explosive calls. During choruses, staccato hooting of different individuals is almost perfectly synchronized so that one individual acts as the "echo" of another, or emits calls at the same moments as another. The calls are given in a steady rhythm of about two per second. The second vocalization, called legato hooting, reaches the same peak frequency, but continues with a relatively long, modulated second syllable. As a consequence, the peak frequency occurs far before the middle of the call, and the vocalization has a less sharp, more melodious quality. Legato hooting also occurs in series, but both the duration per call and the rhythm of emission are irregular.

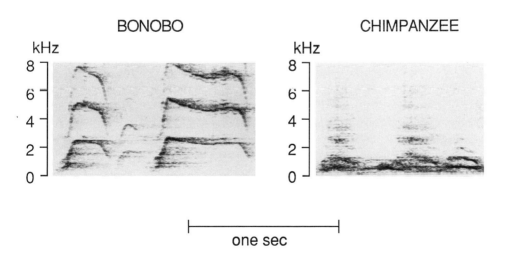

Figure 5. Spectrogram of legato hooting by a single adult female bonobo compared to that of a pant-hoot concert by several female chimpanzees simultaneously. The bonobo's hooting calls start at about the same pitch but then far exceed the pitch of the average chimpanzee hooting, resulting in melodious, gibbon-like sounds. The chimpanzee vocalizations were recorded at the Field Station of the Yerkes Primate Center.

Figure 6. Three bonobos respond with high hooting to the animal caretaker. On the far right, an adolescent male; the other two animals are adult females.

The two forms of high hooting occur in situations similar to those of the pant-hoot choruses of captive chimpanzees, for example, as a group response to anything that excites the bonobos (e.g., the arrival of food) or upsets them (e.g., a visit of the vet). They usually direct the vocalization to the source of disturbance or excitement rather than to one another (Fig. 6). A functional connection between the hooting of the two species is evident in the Frankfurt Zoo (Jordan, 1977) and at the Field Station of the Yerkes Primate Center (pers. obs.) where bonobos and chimpanzees, living in separate enclosures, respond to each other's hooting.

Low hooting by the chimpanzee is louder, deeper and, at least in the adult male, much more drawn out than in the bonobo. It has a challenging, hostile component if accompanied by display behavior. I did not observe this during low hooting in the bonobos. Bonobos seem to have evolved a specific type of hooting, a modification of legato hooting, for agonistic confrontation. This so-called *contest hooting* is softer, tonally flatter, and monosyllabic, but does reach the same pitch and sounds like a tense form of legato hooting. Contest hooting series have a rhythm of one or two calls per second; a monotonous "wee-wee-wee." In contrast to high hooting, which is mostly undirected, the performer always orients to another individual and gives some form of display, usually a rocking or swaying movement in the same rhythm as the vocalization. Teeth-baring is quite common in the middle of the call, but in between calls there is usually no particular expression. Mori (1983) probably referred to the same behavior when describing display hooting by bonobos as "hui-hui-hui" (in contrast to the chimpanzee's "hoo-hoo-hoo"), adding that vocalizers showed defensive expressions.

Only the adult male was heard to make these sounds. It served as a conspicuous warming up for and warning of an incipient attack. Charges followed the display more than 80% of the time. Occasionally, the target individual responded to contest hooting with other sounds, such as peep-yelps, greeting grunts, or screaming. In particular, when peep-yelps were the response to the aggressor's hooting, a sort of dialogue was formed, going rapidly back and forth between the two antagonists. Similarly, Jordan (1977, p. 113) described so-called *Quiekduelle* between captive adult males engaged in a dominance struggle: ". . . the high-pitched rhythmic squeaks of one animal often turn into a direct personal confrontation because of the 'answers' by a second animal during intervals between sounds."

This dialogue aspect of vocal antagonism is unknown in chimpanzees. While bonobos tend to synchronize their calls during staccato hooting, directed outside the group, they may alternate calls in a remarkably high tempo during confrontations within the group. Apart from the instances of contest hooting mentioned above, a similar exchange occurred once with scream vocalizations. When the adult and the oldest adolescent male were first introduced to each other, they gave rasping screams in perfect alternation for six minutes. In this case a mixture of fear and attraction seemed the predominant motivation. The vocal exchange went together with invitational gestures from a distance (i.e., genital presents and begging hand gestures) and ended with a ventro-ventral embrace with mutual penis rubbing.

TRADITIONS AND GAMES

When, in the Arnhem Zoo, one of the adult male chimpanzees had been bitten on his hand and stumbled around supporting himself on a bent wrist, several youngsters started imitating him. Soon the wrist-walk was incorporated into their games. It was used to approach playmates in a funny way, or as a self-handicap during pursuits. Whereas the male's hand had healed within weeks, the juveniles showed the pattern for at least one year. Köhler (1925) described similar locomotory fashions among his juvenile chimpanzees. Propagation of innovative behaviors by means of observational learning is probably widespread among primates and may, if it spans generations, be termed cultural propagation. Such traditions are of special relevance if they concern feeding habits or tool-use techniques with survival value, but they may also concern communication patterns and thus create the primate's equivalent of a local dialect (e.g., Kawai, 1965; Goodall, 1973; McGrew and Tutin, 1978; Nishida et al., 1983; Kummer and Goodall, 1985).

My study in San Diego did not last long enough to decide whether certain unusual behaviors represented fads or stable traditions, but since they were never reported for other bonobos, captive or wild, I assume at least that they developed only in this colony.

Hand-clapping. Apes in zoos often develop hand-clapping as a way of drawing attention from the public or their keepers. The bonobos in San Diego, however, also used the same behavior among themselves, together with variations such as clapping the feet together, clapping a hand and a foot together, or rhythmically hitting the ground or beating their chest with one hand. These behaviors occurred mainly during grooming sessions (Figs. 7a, 7b). One individual would approach another, softly beat his or her chest three times, clap hands twice in front of the other's face, and start the grooming session. In between this meticulous work, the groomer would show some more clapping or rhythmically stamp a foot on the ground. Some individuals clapped quite loudly, providing their grooming bouts with an acoustical element. The behavior was very common in groomers but rare in groomees.

Whether these peculiar hand and foot movements have any specific meaning is doubtful. At first sight it may look like sophisticated gestural communication – comparable to that of apes trained in American Sign Language – but there was no response except for a passive acceptance of the grooming that came with it. The gestures seemed to be an expression of utmost concentration and enthusiasm on the part of the groomer. Once, a juvenile female discovered an injury on another's leg and immediately intensified the hand-clapping and chest beating before starting to clean it. I believe the use of this behavior is best compared with that of lipsmacking in monkeys, or tooth clacking and spluttering in chimpanzees. Another reason to see the behavior as a sign of concentration and devotion to a particular job is that a recipient was not required. The gesturing also alternated with autogrooming and nestbuilding.

The rhythmic hand and foot movements could very well have been introduced by humans. Ninety-eight and a half percent of the 673 observations concerned seven of the ten bonobos. The nonperformers were the only two individuals not born in San

Figures 7a and 7b. An adolescent male bonobo alternates his grooming of an adult female with audible hand clapping.

Diego and the only second-generation bonobo. The seven performers had two points in common: all were offspring of the zoo's wild-born founding pair (which was not involved in the study), and all had spent a period early in life in the zoo nursery. If we exclude the unlikely possibility of a genetic basis for hand-clapping, the behavior might have started under the influence of people in the nursery. The one second-generation individual was a two-year-old infant raised in the group by her mother. This individual did not show the behavior, but also rarely groomed. She did, however, show chest beating, not with both hands and with the fast rhythm of a gorilla, but a loose tapping of the chest during solitary charges. Whatever the origin of these behaviors, the bonobos of San Diego clearly have developed several interesting variants on rhythmic clapping, and have integrated these gestures into their everyday life.

Funny faces. Initially, my list of facial expressions in bonobos was getting quite long as I kept adding new descriptions. Almost every observation session on the group of four juveniles brought new expressions, each one more fantastic than the previous ones. The youngsters would stick out their tongues, curl them up, and wildly shake their heads. Or, they would move the lower jaw rapidly up and down while baring the teeth. Sucking in the cheeks or poking a finger into the opened mouth were also popular. After a while it became evident that these were not species-specific stereotypical expressions, and I labelled them "funny faces" (for photographs, see de Waal, 1986a).

The facial contortions were rarely aimed at a partner. Performers were either sitting alone or in a group without much social orientation. Occasionally two individuals tickled each other, both staring in the distance and continuously changing their facial expressions. The apparent absence of strong emotions and the irrelevance of social partners mean that the production of funny faces is best regarded as a solitary game. This game has not been reported for other bonobos, either in the wild or in captivity. The only reference to a similar game in primates concerns gorillas, which are said to "purposefully make novel, nonemotional faces to themselves in play" (Chevalier-Skolnikoff, 1982, p. 337).

Blindman's bluff. This was a daily game in the group of juveniles. With closed eyes the apes stumbled through the climbing structure, high above the ground. The sophistication of the game depended on the age of the players. The two oldest juveniles usually held their eyelids closed with the thumb and index finger of one hand. The individual next in age simply placed a whole arm over his face (Fig. 8), whereas the youngest poked one finger in one eye, closing and opening the other eye depending on how confident he felt. While he was clearly cheating, the others played the game seriously. This was evident from the many times they almost lost balance or bumped into objects or one another.

This game proves that bonobos are capable of self-imposed rules, as if telling themselves "I'm not allowed to look." Except in human children, I have never seen blindman's bluff played with such dedication and concentration. Yet the game is not unknown in other ape species, or in monkeys either (e.g., Harrisson, 1961; Rensch, 1973; Thierry, 1984).

Figure 8. A juvenile male bonobo sits down after having stumbled around for several minutes with an arm over his eyes. He does not remove his arm immediately. Is he guessing where he is?

SEXUAL BEHAVIOR

Because sexual activities are conspicuous and variable in bonobos, they have attracted quite a number of investigations. In view of the extremely close relation between the genera *Pan* and *Homo* (e.g., Sibley and Ahlquist, 1984), these studies are of great theoretical interest in connection with human evolution. Current scenarios try to link the reduced importance of the menstrual cycle and the richness of sexual practices in our species to the heterosexual pair-bond, food sharing, and even bipedalism (Morris, 1967; Lovejoy, 1981; Fisher, 1983). Such evolutionary models may particularly benefit from research into bonobo sexuality as these apes exhibit crucial elements traditionally considered uniquely human.

Qualitative studies on European zoo collections of bonobos first revealed the bonobo's sexual peculiarities such as frequent face-to-face copulation; a frontally oriented vulva facilitating such positions; pseudo-copulations between males, and especially between females; and a facilitating effect of food provisioning on sexual activities (Tratz and Heck, 1954; Rempe, 1961; Kirchshofer, 1962; Hübsch, 1970; Jordan, 1977). The first systematic data on sexual behavior of captive bonobos were collected by Savage-Rumbaugh et al. (1977) and Savage-Rumbaugh and Wilkerson (1978). Their observations supported the above reports. Savage-Rumbaugh et al. (1977) further claimed that bonobos use "iconic" communication gestures to negotiate mounting positions. As yet, such gestures have not been reported by other investigators.

164

Field studies in Wamba and the Lomako Forest, both in Zaire, have shown that the rich sexual repertoire of captive bonobos is not a product of the artificial environment. Ventro-ventral copulations, which are virtually absent in chimpanzees, constitute between 26% and 38% of the heterosexual copulations observed at these field sites (Kano, 1980; Thompson-Handler et al., 1984). Also, isosexual contacts are quite common in the wild. Characteristic of the species are lateral genito-genital rubbing movements between females mounted in a ventro-ventral position. This pattern was termed GG-rubbing by Kuroda (1980).

De Waal (1987) provides a detailed report on sexual and erotic interactions in the San Diego bonobo colony. Six main types of interaction were distinguished.

Ventro-ventral mount (n = 420). This was the most common mounting pattern. In heterosexual pairs it usually occurred with the male in the active role, on top of the female, but exceptions did occur. In these cases the female lowered herself over the male, manually inserting his penis. Among females, the ventro-ventral mount differed from such contact in other sex combinations in that (a) one female usually lifted the other off the ground, the other clinging with arms and legs around her (Fig. 9),

Figure 9. The typical GG-rubbing posture between two adult females, the older female carrying the younger one. This sexual pattern is unique to the bonobo species.

(b) pelvic movements were made by both partners, and (c) the movements were laterally directed. The males' equivalent of this GG-rubbing posture is a ventro-ventral mount or embrace with mutual penis rubbing (i.e., one or both males making thrusting movements with their penises touching). Perhaps this behavior corresponds with Kano's (this volume) observation of "penis fencing."

Ventral-dorsal mount (n = 156). This pattern is similar to the mounting pattern of chimpanzees, except that a bonobo female, with her ventrally directed vulva, has to lift her abdomen slightly off the ground to allow intromission.

Opposite mount (n = 23). The two partners face in opposite directions, rubbing their genitals or buttocks together. This pattern occurred especially between the two adult females. While one female lay on her back, the other stood quadrupedally over her, with her back turned, while initiating GG-rubbing.

Genital massage (n = 39). Rhythmic manual stimulation of a partner's genitals occurred almost exclusively among the adult and adolescent males.

Oral sex (n = 17). In all cases, oral stimulation of a partner's genitals was classifiable as fellatio. One male would thrust his erect penis in his partner's face, after which the partner would take the penis in his or her mouth.

Mouth kiss (n = 43). One partner places his or her opened mouth over that of the other. In about one-quarter of the instances, extensive tongue-tongue contact was observed; in the other instances such contact could not be excluded. Like oral sex, mouth kissing was virtually limited to the juvenile group.

Sexual and erotic interactions were not more common between individuals capable of reproduction than in other combinations. De Waal (1987) estimates that over three-quarters of the sexual and erotic activity in the San Diego colony served no direct reproductive function whatsoever. On the other hand, this activity did not conflict with reproduction, as males were never observed to achieve intromission or ejaculate during contacts with partners other than mature females. Both the variety of sexual practices and their distribution over all combinations of age and sex classes are very different from the sex life of chimpanzees, which seems more oriented toward repro-duction.

CONTEXTUAL ANALYSIS

To understand which situations elicit sociosexual behavior, two analyses were conducted: one general analysis, treated below, and a more specific one, treated in the next section. The general analysis sought to determine associations between 50 behavior patterns and 40 context types. Contexts were defined by external events, the apparent objective of the interactants, and ongoing social behavior. The overall matrix included 6,857 observed combinations of behavior and context. The observed fre-quency of a particular combination was compared with the frequency expected on the basis of the matrix's marginal totals. The discrepancy between observed (o) and expected (e) frequencies was expressed in a "standardized residual," $s.r. = (o-e)/\sqrt{e}$ (Everitt, 1977). A behavior pattern was considered to be associated with a particular context if the combination had been observed at least five times and reached an s.r. value of at least plus two. Thus defined, less than 8% of all possible combinations of behaviors and contexts showed an association. If the matrix of Table 1 contains a higher proportion of associations, this is because this 9 x 13 selection from the large 40 x 50 matrix serves to illustrate particular results (i.e., it is a nonrandom selection). The s.r. values in Table 1 are based on calculations involving the overall matrix (for the complete analysis, see de Waal, 1988).

Inspection of the context types in Table 1, from left to right, reveals that although high hooting was associated with feeding time it was not linked to interactions specifically related to food. Interactions over food typically involved sexual mounts, the silent pout, silent teeth-baring, embracing, and self-caressing of a nipple. Mounting also occurred relatively frequently when the apes were just released from their night quarters (i.e., just out) and during introductions of new group members (i.e., intros).

In agonistic contexts (i.e., aggression given or received), mounting and other contact behaviors were uncommon. In the post-conflict context, however, these behaviors occurred in the following order of strength of association: holding out a hand in a begging gesture, massage of the partner's genitals, lateral embracing, and mounting. Punching the partner with a fist on the chest or forehead (interpretable as a form of restrained assertiveness) and self-caressing of a nipple (a form of self-reassurance) were also relatively frequent after agonistic encounters. Finally, Table 1 shows that the bonobos responded to alarming or disturbing events caused by people inside or outside their facility with high-hooting vocalizations, and that the mouth kiss was associated with play.

Table 1. Standardized residuals (s.r.) indicating the association between behavior patterns (left) and particular contexts (top); s.r. $= (o-e) / \sqrt{e}$, where o is the observed frequency of a behavior in a particular context and e is the expected frequency of the pattern. Only positive s.r. values for which $o \geq 5$ are presented. For further information on the standardized residual and behavioral terms, see text.

Behavior	Feeding		Just out	Intros	Aggression		Post conflict	Disturbed	Play
	Time	Interact			Given	Received			
Silent pout		4.5				1.1			
Silent teeth-baring	1.6	4.5		0.4		5.6	1.7		0.2
Screaming			4.6	2.8		22.2			
High hooting	6.4		9.3	6.3				26.8	
Begging gesture		1.8					5.6		
Self nipple		4.2				1.7	7.3	1.3	
Charging display	0.9		6.8		23.0			0.1	
Lateral embrace		3.3					3.8		
Ventral embrace		0.6		1.9			1.8		
Punching					3.4		8.6		
Mouth kiss							1.0		9.4
Mounting	4.7	2.5	3.1	10.7			3.1		
Genital massage							5.3		

The conclusion of this selective review of associations is that sexual mounting occurred in a variety of tense "agitation" contexts, excluding, however, agonistic episodes and external disturbances. Genital massage and embracing were associated with one or two such contexts only, whereas mouth kissing seemed to be play related.

TENSION REGULATION

The relation between food and sex is quite conspicuous in bonobos. It is not only evident in captivity, when the keeper arrives with their meal, but also in the wild when bonobos enter a fruit tree, or are provided with sugar cane by field workers (Kano, 1980; Thompson-Handler et al., 1984; Kuroda, 1980, 1984; Mori, 1984). Observations at the provisioning site in Wamba have led to a specific hypothesis concerning

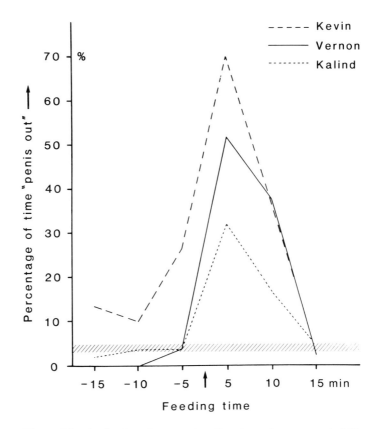

Figure 10. At 5-min point samples, "penis out" was recorded if a male's penis was unsheathed over more than half its length. Usually this meant that the male had an erection. Samples taken around feeding time are compared with samples at other times of the day, represented as a grey area. The adult male (Vernon) and both adolescent males (Kevin and Kalind) showed a significant increase in erections at feeding time.

Figure 11. Kevin on the lookout for the keeper with his meal.

the function of food-related sexual behavior. Kuroda (1980, p. 190) states that GG-rubbing among females "works to ease anxiety or tension and to calm excitement" and "thus to increase tolerance, which makes food sharing smooth." An element of calculation is even suggested for sexual contacts of estrus females with male posses-sors of food: "numerous observations suggest that in these instances the female's purpose was to obtain food, not simply to copulate." Presenting and copulation buffer the food interaction and render the males more tolerant (Kuroda, 1984, p. 318). Kano (1980, p. 156), referring to similar sequences, wrote that "copulations had the effect of flattering or appeasing the males."

There is a simpler explanation for a relation between food and sex. The sudden availability of attractive food may induce a state of arousal, called food excitement, which then "sparks over" into sexual arousal. Thus, the mere sight of food causes penile erections in male bonobos. One male, Kevin, showed erections well before feeding time if he had seen the keepers begin food preparation in the kitchen (Figs. 10 and 11). The question is, however, whether this explanation in terms of arousal states is sufficient.

The two possible causal factors – food excitement and social tension – are confounded at feeding time. If social tensions have an independent effect on sexual activity, we expect this effect to occur also when tensions are unrelated to food. In other words, it is critical to investigate how confrontations in the absence of food affect the occurrence of sociosexual behavior. As yet, field data cannot resolve this point. Aggressive behavior is rare in the unprovisioned situation, and, if it does occur, it is usually when parties meet at a natural food source (Kano and Mulavwa, 1984; Badrian and Badrian, 1984).

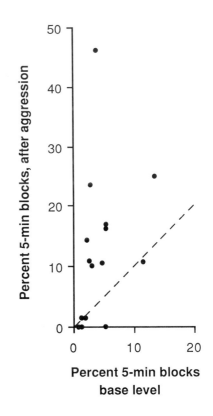

Figure 12. Aggressive incidents are followed by an increase in nongrooming interactions between the opponents, i.e., genital presentations, sexual mounts, erotic contacts, touch, and embrace. Frequency is measured as the percentage of 5-min time blocks during which a dyad engaged at least once in one of these behaviors. Each dot represents one dyadic relationship. Compared are periods of three 5-min blocks immediately following aggression unrelated to food, and a base level when there was no food and no aggressive behavior. The dotted line indicates the expected relation if previous aggression had no effect on the above behavior patterns. Data from de Waal (1987).

The data in Table 1 already indicate that sexual behavior increased after aggressive episodes. In this analysis, however, data on all aggression, including aggression at feeding time, were taken together. In addition, these are pooled observations; there was no distinction between different dyadic combinations of individuals. De Waal (1987) conducted a more specific analysis for each dyadic relationship, comparing the rate of behavior before and after food provisioning, as well as before and after aggressive incidents, with the base level at other times of the day.

It was found that grooming was suppressed at feeding time. Sexual mounts, erotic behaviors (e.g., genital massage), and other intensive body contacts (e.g., embrace, touch), however, increased dramatically both in the presence of food and between former adversaries within 15 minutes after conflicts unrelated to food. The latter increase fits the *reconciliation* concept developed and investigated by de Waal and van Roosmalen (1979), de Waal and Yoshihara (1983), and de Waal (1986b); that is, it suggests that bonobos actively repair disturbed relationships by means of reassuring body contact. Peacemaking was most predictable between one of the adolescent males and the infant female. They frequently played together, but had many quarrels as well. Normally, these two individuals engaged in sexual or other intensive contact during 3.8% of the 5-minute observation blocks. They reached a rate of 46.2% in periods following aggression between them, a 12-fold increase. As Figure 12 shows, this change was representative for the majority of dyads. Nongrooming contacts increased in nine dyads, decreased in one dyad, and were virtually unaffected by previous aggression in five dyads (Wilcoxon test, p < 0.02, two-tailed).

In chimpanzees, kissing is by far the most characteristic conciliatory contact between former antagonists (de Waal and van Roosmalen, 1979), while mounting as a form of reassurance is uncommon. In terms of frequency, mounting and mating never reached a position among the top ten reconciliation behaviors in our studies of the Arnhem chimpanzee colony. Kissing, on the other hand, was rare during reconciliations observed in the bonobos. Apart from this morphological contrast between the reconciliation behavior of the two *Pan* species, the study in San Diego indicated major differences in the rate and initiation of peacemaking. Compared to the Arnhem chimpanzees, the San Diego bonobos reconciled a greater proportion of their conflicts, and these reunions were initiated more often by the aggressors (de Waal, 1987). It should be kept in mind, however, that the ape colonies of the Arnhem Zoo and the San Diego Zoo live under very different conditions, which may affect these variables.

AGGRESSION

It is my distinct impression that bonobos have their emotions better under control than chimpanzees. This temperamental difference is hard to measure, but I never observed in bonobos, for instance, the occasional temper tantrums or violent displays during which chimpanzees seem to lose all inhibitions. Of the more than 500 aggressive acts observed in the San Diego colony, 6.4% involved biting, and minor injuries resulted six times. After biting another bonobo, aggressors usually returned to the victim to inspect the spot where the teeth had been put. The lack of hesitation with which they took the finger, toe, or other bitten body part in their hands indicated that they remembered in detail what they had done. If an injury had resulted, the aggressor would lick and clean it.

Severe violence has, until now, not been observed in the species, either in captivity or in the wild. Jordan (1977) described a tense process of dominance reversal between two captive adult males (father and son) which took place without injuries except for a few scratches. When the adult and the oldest adolescent male of the San Diego colony were united after two years of physical separation (during which time they did have visual and vocal contact), they did not fight (see above, Vocal Communication). However, the next morning the younger male had a canine gash in his upper lip, not a really serious injury, but enough to stop his sexual advances to the cycling female in the presence of the dominant male. It remains to be seen whether the picture of bonobos as a peaceful species holds up after we know more about them. Chimpanzees used to have the same flower-child image until the late 1970s, after which observations of violence and cannibalism thoroughly disturbed this picture (Goodall et al., 1979; Nishida et al., 1985; de Waal, 1986c).

Nevertheless, on the basis of our present knowledge it seems a plausible hypothesis that aggressive behavior is more restrained and more effectively buffered in the bonobo than in the chimpanzee. Also, the style of peacemaking and tension regulation is different. Bonobos use overtly sexual and erotic behaviors at moments when chimpanzees merely kiss and embrace. This difference suggests that close and tolerant heterosexual relations have been of greater importance in the bonobo's evolutionary history than in the chimpanzee's. After sexual behavior had become established as a mechanism of tension regulation between the sexes, it may have been adopted as a mechanism among males, and especially among females. Alternative evolutionary scenarios are possible, but emphasis on heterosexual relationships seems the most logical route for the introduction of sex as a reassurance behavior.

Perhaps related to the lower intensity of agonistic confrontations in bonobos is the virtual absence of so-called side-directed behavior, that is, behavior shown by agonistically involved apes towards non-opponents. In chimpanzees, this is an extremely well-developed category of behavior. De Waal and van Hooff (1981) distinguished seven different functional types, ranging from recruitment of supporters to seeking reassurance from a third party. Only rarely did I notice such behaviors among the bonobos. Holding out a hand, for example, is a common begging gesture in both species, but whereas chimpanzees frequently show it to potential supporters when they are in trouble, the bonobos never did so.

One might argue that the smaller group size of the San Diego bonobos, compared with the Arnhem chimpanzee colony, may have made such strategies superfluous. I did observe the merging of two bonobo subgroups of three individuals each, however. I am convinced that six chimpanzees would have thoroughly tested out the new constellation of alliances, performing all kinds of side-directed communication in the process. Although the six bonobos did show a few interesting new coalitions, their agonistic strategies struck me as remarkably simple and straightforward. The same can be said for the display behavior of males, which is unspectacular compared to the elaborate shows of strength and acrobatics by male chimpanzees.

One aspect of agonistic behavior, however, may be better developed in the bonobo than in the chimpanzee. This is the rapid vocal dialogue between male antagonists. The observations by Jordan (1977) and me (see above, Vocal Communication) are best regarded as a first tentative suggestion of the existence of this phenomenon. From the perspective of conflict resolution, this interaction pattern is highly intriguing. Agonistic vocalizations probably contain information on the preparedness to attack, to defend oneself, to flee, to submit, or to reconcile. The vocal exchanges may, therefore, serve to avoid physical confrontations by helping the antagonists predict the response to an approach or a charge.

CONCLUSION

No doubt the variety in sexual behaviors and the nonreproductive functions of sex in bonobos represent a principal difference from the other *Pan* species. The present study lends quantitative support to suggestions by field workers that food-related sex serves to reduce competition and aggression. The simplest explanation for food-related sex, namely that it is due to a shift in arousal state, appears insufficient. The study on the San Diego bonobos demonstrates that sexual and erotic behaviors also increase after conflicts that are unrelated to food. In other words, the presence of food is not a prerequisite for sociosexual behavior, and the causal factor most elegantly explaining both measured increases is interindividual tension.

ACKNOWLEDGMENTS

I am grateful to the Zoological Society of San Diego for having allowed a study of their unique bonobo colony. The research was made possible by National Geographic Society Grant No. 2732-83 to the author and National Institutes of Health Grant No. RR00167 to the Wisconsin Regional Primate Research Center. I thank the curators, scientists, and animal care staff in San Diego for their pleasant cooperation and encouragement, especially Diane Brockman, Dr. Kurt Benirschke, Gale Foland, Mike Hammond, Fernando Covarrubias, and Joe Kalla. Data analysis in Madison was done with the expert assistance of Katherine Offutt. I am grateful to the Yerkes Primate Center for permission to record vocalizations of chimpanzees, Lesleigh Luttrell for the spectrographic analysis, and Dr. Charles Snowdon for making available his spectrographic equipment. I thank Mary Schatz and Jackie Kinney for typing the manuscript, Bob Dodsworth for excellent darkroom work with my negatives, and Linda Endlich for drafting the illustrations. This is publication No. 26–016 of the Wisconsin Regional Primate Research Center.

REFERENCES

Badrian, A., and N. Badrian. 1984. Social organization of *Pan paniscus* in the Lomako Forest, Zaire. In R. Susman, ed., *The Pygmy Chimpanzee,* pp. 325–346. New York: Plenum Press.

Becker, C. 1983. Sozialspiel in einer gemischten Gruppe Orang-Utans und Bonobos, sowie Spielverhalten aller Orang-Utans im Kölner Zoo. *Z. Kölner Zoo* 26:59–69.

Chevalier-Skolnikoff, S. 1982. A cognitive analysis of facial behavior in Old World monkeys, apes and human beings. In C. Snowdon, C. Brown, and M. Peterson, eds., *Primate Communication,* pp. 303–368. Cambridge: Cambridge Univ. Press.

Coolidge, H. 1933. *Pan paniscus:* Pygmy chimpanzee from south of the Congo River. *Am. J. Phys. Anthropol.* 28:1–57.

Everitt, B. 1977. *The Analysis of Contingency Tables.* London: Chapman and Hall.

Fisher, H. 1983. *The Sex Contract.* New York: Quill.

Goodall, J. van Lawick. 1968. The behaviour of free-living chimpanzees in the Gombe Stream Reserve. *Anim. Behav. Monogr.* 1(3):161–311.

———. 1973. Cultural elements in a chimpanzee community. In E. Menzel, ed., *Pre-cultural Primate Behavior,* pp. 144–184. Basel: Karger.

Goodall, J. 1986. *The Chimpanzees of Gombe: Patterns of Behavior.* Cambridge, MA: Belknap.

Goodall, J., A. Bandora, E. Bergman, C. Busse, H. Matama, E. Mpongo, A. Pierce, and D. Riss. 1979. Intercommunity interactions in the chimpanzee population of the Gombe National Park. In D. Hamburg and E. McCown, eds., *The Great Apes,* pp. 13–53. Menlo Park, CA: Benjamin/Cummings.

Harrisson, B. 1961. A study of orang-utan behavior in the semi-natural state. *Int. Zoo Yearbook* 3:57–68.

van Hooff, J. 1967. The facial displays of the Catarrhine monkeys and apes. In D. Morris, ed., *Primate Ethology,* pp. 7–68. London: Weidenfeld Nicolson.

_____. 1972. A comparative approach to the phylogeny of laughter and smiling. In R. Hinde, ed., *Non-Verbal Communication,* pp. 209–241. Cambridge: Cambridge Univ. Press.

_____. 1973. A structural analysis of the social behaviour of a semi-captive group of chimpanzees. In M. Cranach and I. Vine, eds., *Expressive Movement and Non-Verbal Communication,* pp. 75–162. London: Academic Press.

Hübsch, I. 1970. Einiges zum Verhalten der Zwergschimpansen *(Pan paniscus)* und der Schimpansen *(Pan troglodytes)* im Frankfurter Zoo. *Zool. Garten* 38:107–132.

Jordan, C. 1977. Das Verhalten Zoolebender Zwergschimpansen. Unpubl. dissertation. Frankfurt: Goethe University.

Kano, T. 1980. Social behavior of wild pygmy chimpanzees *(Pan paniscus)* of Wamba: a preliminary report. *J. Human Evol.* 9:243–260.

Kano T., and M. Mulavwa. 1984. Feeding ecology of the pygmy chimpanzees *(Pan paniscus)* of Wamba. In R. Susman, ed., *The Pygmy Chimpanzee,* pp. 233–274. New York: Plenum Press.

Kawai, M. 1965. Newly acquired pre-cultural behavior of a natural troop of Japanese monkeys on Koshima Island. *Primates* 6:1–30.

Kirshshofer, R. 1962. Beobachtungen bei der Geburt eines Zwergschimpansen *(Pan paniscus,* Schwarz 1929) und einige Bemerkungen zum Paarungsverhalten. *Z. Tierpsychol.* 19:597–606.

Köhler, W. 1925. *The Mentality of Apes.* London: Routledge and Kegan Paul.

Kummer, H., and J. Goodall. 1985. Conditions of innovative behaviour in primates. *Phil. Trans. R. Soc. Lond.* 308:203–214.

Kuroda, S. 1980. Social behavior of the pygmy chimpanzees. *Primates* 21:181–197.

_____. 1984. Interaction over food among pygmy chimpanzees. In R. Susman, ed., *The Pygmy Chimpanzee*, pp. 301–324. New York: Plenum Press.

Learned, B. 1925. Voice and "language" of young chimpanzees. In R. Yerkes and B. Learned, eds., *Chimpanzee Intelligence and Its Vocal Expressions,* pp. 57–157. Baltimore: Williams Wilkins.

Lovejoy, C. 1981. The origin of man. *Science* 211:341–350.

Marler, P., and L. Hobbett. 1975. Individuality in a long-range vocalization of wild chimpanzees. *Z. Tierpsychol.* 38:97–109.

Marler, P., and R. Tenaza. 1977. Signaling behavior of apes with special reference to vocalization. In T. Sebeok, ed., *How Animals Communicate,* pp. 965–1033. Bloomington: Indiana Univ. Press.

McGrew, W., and C. Tutin. 1978. Evidence for social custom in wild chimpanzees? *Man* 13:234–251.

Mori, A. 1983. Comparison of the communicative vocalizations and behaviors of group ranging in Eastern gorillas, chimpanzees and pygmy chimpanzees. *Primates* 24:486–500.

_____. 1984. An ethological study of pygmy chimpanzees in Wamba, Zaire: a comparison with chimpanzees. *Primates* 25:255–278.

Morris, D. 1967. *The Naked Ape.* New York: McGraw-Hill.

Napier, J. 1975. The tree of evolution. In V. Goodall, ed., *The Quest for Man,* pp. 57–78. London: Phaidon.

Nishida, T., M. Hiraiwa-Hasegawa, T. Hasegawa, and Y. Takahata. 1985. Group extinction and female transfer in wild chimpanzees in the Mahale National Park, Tanzania. *Z. Tierpsychol.* 67:284–301.

Nishida, T., R. Wrangham, J. Goodall, and S. Uehara. 1983. Local differences in plant-feeding habits of chimpanzees between the Mahale Mountains and Gombe National Park, Tanzania. *J. Human Evol.* 12:467–480.

Patterson, T. 1979. The behavior of a group of captive pygmy chimpanzees *(Pan paniscus)*. *Primates* 20(3):341–354.

Rempe, U. 1961. Einige Beobachtungen an Bonobos *(Pan paniscus,* Schwarz 1929). *Z. Wissensch. Zool.* 165:81–87.

Rensch, B. 1973. Play and art in apes and monkeys. In E. Menzel, ed., *Pre-cultural Primate Behavior,* pp. 102–123. Basel: Karger.

Rijksen, H. 1978. *A Field Study on Sumatran Orang-utans* (Pongo pygmaeus abelii, *Lesson 1827): Ecology, Behavior and Conservation.* Wageningen (Netherlands): Veeneman.

Savage-Rumbaugh, S., and B. Wilkerson. 1978. Socio-sexual behavior in *Pan paniscus* and *Pan troglodytes:* a comparative study. *J. Human Evol.* 7:327–344.

Savage-Rumbaugh, S., B. Wilkerson, and R. Bakeman. 1977. Spontaneous gestural communication among conspecifics in the pygmy chimpanzee *(Pan paniscus).* In G. Bourne, ed., *Progress in Ape Research,* pp. 97–116. New York: Academic Press.

Schwarz, E. 1929. Das Vorkommen des Schimpansen auf den linken Kongo-Ufer. *Rev. Zool. Bot. Afr.* l6:425–426.

Sibley, C., and J. Ahlquist. 1984. The phylogeny of the Hominoid primates, as indicated by DNA-DNA hybridization. *J. Mol. Evol.* 20:2–15.

Thierry, B. 1984. Descriptive and contextual analysis of eye-covering behavior in captive rhesus macaques *(Macaca mulatta). Primates* 25:62–77.

Thompson-Handler, N., R. Malenky, and N. Badrian. 1984. Sexual behavior of *Pan paniscus* under natural conditions in the Lomako Forest, Equateur, Zaire. In R. Susman, ed., *The Pygmy Chimpanzee,* pp. 347–368. New York: Plenum Press.

Tratz, E., and H. Heck. 1954. Der afrikanische Anthropoide "Bonobo," eine neue Menschenaffengattung. *Saugetierk. Mitteilungen* 2:97–101.

de Waal, F. 1982. *Chimpanzee Politics.* London: Jonathan Cape.

———. 1986a. Imaginative bonobo games. *Zoonooz* 59:6–10.

———. 1986b. Conflict resolution in monkeys and apes. In K. Benirschke, ed., *Primates – The Road to Self-sustaining Populations,* pp. 341–350. Berlin: Springer Verlag.

———. 1986c. The brutal elimination of a rival among captive male chimpanzees. *Ethol. Sociobiol.* 7:237–251.

———. 1987. Tension regulation and nonreproductive functions of sex in captive bonobos *(Pan paniscus). Nat. Geogr. Research.* 3:318–335.

———. 1988. The communicative repertoire of captive bonobos *(Pan paniscus),* compared to that of chimpanzees. *Behaviour* 106:183–251.

de Waal, F., and J. van Hooff. 1981. Side-directed communication and agonistic interactions in chimpanzees. *Behaviour* 77:164–198.

de Waal, F., and A. van Roosmalen. 1979. Reconciliation and consolation among chimpanzees. *Behav. Ecol. Sociobiol.* 5:55–66.

de Waal, F., and D. Yoshihara. 1983. Reconciliation and redirected affection in rhesus monkeys. *Behaviour* 85:224–241.

THE SEXUAL BEHAVIOR
OF PYGMY CHIMPANZEES

Takayoshi Kano

Sexual behavior, or more precisely copulatory behavior, is indispensable to reproduction. Usually, however, sexual behavior has no immediate meaning for the survival of any individual. In most animal species, therefore, females become receptive towards males only when they are likely to conceive, so as to minimize the energy and risks of copulation. Humans are exceptional; their heterosexual copulation functions not only for the production of offspring but also for the formation and maintenance of male-female pairs. The aim of this paper is to show that pygmy chimpanzees resemble humans in their sexual behavior. Their copulatory and pseudo-copulatory behaviors have purely social functions as well as reproductive functions. The data presented in this preliminary paper were collected at Wamba, in the central Zaire basin, between 1977 and 1983.

PSEUDO-COPULATORY BEHAVIORS

Besides heterosexual copulation, sexually mature (adolescent and adult) pygmy chimpanzees have four kinds of pseudo-copulatory behaviors which involve genito-genital (GG) contacts: genito-genital (GG) rubbing, mounting, rump contact, and penis fencing.

GG-rubbing was primarily restricted to females (Table 1), who, on the ground or in trees, embraced each other ventro-ventrally and swung their pelves sideways so that the tips of the genitals (frenulum clitoridis) rubbed each other (Kano, 1980; Kuroda, 1980). In many cases, both females rubbed the genitals silently, but sometimes one or both participants emitted vocal sounds similar to copulatory screams. GG-rubbing was usually preceded by a kind of sexual display: one female stood up bipedally and extended her hand towards the partner to touch her. In this way, she solicited the other for GG-rubbing. This behavior pattern is, in behavioral morphology, similar to the courtship display exhibited by a male towards a receptive female. This solicitation for GG-rubbing was directed more frequently from the younger to the older than the other way around. Therefore, GG-rubbing may be interpreted as reassurance behavior between females. With few exceptions, GG-rubbing was not preceded or followed by agonistic interactions.

Table 1. Genito-genital contacts (1978 and 1979 under provisioning).

	Mounting	Rump contact	GG-rubbing
Male – male	103	31	2*
Male – female	41	12	0
Female – female	5	0	318
Total	149	43	320

* Penis fencing.

The most frequent GG contacts between males were mounting and rump contact (Table 1). These often followed chasing or attacking. However, there were also many cases of peaceful mounting or rump contact which were not accompanied by agonistic behavior but were preceded by one partner's solicitation, analogous to the heterosexual courtship display. In male-male mounting, the dominant more often mounted the subordinate. Both mounting and rump contact were usually accompanied by quick rhythmical pelvic thrusts, and screams or grimacing by one or both participants.

Mounting and rump contact were not restricted to interactions between males (Table 1). Male to female, female to male, and female to female mountings also occurred, although less frequently. Rump contact between male and female also occurred. Heterosexual mounting and rump contact often occurred when the female was not showing maximum swelling.

The behavior that is called "penis fencing" in our research group may be interpreted as male-male genital rubbing, in which each partner's erect penis rubs the other's erect penis while the partners hang face to face from a branch. This is a very rare behavior: I have recorded only two cases of penis fencing among sexually mature males (Table 1).

The above mentioned pseudo-copulatory behaviors may be classified as sexual behavior in the sense that these involve the contact of external sexual organs. However, these behaviors were frequent not in the resting periods, when individuals could freely choose their spatial position, but in the feeding periods, when they were forced to shorten inter-individual distances. This fact seems to indicate that these behaviors function to ease inter-individual tensions, no matter how erotic they appear by our standards.

HETEROSEXUAL COPULATION

Immature pygmy chimpanzees were frequently involved in heterosexual copulations. In this paper, however, I will report not about such immature copulations, which are not reproductive, but only about copulations in which both partners are sexually mature. Though some adolescents may be sterile, copulations between adolescents, or between an adolescent and an adult, are included in the mature copulations, since they perform adult-type copulations. I recorded 1,209 cases of mature copulations (simply copulations hereafter).

About 40 behavioral elements accompanying copulations were recorded (Table 2). The elements restricted to males were those included in the courtship display, as well as mounting, intromission, and thrusting. The element restricted to females was presenting. Some of the courtship elements were identical to those of common chimpanzees. Among pygmy chimpanzees, however, a more prominent behavior appeared to be one in which the male stood bipedally and stretched hands toward the female. Aggressive elements such as hair erection and glaring reported for common chimpanzees (Goodall, 1968) were rare. Neither leaf clipping nor pelvic thrusting (Nishida, 1979), which young and low-ranking common chimpanzee males furtively displayed, avoiding the sight of higher-ranking males, has been recorded among Wamba pygmy chimpanzees.

A total of 139 copulations was recorded between October 1984 and February 1985. In 101 of those copulations, the total mating sequence was observed.

Courtship was observed in 55 out of these 101 copulations (54%). However,

Table 2. Copulatory behavior elements.

PRE-COPULATORY		COPULATORY		POST-COPULATORY	
Advance	(M, F)	Intromission	(M)	Sit	(M, F)
Approach	(M, F)	Mounting	(M)	Walk away	(M, F)
Stare	(M, F)	Thrusting	(M)	Run away	(F)
Begging food	(F)	Female postures	(F)	Grooming	(M, F)
Sharing food	(M)	clinging			
Grooming	(M, F)	sitting			
Retreat	(M)	crouching (flat			
Follow	(M, F)	prone, raising			
Step away	(M)	buttocks)			
Walk away	(M, F)	latero-ventral			
Male display	(M)	quadrupedal			
bipedal standing		standing			
sitting		hanging			
branch shake		Grimace	(M, F)		
reach out		Scream or squeal	(F)		
penile erection		Touching scrotum			
rocking		with hand or foot	(F)		
touching females					
Presenting	(F)				
dorso-ventral					
ventro-ventral					

Notes: M = males, F = females.

courtship display does not necessarily indicate that a male initiated the mating sequence. Before courtship display, it was often observed that an estrous female approached a male so that he noticed her swollen sexual skin, and she responded immediately to his slightest courtship display by running up and presenting to him. Thus some female approaches may be considered as the initiation of a mating sequence through the solicitation of a male's display. In my judgement, 63 (62%) of the total 101 copulations were initiated by a male's courtship displays or approaches, 26 (26%) were initiated by a female's approaches or solicitations, and in 12 cases it was impossible to judge which sex initiated.

All observed copulations were preceded by a female presenting; no forced copulation was observed. A ventro-ventral present was followed by a ventro-ventral copulation, and a dorso-ventral present was followed by a dorso-ventral copulation; in one set of 940 recorded copulations, 138 (14.7%) were ventro-ventral. The ventro-ventral copulation seemed to be preferred by females more than by males. Sometimes males did not react positively to ventro-ventral presenting, while they almost always reacted to dorso-ventral presenting by mounting. In addition, there were 20 cases recorded in which during dorso-ventral copulation the female separated herself from the male and again embraced the male ventro-ventrally, and consequently the copulation was switched to the ventro-ventral. But there was no case of switching from ventro-ventral to dorso-ventral copulation.

The mean duration of copulatory junction of Wamba pygmy chimpanzees was 15.3 seconds (range: 1–52; n = 121). This average is closest, among the great apes, to that for common chimpanzees (Table 3). However, although short, the duration was longer than in common chimpanzees. During the junction, the male incessantly thrust his pelvis. The mean number of thrusts was 2.7 times per second and 43.8 times per junction.

In common chimpanzees, three mating patterns are recorded: opportunistic mating, consortship, and male possessive behavior (Tutin, 1979). In Wamba pygmy chimpanzees, all observed copulations were opportunistic. This seems to suggest that pygmy chimpanzee males are sexually less competitive than common chimpanzee males. In a set of 515 copulations, interruption of copulation by an individual other than the mating pair was observed for 34 cases (7%). In 27 cases, an adult male interrupted, and in seven cases an adult female did so. Without exception, the interrupting male was dominant over the mating male.

Copulations, as well as pseudo-copulations such as GG-rubbing, rump contact, or mounting, more often appeared to stimulate others sexually than to induce aggression. It was frequently observed that a sexual interaction induced another involving different chimpanzees, or that one of the participants who had just finished a sexual contact was immediately solicited sexually by another. In four cases, one or two individuals joined a mating couple by mounting or by thrusting their buttocks against them.

A juvenile or infant often interfered in adult copulations by clinging to one individual of the mating pair and screaming. Such immature individuals were not limited to the offspring of the mating female. Their screams were not due to fear or

Table 3. Copulatory activity of great apes.

		Duration of copulation (sec.)			Number of pelvic thrusts			Reference
		Mean	Range	n	Mean	Range	n	
Pygmy chimpanzee	Wild	15.3	1-52	121	43.8	14-78	16	This study
	Captive	20			25			Patterson[1]
Common chimpanzee	Wild	7	3-35	45	8.8	3-30	1084	Tutin and McGinnis[2]
	Captive	8	5-14					Yerkes[3]
								Yerkes and Elder[4]
Gorilla	Wild	96	30-310	11	27.5	6-52.5	8	Harcourt et al.[5]
	Captive	52.5	20-110	11				Hess[6]
Orangutan	Wild	648	180-1680					Galdikas[7]
	Captive	900						Nadler[8]

1. Patterson (1979)
2. Tutin and McGinnis (1981)
3. Yerkes (1939)
4. Yerkes and Elder (1936)
5. Harcourt et al. (1981)
6. Hess (1973)
7. Galdikas (1981)
8. Nadler (1977)

antagonism against the mating male, but seemed to be an expression of sexual excitement. This is supported by the fact that after the copulation the immature individual often participated in a sexual interaction with one of the partners.

FEMALE SEXUAL STATE AND OCCURRENCE OF COPULATION

The occurrence of copulation is greatly dependent on female sexual state. Seventy-eight percent of the copulations took place when the females showed maximal or near-maximal tumescence of their sexual skins.

In pygmy chimpanzees, the size and cycle of tumescence changes according to the female's age. Up until six years of age, the vulva is very small and shows no swelling. Around seven years of age, at the beginning of adolescence, the vulva begins to show slight tumescence, but the size is still small. Throughout adolescence, the vulva increases its size with age. In this stage, the shape is largely semi-spherical and resembles that of common chimp females. Characteristic of pygmy chimpanzee females is the oblong sexual skin which stretches from dorsal to ventral side of the vulva. This characteristic gradually becomes prominent after first parturition.

No adolescent female showed a regular tumescence-detumescence cycle. They almost always maintained the level of maximum or near-maximum tumescence, and even if their genital organs became deflated, they usually recovered the former swelling after only a few days. As adolescent females were almost always sexually receptive, they frequently copulated at any time in their irregular cycles. However, pregnant adolescent females seemed to be exceptions to this rule. A pregnant adolescent female began to show shrinkage of sexual skin two months before parturition, though she continued to copulate up until three weeks before parturition.

Adult females show regular tumescence-detumescence cycles. According to Furuichi (1987), the mean cycle length of Wamba pygmy chimpanzee females is 42 days (range: 37–49 days). The duration of maximal tumescence varied, depending on the individual female and on her age. One multiparous female continued her estrus for at least 50 days; on the other hand, some females showed maximal swelling for only a few days, while in many females the state of maximal and near-maximal tumescence lasted more than 20 successive days. Thus the average duration of estrus in adult pygmy chimpanzee females is longer than that of common chimpanzees (9.6 days, Tutin, 1979; 12.5 days, Hasegawa and Hiraiwa-Hasegawa, 1983).

In Wamba pygmy chimpanzee females, the anestrous period after parturition, or postpartum acyclicity, appeared to be much shorter than that of common chimpanzees (30 months for primiparous and 48 months for multiparous females) (Goodall, 1983). Some of the females with infants younger than one year old resumed maximal tumescence and participated in heterosexual copulations. Since, in Wamba, lactation lasts until the infant becomes about four years of age, the resumption of swelling is unrelated to lactation. Swelling may also be unrelated to menstruation, since a pygmy chimpanzee mother with a dependent offspring does not, as a rule, give birth again for five to six years (Kano, 1987). This birth interval is almost the same as that of common chimpanzees (5 to 5.5 years, Goodall, 1983). That means the mother is probably in postpartum lactational amenorrhea while showing cyclic sexual swelling and receptivity. Consequently, the proportion of sexually receptive females becomes fairly high in a group of pygmy chimpanzees. In almost all parties of pygmy chimpanzees I observed, there was at least one estrous female, and in 50–100% of parties with more than 11 members, at least one heterosexual copulation (Kano, 1982).

CONCLUSION

It is very difficult to compare appropriately the frequencies of copulation among different species of different group size, particularly when recorded by different observers under a variety of observational conditions. It can be safely said, however, that pygmy chimpanzees are the most sexual of the apes and are engaged in heterosexual copulation more frequently than any other living anthropoid apes (Kano, 1987). This is probably related to the elongated duration of female estrus. However, this frequent sexual activity does not mean a high reproduction rate in pygmy chimpanzee groups. Very few copulations lead to pregnancy, partly because females go into estrus during adolescent sterility or postpartum amenorrhea, and partly because the period of maximal tumescence is elongated before and/or after ovulation in each cycle.

In Wamba pygmy chimpanzee groups, copulations occurred in the same context as that in which pseudo-copulatory behaviors occurred. Copulations most frequently occurred immediately after strong social stimuli such as arrival at a large natural food source or at the artificial feeding site, reunion of a party unit or group, or detection of another unit group nearby; and their incidence decreased rapidly after the first enthusiasm. The contexts and the declining pattern are almost identical to those for pseudo-copulation behaviors (i.e., GG-rubbing, mounting, and rump contact) (Kano, 1987). This strongly suggests that copulation has the same social function as that of the other GG contacts, that is, easing tensions between individuals.

Pygmy chimpanzees have a more or less fluid society like common chimpanzees. In pygmy chimpanzees, almost all temporary parties include all age-sex classes at the same time, though size is changeable. In other words, mixed-type parties are predominant (e.g., Kano, 1982). In common chimpanzees, on the other hand, parties of various age-sex compositions are often formed: mixed, adult, mother, and male parties, and temporary lone individuals (e.g., Goodall, 1968). Since most adult female pygmy chimpanzees have dependent offspring, the prominence of mixed parties means that, in pygmy chimpanzee groups, the coexistence of males and females is profoundly realized throughout the reproductive cycle. This sociological feature of pygmy chimpanzees is probably related to the frequent occurrence of copulation based on the nearly continuous sexual receptivity of females. In pygmy chimpanzees, copulatory behavior between sexually mature individuals may be not only reproductive behavior but also, or more importantly, social behavior.

REFERENCES

Furuichi, T. 1987. Sexual swelling, receptivity and grouping of wild pygmy chimpanzees at Wamba, Zaire. *Primates* 28(3):309–318.

Galdikas, B. M. F. 1981. Orangutan reproduction in the wild. In C. E. Graham, ed., *Reproductive Biology of the Great Apes, Comparative and Biomedical Perspectives*, pp. 281–299. New York: Academic Press.

Goodall, J. van Lawick. 1968. The behaviour of free-living chimpanzees of the Gombe Stream Reserve. *Anim. Behav. Monogr.* 1:161–311.

Goodall, J. 1983. Population dynamics during a fifteen-year period in one community of free-living chimpanzees in the Gombe National Park, Tanzania. *Z. Tierpsychol.* 61:1–60.

Harcourt, A. H., K. J. Stewart, and D. Fossey. 1981. Gorilla reproduction in the wild. In C. E. Graham, ed., *Reproductive Biology of the Great Apes, Comparative and Biomedical Perspectives*, pp. 265–278. New York: Academic Press.

Hasegawa, T., and M. Hiraiwa-Hasegawa. 1983. Opportunistic and restrictive matings among wild chimpanzees in the Mahale Mountains, Tanzania. *J. Ethol.* 1:73–85.

Hess, J. P. 1973. Some observations on the sexual behaviour of captive lowland gorillas. In R. P. Michael and J. H. Crook, eds., *Comparative Ecology and Behaviour of Primates*, pp. 507–581. London: Academic Press.

Kano, T. 1980. Social behavior of wild pygmy chimpanzees *(Pan paniscus)* of Wamba: a preliminary report. *J. Human Evol.* 9:243–260.

_____. 1982. The social group of pygmy chimpanzees *(Pan paniscus)* of Wamba. *Primates* 23: 171-188.

_____. 1987. A population study of a unit group of pygmy chimpanzees of Wamba, with a special reference to the possible lack of intraspecific killing. In Y. Ito, J. L. Brown, and J. Kikkawa, eds., *Animal Societies: Theories and Facts,* pp. 159–172. Tokyo: Japan Scientific Societies Press.

_____. In press. The social organization of pygmy chimpanzees and common chimpanzees: similarities and differences. In S. Kawano, J. H. Connell, and T. Hidaka, eds., *Evolution and Coadaptation in Biotic Communities.* University of Tokyo Press.

Kuroda, S. 1980. Social behavior of the pygmy chimpanzees. *Primates* 21(2):181–197.

Nadler, R. D. 1977. Sexual behavior of the chimpanzee in relation to the gorilla and orang-utan. In G. H. Bourne, ed., *Progress in Ape Research,* pp. 191–206. New York: Academic Press.

Nishida, T. 1979. The leaf-clipping display: a newly discovered expressive gesture in wild chimpanzees. *J. Human Evol.* 9:117–128.

Patterson, T. 1979. The behavior of a group of captive pygmy chimpanzees *(Pan paniscus).* *Primates* 20(3):341–354.

Tutin, C. E. G. 1979. Mating patterns and reproductive strategies in a community of wild chimpanzees *(Pan troglodytes schweinfurthii). Behav. Ecol. Sociobiol.* 6:29–38.

Tutin, C. E. G., and P. R. McGinnis. 1981. Chimpanzee reproduction in the wild. In C. E. Graham, ed., *Reproductive Biology of the Great Apes, Comparative and Biomedical Perspectives,* pp. 239–264. New York: Academic Press.

Yerkes, R. M. 1939. Social dominance and sexual status in the chimpanzee. *Human Biol.* 11:78–111.

Yerkes, R. M., and J. H. Elder. 1936. Oestrus, receptivity, and mating in the chimpanzee. *Comp. Psychol. Monogr.* 13:1–39.

DEVELOPMENTAL RETARDATION
AND BEHAVIORAL CHARACTERISTICS
OF PYGMY CHIMPANZEES

Suehisa Kuroda

INTRODUCTION

One of the important aspects of human evolution is the acquisition of paedo-morphic traits and retardation of growth (Gould, 1977). From a morphological perspective, this could hardly be denied, and behavioral and sociological changes connected with such physical changes have been hypothesized by many authors. The essential part of this hypothesis is familiar. The prolonged maturation period enabled behavioral flexibility, strengthened the mother-offspring tie, and involved males in child-rearing, all of which increased the complexity of social organization. This seems almost self-evident, but it has rarely been demonstrated by actual observation of living animals.

In this aspect, the pygmy chimpanzee *(Pan paniscus)* is an interesting subject because H. Coolidge (1933) stated that this ape is a paedomorphic species when compared to the common chimpanzee *(Pan troglodytes)*, and the pygmy chimpanzee seems to preserve some immature behavioral patterns after reaching adulthood (Kuroda, 1979, 1980).

Coolidge's theory has been rejected by some researchers, but more detailed evidence has been presented by Shea (1983, 1984). Broadly stated, adult pygmy chimps have hindlimbs of adult common chimps, trunks and forelimbs of subadults, and skulls of juveniles (Shea, 1983, 1984). This mosaic state of the pygmy chimp's paedomorphosis (since we do not know which is the prototype, there is the converse prospect of the common chimpanzee's hypermorphosis) has led to further controversy about Coolidge's theory. On the other hand, with the progress of socio-behavioral studies on pygmy chimps in natural habitats and in captivity, it has become clearer that between the two chimp species, there is a great difference which is mainly attributed to the development and diversity of sexual behaviors and to the morphological and "behavioral" paedomorphosis in the pygmy chimps. Doubtless the study of these characteristics in the pygmy chimp will bring not only understanding of this ape but also insight into the reconstruction of the process of human evolution.

In this paper, I will first distinguish the type of heterochronic process which resulted in the pygmy chimp's paedomorphosis, and second, I will discuss the retardation of infant development and the behavior and social structure which closely relates to this retardation.

HETEROCHRONIC PROCESS

Shea (1984) suggested that the pygmy chimp's paedomorphosis is caused by rate hypomorphosis (Gould, 1977). That is, the two chimp species have the same duration of growth, but the delays in the speed of growth in the pygmy chimp result in its paedomorphosis. The following data support Shea's hypothesis.

Female pygmy chimps at Wamba come into their first estrus at the age of eight or nine years, and then they leave their natal groups. The age of females who transfer into E-group, our most intensively studied group, is difficult to determine, but we estimate it to be 13 to 15 years when they give birth for the first time (13 years: one case; 14 years: three cases; 15 years: two cases). It has been observed that two females over 15 years old, in different unit groups, had not given birth, so the average age will be higher. Therefore, the age of sexual maturity of the female pygmy chimps in the natural habitat is the same as that of female common chimps at Gombe and Mahale. In captivity, the pygmy chimp's first birth occurs at nine years (n = 3, San Diego Zoo), which also does not seem different from the common chimpanzee case under the same conditions. According to the records of Japanese zoos, the minimum age for first births among common chimps is seven years, while ten to eleven years proved to be the most frequent age (n = 106).

There is no observation record of the male pygmy chimp's first ejaculation period at Wamba. The age at which males reach sexual maturity has been estimated to be approximately nine years, when body size and testes begin to enlarge markedly. This estimated age is close to the period of sexual maturity of male common chimps in natural habitats.

Both female and male pygmy chimps at Wamba attain adult body size at the age of 14 to 16 years, which is approximately equivalent to data on weight increases of common chimps from Mahale (Uehara and Nishida, 1987). We therefore conclude that there is no time difference in maturation for common chimps and pygmy chimps and that the pygmy chimp's paedomorphosis results from rate hypomorphosis. The small sexual dimorphism in the pygmy chimps can also be attributed to truncation of maturation by rate hypomorphosis.

RETARDATION OF INFANT DEVELOPMENT

According to data from the San Diego Zoo, the pygmy chimp gestation period is 255 days (n = 1), which is within the range of the common chimpanzee (Schultz, 1940). The weight of newborn pygmy chimps is about 80% that of newborn common chimps (1.27 kg, n = 4, San Diego Zoo). This is the same ratio of weight between adult females of both species.

I obtained a male infant who had died just after being caught near Wamba (Fig. 1, Fig. 2), whose weight was 1.3 kg; its deciduous teeth had not yet appeared, but the frontal gums were hard. Considering the overall appearance of the animal, his age was estimated at around two months, and his weight is in accord with data of newborns in the zoo. The infant's arms and legs looked too slender, but young infants in general have such slender limbs. Newborns and infants of one year are not very different in body size or slenderness.

It appears that the weight of pygmy chimp infants in natural habitats increases slowly. In another case, I cared for a female infant right after she was caught. Her weight was 3.0 kg, and the M1's of the lower jaw had almost reached the occlusal plane. She was in good health and could eat natural foods; her age, based on body size and tooth condition (applying the tooth standard for common chimps, Schultz, 1940; Nissen and Riesen, 1964) was estimated to be nearly three years.

Records of the San Diego Zoo show that a healthy, breast-fed two-year-old pygmy chimp weighed 3.15 kg (though artificially fed animals grew more rapidly, Table 1). From these data, we can conclude that a two- to three-year-old pygmy chimp in a natural habitat would also weigh no more than 3.0 kg, only 2.3 times that of its birth

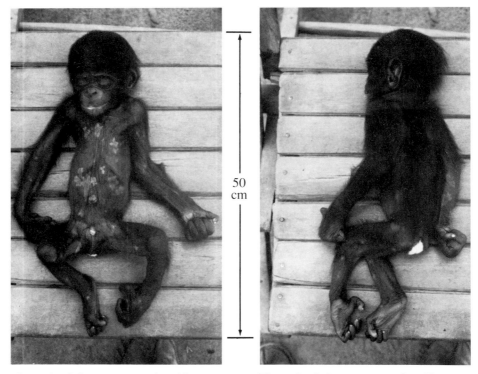

50
cm

Figure 1. Infant, two months old.

Figure 2. Infant, two months old. Rump skin is bare; white tuft grows later in the pygmy chimpanzee than in the common chimpanzee.

Figure 3.   One-year-old infant.

Figure 4.   Three-year-old infant (female).

weight. On the other hand, a common chimp of three years and three months and another of three years and nine months weighed 8.5 kg and 7.5 kg respectively (Uehara and Nishida, 1987), which is more than four to five times its newborn weight. From the age of three to four, the pygmy chimp's weight does not seem to increase much (Fig. 4). Therefore, we can say that the pygmy chimp grows considerably slower in the natural habitat than the common chimp (Table 1).

Table la. Weight increase (kg) of *Pan paniscus* infants.

| | San Diego Zoo [1] | | | | | | Wild caught | |
| | Artificially fed | | | | Breast fed | | | |
Month	Kalindo	Kakowet	Leslie	Lana	n = 1	n = 1	Inf (M)	Juv (F)
0	1.43[a]	1.18[b]	1.23[c]	1.48[d]				
1	1.83	1.58	1.60	1.83				
2	2.38			2.26			1.3	
3	2.79	2.53	2.26					
4		2.84						
5		3.04						
6		3.35						
7								
8								
9			3.65					
10			4.15					
11	5.51							
12						3.0[2]		
•								
14				5.62				
•								
17				5.82				
•								
19				6.41				
•								
22	7.80							
•								
24						3.15		
•								
26			7.88					
•								
35								3.0
•								
39								
•								
45								

Sources: 1. Data from JoAnn Thomas
2. Data from M. Hammond

a. 5 days after birth
b. 4 days after birth
c. 7 days after birth
d. 3 days after birth

Table 1b. Weight increase (kg) of *Pan troglodytes* infants.

	Asahiyama Zoo [1]	Kyoto University PSI [2]		Schultz [3]	Wild [4]
	Breast fed	Breast fed	Artificially fed		
	n = 1	Leo (M)	Popo (F)	n = 6	n = 1
Month					
0				1.58	
1		1.5	1.8		
2	2.2	2.0	2.7		
3	2.7	2.2	3.4		
4	3.0	2.3	3.9		
5	3.3	2.4	4.4		
6	3.5	2.5	4.7		
7	3.6	2.5	5.0		
8	3.8	2.7	5.5		
9	4.2	2.7	5.6		
10	4.4	3.0	5.7		
11	4.7	3.7	6.2		
12	5.1	4.0	6.5		
•					
14			7.0		
•					
17					
•					
19					
•					
22					
•					
24					
•					
26					
•					
35					
•					
39					8.5
•					
45					7.5

Sources: 1. M. Kosuga (pers. comm.)
2. Kumazaki (1983)
3. Schultz (1940)
4. Uehara and Nishida (1987)

This retardation of growth corresponds to the retardation of motor and social development. Pygmy chimp infants of three months and under never leave their mothers. At six months, pygmy chimps are rarely seen more than one meter away from their mother, and their locomotion is still unsteady. Common chimp infants, on the other hand, can be seen parting from their mothers and approaching unrelated individuals at an age of six months. By approximately ten months, pygmy chimps can be seen slowly climbing narrow branches and touching siblings and unrelated individuals. Still, if an infant moves more than three to four meters away from its mother, the mother will bring the infant back to her side. When they reach one year of age, they can walk quadrupedally only very slowly and for just a few meters. Two-year-old infants begin to approach and play with other infants, juveniles, and adults, but they cannot move as actively as their elders before the age of three years. Three-year-old and older infants often move more than ten meters from their mothers, but they never go so far that they cannot come back immediately when something strange happens. In most cases (about 80% of two-year-old infants), mothers convey infants in a ventral clinging position. Three-year-old and older infants usually ride on their mother's back during long-distance traveling.

The infant under one year of age practices mouthing of solid food, but never eats it. Infant pygmy chimps reach the weaning period by the age of four to five years. Even though infants of both species are weaned at approximately the same time, the common chimp begins to eat solid food and to depend less on its mother at a much earlier age.

PAEDOMORPHOSIS, RETARDATION OF DEVELOPMENT, AND SOCIAL BEHAVIOR

Pygmy chimps have characteristic behaviors which support the notions of paedomorphosis and retardation of motor and social development. These behaviors are keys to understanding pygmy chimpanzee society and to considering the process of human evolution.

Food sharing is one such behavior. Pygmy chimp mothers share food with their offspring longer than common chimp mothers do, and they share food even with fully adult sons (Kuroda, 1984). Although there are not many observed cases (n = 7), pygmy chimp mothers put food into the mouths of infants of about one year of age without their begging. This behavior has never been seen in any other nonhuman primate and suggests one source of the evolution of voluntary food-sharing. For the pygmy chimps themselves, this behavior would promote the infants' weaning.

The mothers' strong and continuous attention to their tottering infants is also peculiar to the pygmy chimps. That is, the mothers bring infants back whenever they are too far away or appear to be going into a dangerous situation. According to a comparative study of two chimp species at Yerkes, pygmy chimp mothers pay closer and more continuous attention to their infants than common chimp mothers do, and pygmy chimp mothers seem to understand the infants' situations very well (Savage-

Rumbaugh, 1984). When infants scream, pygmy chimp mothers can quickly assess the situation and respond in a variety of ways – either by embracing the infant, attacking the threatener, embracing the infant and appeasing the threatener, or leaving the infant alone. However, the common chimp mother's reaction to the infant's scream may be one of two patterns: either attack its neighbors or embrace the infant and escape (Savage-Rumbaugh, 1984). These differences constitute one example of the flexibility of pygmy chimp behavior. Others follow.

High tolerance among individuals is also an important trait of pygmy chimp society. In common chimpanzees, juveniles begin to be threatened by adult males just after weaning (four to five years). By contrast, pygmy chimp juveniles are not threatened by adult males until they reach puberty (eight years); pygmy chimps have a longer duration of immature morphology and a longer period of toleration by adult males. In addition, adult male pygmy chimps exhibit much less aggression among themselves than common chimps do; they never kill or even seriously injure each other. In general, pygmy chimp society members have a higher tolerance for each other.

Strong bonding between mothers and sons is also a special pygmy chimp characteristic and is probably derived from the long duration of immaturity. Even after weaning, juveniles often play "baby" by rocking, whimpering, and having temper tantrums. This is interpreted as the juveniles' resistance to the decrease in contact with the mother. There is a strong possibility that male juveniles play "baby" more often and persistently (male: seven cases; female: two cases). Even six-year-old male juveniles (n = 2) exhibited this behavior.

In the case of female juveniles, the distance between mother and daughter during foraging gradually increased after the age of six to seven years, although they could always be seen in the same foraging party. When the party began long-distance traveling, six- to seven-year-old female juveniles frequently started moving only after the mothers left. When they reach about nine years of age, daughters emigrate to other unit groups. However, sons, even at six or seven years of age, can be seen very close to their mothers. Even if they cannot be found near their mothers, they return when long-distance traveling begins. When they reach adulthood, males still forage in the same party as their mothers, unlike the male common chimps that begin foraging in a different party from adolescence onward.

In pygmy chimpanzees, the most stable group is the matrilineal group which consists of a mother and her offspring, including adult sons. The reasons for the high frequency of mixed groups among pygmy chimps, even in small-sized parties (Kuroda, 1979), is explained by this matrilineal grouping. This matrilineal group appears similar to a nuclear family because of its composition and stability, but copulation does not occur in such a group. Pygmy chimp groups are based on three types of bonding: matrilineal bonding, interfemale bonding which is formed through GG (genito-genital) rubbing (Kano, 1980; Kuroda, 1980), and male-female bonding.

The strong bond between mother and adult son can also be seen in the protective behavior of the mother and the dependency of the son on the mother. This greatly affects male ranking order. When the son fights, his mother attacks the antagonist or attacks or appeases its mother. The son retreats even when his mother is attacked. Pygmy chimp males and females have almost equal power both socially and physically; the son of a dominant female tends to be highly ranked. In one case, a young adult male won a fight with three top-ranking males, supported by his mother who was the most dominant in E-group. He also had been the highest ranked in a subgroup which included five adult males. However, when the mother's health failed and she lost her high status, the son's rank also fell suddenly in the whole E-group as well as in the subgroup. The empty high-status position in the subgroup was taken over by the son of the new most-dominant female, who had just reached adulthood. (This subgroup fissioned from E-group during the event.) Since female power affects the male ranking order, the pygmy chimp ranking system is unstable relative to that of the common chimps, which is determined only by male power.

The rate hypomorphosis process of pygmy chimps decreases sexual dimorphism both physically and behaviorally. Even an alpha male, when attacked by an adult female, always retreats. Sometimes an adult female suddenly attacks a male, for no apparent reason. Females make alliances to co-attack others, yet males, even siblings, never do this. From this, we can say that the pygmy chimp society is highly female-centric.

One of the most important paedomorphic characters is the ventral positioning of the female genitals (Bolk, 1926), which, in pygmy chimps, enables GG-rubbing between females. This behavior decreases tension between females and makes it more possible for females to form alliances with each other. In particular, this behavior functions critically when young females are transferring into a new unit group. They repeat this behavior with resident females to facilitate acceptance.

High frequency of face-to-face postures and eye contact in pygmy chimp interactions is also considered to be a result of prolonged immaturity. These are the bases of the development of their complicated communication system. Thus, pygmy chimp paedomorphosis is characterized by rate hypomorphosis, of which the most important aspects are the retardation of physical growth and behavioral development. This is the source of their frequent food-sharing, behavioral flexibility, complicated communication, matrilineal groupings, female-centric society, and highly tolerant interaction patterns.

In conclusion, we can gain much insight into human evolution from the study of the pygmy chimpanzee's behavioral development, mother-infant relations, female-female interactions, and sexual behavior. These findings present a contrast to studies of the common chimpanzee which describe a male-dominant society with male alliance and strong sexual competition, advanced feeding techniques, and predation which implies the evolution of an open-land adaptation.

ACKNOWLEDGMENTS

I am indebted to The Zoological Society of San Diego and the Asahiyama Zoo Park for permission to use their unpublished data. I wish to thank Ms. E. O. Vineberg for kind help translating this paper into English.

REFERENCES

Bolk, L. 1926. On the problem of anthropogenesis. *Proc. section sciences Kon Akad. Wetens.* Amsterdam 27:329–344.

Coolidge, H. J. 1933. *Pan paniscus:* Pygmy chimpanzee from south of the Congo River. *Am. J. Phys. Anthropol.* 18(1):1–57.

Gould, S. J. 1977. *Ontogeny and Phylogeny.* Cambridge, Massachusetts: Harvard Univ. Press.

Kano, T. 1980. Social behavior of wild pygmy chimpanzees *(Pan paniscus)* of Wamba: a preliminary report. *J. Human Evol.* 9:243–260.

Kumazaki, K. 1983. Artificial rearing of a chimpanzee. *Monkey* 27(3,4):40–45.

Kuroda, S. 1979. Grouping of the pygmy chimpanzees. *Primates* 20(2):161–183.

_____. 1980. Social behavior of the pygmy chimpanzees. *Primates* 21(2):181–197.

_____. 1984. Interaction over food among pygmy chimpanzees. In R. L. Susman, ed., *The Pygmy Chimpanzee: Evolutionary Biology and Behavior,* pp. 301–324. New York: Plenum.

Nissen, H. W., and A. H. Riesen. 1964. The eruption of the permanent dentition of chimpanzee. *Am. J. Phys. Anthrop.* 22:285–294.

Savage-Rumbaugh, E. S. 1984. *Pan paniscus* and *Pan troglodytes.* In R. L. Susman, ed., *The Pygmy Chimpanzee: Evolutionary Biology and Behavior,* pp. 395–413. New York: Plenum.

Schultz, A. H. 1940. Growth and development of the chimpanzee. Publ. 525. *Contrib. Embryol. Carneg. Inst.* 28:1–63.

Shea, B. T. 1983. Paedomorphosis and neoteny in the pygmy chimpanzee. *Science* 222:521–522.

_____. 1984. An allometric perspective on the morphological and evolutionary relation ships between pygmy *(Pan paniscus)* and common *(Pan troglodytes)* chimpanzees. In R. L. Susman, ed., *The Pygmy Chimpanzee: Evolutionary Biology and Behavior,* pp. 89–130. New York: Plenum.

Uehara, S., and T. Nishida. 1987. Body weights of wild chimpanzees *(Pan troglodytes schweinfurthii)* of the Mahale Mountains National Park, Tanzania. *Amer. J. Phys. Anthrop.* 72:315-321.

SOCIAL ORGANIZATION OF PYGMY CHIMPANZEES

Frances J. White

INTRODUCTION

The family of primates that are most closely related to humans are the Pongidae, or great apes. There are four species of great apes living today: the orangutan, the gorilla, and two species of chimpanzee – the more familiar common chimpanzee and the pygmy chimpanzee or bonobo, *Pan paniscus*. The name "pygmy" is not an accurate description of *Pan paniscus*. Pygmy chimpanzees are not smaller than the best-studied subspecies of common chimpanzee, *P. troglodytes schweinfurthii*. Pygmy chimpanzee males weigh approximately 45 kilograms, with a range that extends from 40 to 60 kilograms for a large male. Females weigh less, averaging 33 kilograms with a range from 30 to 40 kilograms (Jungers and Susman, 1984).

Between July 1983 and July 1985, I spent a total of 20 months in central Zaire at the Lomako Forest Pygmy Chimpanzee Project field site. The Lomako site trail system covers an area of about 40 square kilometers of mainly climax evergreen forest (White, 1986). The major part of the study site is high, polyspecific evergreen forest. This is tall forest with a canopy of 30 to 40 meters, with large emergents reaching 60 meters. The forest is dissected by many small streams, each of which is flanked by an area of low swamp forest and then a characteristic slope forest. There are areas of second growth that represent areas cut in the 1920s, and they are typically low and contain high densities of the tangled form of *Haumania* vine. The pygmy chimpanzees eat the center of these vines on the ground, but otherwise the majority of their food is fruit and young leaves.

We are able to recognize individual pygmy chimpanzees to varying degrees. The easiest parameter to determine is the sex. Males are generally the hardest to tell apart although there are differences in scrotum color. Females, when adult, often have infants or juveniles with them, and their identification is aided by the idiosyncratic nature of the sexual swellings. We can often recognize individuals by their faces, many of which are very distinctive. Some animals even have deformed faces. These are rare, and we do not know the exact cause in each case.

Our understanding of the social organization of this rare ape species is rapidly increasing as the Lomako Forest study population gradually becomes more habituated to observers (White, 1988). This chapter presents a summary of recent studies of the social organization of the Lomako Forest pygmy chimpanzees, and makes some direct comparisons with the social organization of the common chimpanzees. This chapter also discusses recent work on the possible ecological correlates of the major differences in the social organizations of the common and pygmy chimpanzee.

SOCIAL ORGANIZATION

There are many features of an animal's social organization. The most commonly considered aspects are the observed social interactions and the way individuals associate in groups or parties. Social organization can also be determined from the way individuals associate in parties while feeding, as measured by their nearest neighbor choice.

Social Structure and Observed Interactions

Studies of social organizations ultimately depend on observations of the way in which individuals interact in a social situation. The different interactions between two individuals reflect their relationship to each other, and the social structure of sets of individuals is characterized by the context, quality, and patterning of the constituent relationships (Hinde, 1976, 1979).

The diverse social structures of primates in general, and great apes in particular, can be described by the distribution of interactions between the various age and sex classes. Thus, common chimpanzee social organization is typified by frequent affiliative behavior between males compared to infrequent affiliations among females (Goodall, 1968; Sugiyama, 1968; Nishida, 1979; Pusey, 1979). Gorillas display close relationships between the leading male and the females of a group, but not between the females themselves (Harcourt, 1979a, 1979b, 1979c). Orangutans are notable for the infrequency of any social interactions (Galdikas, 1979; MacKinnon, 1979; Rodman, 1979). The orangutan is essentially solitary (MacKinnon, 1974). Females do not maintain exclusive territories but have different, overlapping feeding ranges (MacKinnon, 1979). Resident males, in contrast, maintain exclusive territories that they defend from other males and that contain the feeding ranges of several females (Rodman, 1973, 1979). Affiliative behaviors within both males and females are rare, and there is a strong intolerance among males (Galdikas, 1979; Horr, 1975; MacKinnon, 1974; Rijksen, 1978; Rodman, 1973).

The African apes, however, are often found in larger social groupings. Gorillas live in cohesive groups that consist of a dominant male, a number of other adult males, adult females, and immatures (Schaller, 1963). As in the common chimpanzee and orangutan, affiliative behaviors between gorilla females are rare (Harcourt, 1979b), as are those between gorilla males (Fossey, 1974; Fossey and Harcourt, 1977). Gorilla groups maintain their cohesiveness despite the lack of interaction within the sexes because of the affiliation between each adult female and the dominant male of the group (Harcourt, 1979a, 1979b, 1979c).

Table 1. Social systems of apes.

Species	Affiliative bonds within each sex		Social system
	Males	Females	
Gibbon	no	no	monogamous
Orangutan	no	no	noyau
Gorilla	no	no	groups
Common chimpanzee	yes	no	communities
Pygmy chimpanzee	no	yes	communities

Note: Noyau social systems are based on separate but overlapping home ranges for each individual.

Common chimpanzee females also rarely show affiliation toward each other (Bygott, 1979). Female common chimpanzees, with their young, live their lives in different, overlapping core areas (Halperin, 1979; Wrangham, 1975, 1979a, 1979b). Unlike gorilla males, however, common chimpanzee males are gregarious and cooperate to defend communal ranges that include the feeding ranges of several females (Goodall et al., 1979; Nishida, 1979). Males, in contrast, frequently show affiliative behavior toward other males and often groom each other (Goodall, 1968).

Preliminary observations on the pygmy chimpanzees of the Lomako Forest (Badrian et al., 1981) and at Wamba (Kano, 1980, 1982; Kuroda, 1979) have shown that pygmy chimpanzees, like common chimpanzees, display a fission-fusion social system in which individuals are found in parties that are flexible in size and composition. These individuals belong to geographically distinct associations or communities. Unlike common chimpanzees, affiliative behaviors between female pygmy chimpanzees are common. More recent work has supported this picture and expanded the affiliative behaviors to include food sharing between females (Badrian and Badrian, 1984; Kuroda, 1984). It has also become apparent that there is a unique behavior of genito-genital (GG) rubbing (Thompson-Handler et al., 1984) that is a mutual homosexual behavior between females.

Each ape, therefore, has a different social structure as reflected by the different bonds between and within the sexes (Table 1). The orangutan, gorilla, and common chimpanzee are characterized by a lack of strong affiliative bonds between the individual females. In contrast, pygmy chimpanzees appear to have a very different social structure, with strong affiliative behaviors between individual females.

It is perhaps not unexpected that among the ape species, with their drastically different body sizes and adaptations to different habitats, there should be markedly dissimilar social systems. However, it is less clear why there should be a major difference between the comparably sized pygmy and common chimpanzees. It is also not clear why the pygmy chimpanzee, unlike all other apes, should appear to show strong affiliation between females.

Recent studies of social interactions among the Lomako Forest pygmy chimpanzees have confirmed these preliminary observations (White, 1986; White, in review). Interactions between individuals were classed as either affiliative or agonistic (aggressive plus submissive behaviors) (White, 1986; White, in review). Affiliative behaviors included grooming and GG-rubbing. Aggressive behaviors were mainly chases but could also include threats, displacements, pursuits, and fights. Interactions were classed according to the sexes of the individuals involved. If the social organization of the pygmy chimpanzees is described in terms of direct observations, affiliation was observed among females and between females and males, but not among males.

Affiliative interactions involving either two females or a male and a female were common, whereas affiliative interactions between two males were infrequent (Table 2a). Differences were not due to males and females interacting affiliatively at different rates. Males, however, showed higher rates of involvement in agonistic interactions than females (Table 2a).

However, pygmy chimpanzees display a fission-fusion social organization. Females spent more time in parties with other females, whereas males often left or joined parties as lone animals (White, 1988). The chance of observing an interaction was dependent both on the numbers of individuals observed during a sighting and on the length of that sighting. Since there are usually more females than males in a party (White, 1988), the chance of seeing an interaction involving two females is greater. It is possible, therefore, that the observed interactions reflected the way individuals associated in parties.

Each sex may also not interact at equivalent rates. An expected frequency of interaction, based on equivalent rates, can be calculated from the time spent observing individuals of either sex. By comparing this expected frequency to the observed frequency for each sex, it was possible to determine if the sexes interacted at equivalent rates.

Each interaction by a male or a female, however, could be with an individual of the same sex or the opposite sex. If individuals interacted at random, the distribution of interactions would have been dependent on the number of each sex present at each sighting and the duration of that sighting.

In each sighting, the chances of an individual interacting with another of the same sex can be expressed as:

$$\frac{(\text{number of individuals of that sex present} - 1)}{(\text{total party size} - 1)}$$

Similarly, the chances of an individual interacting with another of the opposite sex can be expressed as:

$$\frac{(\text{number of individuals of that sex present})}{(\text{total party size} - 1)}$$

If individuals interacted with each sex randomly, the expected number of interactions can be calculated from these chances and the duration of each sighting.

Table 2a. Total number of observed dyad interactions and number of times each sex was observed interacting. Expected numbers are based on equal rates of interacting for each sex.

Dyads	All Interactions Observed	Affiliative Interactions Observed	Agonistic Interactions Observed	Unclassified Interactions Observed
n	196	150	39	7
Male-male	24	12	12	0
Male-female	84	67	13	4
Female-female	88	71	14	3

Individuals	Observed	Expected	Observed	Expected	Observed	Expected
Male	132	138.5	91	106.0	37	27.6
Female	260	254.5	209	194.0	41	50.6
p[1]		n.s.		n.s.		< 0.05

Source: White, in review.

Note: n.s. = not significant.

1. G test for Goodness of Fit; Sokal and Rohlf, 1981.

There are several significant differences between these expected and observed distributions, as well as many nonsignificant differences. Males and females interacted at equivalent rates except that males interacted aggressively more often than expected. Males interacted with females and other males in direct proportion to the presence of females in the party for all types of interactions, with one exception. They interacted aggressively with other males significantly more frequently than they should if such interaction were independent of sex (Table 2b). Females interacted more frequently with males and less frequently with females than expected overall or in affiliative interactions (Table 2b).

Therefore, both females and males interacted more with females because females remained in parties more than males. On the basis of the observed social interactions, the social structure of pygmy chimpanzees can, therefore, be classed as one with a high degree of affiliative behavior among females and between males and females, but not among males. This observed distribution of social interactions, however, appears to be dependent on which individuals form parties. Social interactions, therefore, appear to be reflecting the prior level of organization of party composition.

Party Compositions

Having demonstrated this relationship between the observed social interactions and the way in which individuals are associating in parties, it is necessary to look at the results of studies of party composition (White, 1988; White and Burgman, in review). In the Lomako Forest, there are three separate associations or communities of pygmy

Table 2b. Division of interactions among individuals of same or opposite sex. Expected interactions are based on equal chances of interacting with either sex, given observed party composition.

	All Interactions		Affiliative Interactions		Agonistic Interactions	
	Observed	Expected	Observed	Expected	Observed	Expected
Male with:						
Male	48	41	24	28	24	11
Female	84	91	67	63	13	26
p[1]	n.s.		n.s.		< 0.001	
Female with:						
Male	84	60	67	48	13	9
Female	176	200	142	161	28	32
p[1]	< 0.001		< 0.01		n.s.	

Source: White, in review.
Note: n.s. = not significant.
1. G test for Goodness of Fit; Sokal and Rohlf, 1981.

chimpanzees. In this study, the 17 best-known individuals from the two large communities, called the Hedons and the Rangers, and the 8 consistent members of a small splinter group, known as the Blobs, are used. Of the adults in the Hedon community, 5 were males and 12 were females; of the Rangers, 7 were males and 10 were females. The Blobs consisted of 5 males and 5 females.

Parties of pygmy chimpanzees frequently undergo fission and fusion and thus maintain the same composition for varying lengths of time. It is possible to examine the associations among individuals that show a social organization of this type by using methods developed for Numerical Taxonomy (White, 1986; White and Burgman, in review). Each time the party composition changed by one or more individuals entering or leaving the party, a new record of the individuals present was made. This produced a matrix of the amount of time that individuals spent together in parties. This matrix was used to calculate the following index of similarity between individuals. The amount of time that Chimp 1 and Chimp 2 were present together in a party was divided by the total time that either one was present (Table 3). This is a quantitative analogue of Jaccard's coefficient (standardized between 0 and 1) and gives a measure of the similarity between individuals based on the time that they spent together in parties.

Clustering was calculated using UPGMA (unweighted pair group method using arithmetic averages). The dendrograms of party composition were constructed separately for the two communities and the splinter group. The data from the Hedons were probably the most realistic as this group was the most habituated and was found in a

Table 3. Method of calculating association matrix for party composition.

	Number of minutes	
	Chimp 2 present	Chimp 2 absent
Chimp 1 present	a	b
Chimp 1 absent	c	d

Measure of similarity between individuals
(Quantitative analogue of Jaccard's coefficient)

$$\frac{a}{a + b + c}$$

Source: White and Burgman, in review.

Figure 1. Hedon community. Dendrogram of relative time spent together.

Figure 2. Hedon community. PCA of relative time spent together.

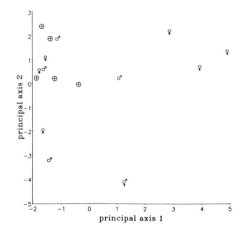

Relative time together (similarity)

Source: White and Burgman, in review.
Note: Proportion of variance explained by axes 1 to 4 = 0.717
♀ = female   ♂ = male   ⊕ = mother

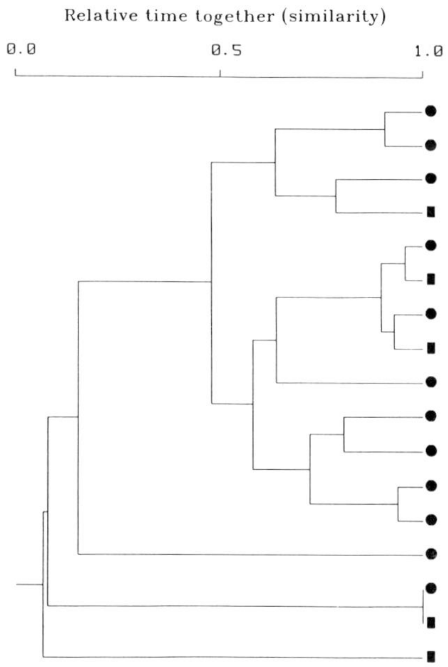

Source: Adapted from White and Burgman, in review.
● = female   ■ = male

wide range of party sizes. The Blobs, as a small splinter group, were confined to showing only small party sizes, and the Rangers were only observed in superabundant food sources and, as party size is related to the size of the food patch (White, 1986; White and Wrangham, 1988), the Rangers always showed unusually large party sizes. These differences are important, as they were reflected in the different structuring of the data. That is, there are different effects when parties are unusually large, whereas having only small party sizes tends to emphasize greatly the underlying structuring in the party composition data.

In the Hedon dendrograms there are subgroups that cluster together (Fig. 1). These subgroups are either males with females, or females with other females. Another way of looking at this same data is by Principal Components Analysis (PCA). The plot of the first two axes (Fig. 2) shows that females fall into two clusters, whereas there are no clusters of males. The dendrogram of the Rangers shows less higher-order structure, but there are still subgroups of males plus females (Fig. 3). The first two axes in this case, however, show an almost random distribution of points (Fig. 4). The

Figure 3. Ranger community. Dendrogram of relative time spent together.

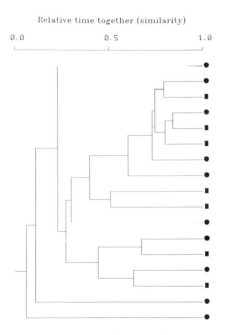

Source: Adapted from White and Burgman, in review.
● = female   ■ = male

Figure 4. Ranger community. PCA of relative time spent together.

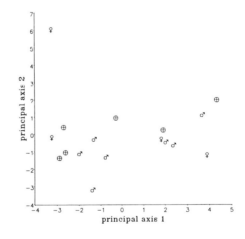

Source: White and Burgman, in review.
Note: Proportion of variance explained by axes 1 to 4 = 0.677
♀ = female   ♂ = male   ⊕ = mother

Figure 5. Blob splinter group. Dendrogram of relative time spent together.

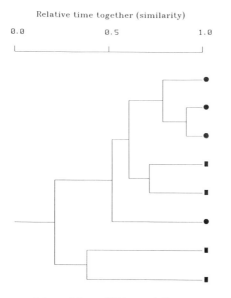

Figure 6. Blob splinter group. PCA of relative time spent together.

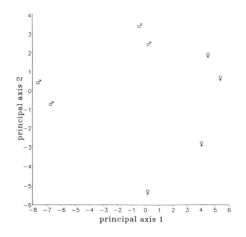

Source: White and Burgman, in review.
Note: Proportion of variance explained by axes 1 to 4 = 0.839
♀ = female   ♂ = male

Source: Adapted from White and Burgman, in review.
● = female   ▪ = male

Blobs have a cluster of females (Fig. 5), and the PCA plot shows linked females and more loosely associated males (Fig. 6). White and Burgman (in review) show that the female cluster is tight and that there is a clear progression from tightly linked females to more loosely linked males.

It is possible to test whether the structure that one observes in data of this type is due to particular factors. This is done with a Mantel test, with significance based on a Monte Carlo simulation of the data (Schnell et al., 1985). The most significant factor in the structure of the party composition data in the Hedons and the Blobs is associations between females (Table 4). However, in the Ranger data, the structure is more dependent on male-female and male-male associations.

Nearest Neighbors

A similar analysis can also be done for feeding proximity data (White, 1986; White and Burgman, in review). The nearest neighbor's identity and distance to a focal animal was recorded at two-minute time points. In this analysis, 5 meters was used as a measure of proximity, as behavioral data have shown that this is the approximate distance of an individual's feeding sphere. The amount of time that individuals spent within 5 meters of each other was converted into a similarity association matrix like that for party composition data. The measure of similarity between individuals was:

$$\frac{X_{i,j} + X_{j,i}}{n_i + n_j}$$

where:

$X_{i,j}$ is the frequency that i was the nearest neighbor within 5 meters of j.

$X_{j,i}$ is the frequency that j was the nearest neighbor within 5 meters of i.

n is the frequency that i and j were focal animals.

This value, therefore, reflected the time that any two individuals were close together as a proportion of the amount of time that either of them was the focal animal being sampled.

The results from Mantel tests on these data (Table 5) were very similar to the results from the Mantel tests of party composition. The structure in the Hedon and

Table 4. Mantel tests of party composition data.

	Community		
	Hedons	Blobs	Rangers
Female-female	0.714	0.964	0.078
Male-male	0.286	0.036	0.922
Male-female	0.300	0.028	0.984*

Source: White and Burgman, in review.
* Denotes significance of $p < 0.05$.

Table 5. Mantel tests of feeding proximity data.

	Community		
	Hedons	Blobs	Rangers
Female-female	0.836	0.984*	0.546
Male-male	0.164	0.016*	0.454
Male-female	0.304	0.028	0.826

Source: White and Burgman, in review.
* Denotes significance of $p < 0.05$.

Blobs data was strongly influenced (in the Blobs, it is statistically significant) by associations between females. In the Rangers, however, the major effect was a male-female association, and there was no male-male effect. This, when taken together with the results from the Mantel test of party compositions, could be an indication that the Ranger males were attracted into the unusually large parties observed, but they entered the parties in order to maintain proximity to females.

In summary, therefore, data on interactions, party composition, and nearest neighbors all indicate that the social organization of pygmy chimpanzees can best be categorized as one in which there is strong affiliation among females and between males and females, but not among males.

ECOLOGICAL CORRELATES OF THE SOCIAL ORGANIZATION

The social organization of pygmy chimpanzees can, therefore, be considered as female-based. This is very different from the male-bonded social system of common chimpanzees. Wrangham (1979a, 1979b) has hypothesized that, since their food occurs in small patches, the high cost of feeding competition prohibits sociality among female common chimpanzees. In a recent comparison of food patches of pygmy and common chimpanzees (White and Wrangham, 1988), it was found that pygmy chimpanzees feed more frequently in large trees than do common chimpanzees (Table 6). Pygmy chimpanzees visit larger patches than common chimpanzees do (Table 7). Feeding competition is, therefore, presumably less in the pygmy chimpanzee. Females can thus afford to associate with other females.

Table 6. Food patch types (percent of all patches).

Patch type	*Pan paniscus*	*Pan troglodytes*
Ground (leaves + pith)	14.8	16.2
Ground (insects)	0.0	2.7
Vines	4.9	17.3
Small trees	15.6	24.9
Large trees	64.8	38.9

Source: White and Wrangham, 1988.
Note: G test value = 28.36, $p < 0.005$.

Table 7. Frequency distribution of patch sizes (percent of all patches).

Patch size estimate (number of chimp-minutes)	*Pan paniscus*	*Pan troglodytes*
1 to 50	38.8	59.1
50 to 100	14.7	19.4
100 to 150	5.4	10.8
150 to 200	7.0	4.3
200 to 250	6.2	2.7
above 250	27.9	3.7

Source: White and Wrangham, 1988.
Note: Kruskal-Wallis H value = 15.19, $p < 0.005$.

ACKNOWLEDGMENTS

I would like to thank John Fleagle for his continuing advice and guidance and Randall Susman and Noel Badrian for introducing me to the pygmy chimpanzees of the Lomako Forest. Many of the results presented here are from productive collaborations with Richard Wrangham and Mark Burgman. I would especially like to thank Nancy Thompson-Handler and Richard Malenky, as it is thanks to them that my time in the forest was both enjoyable and highly productive. This work owes much to the comments and suggestions of Charlie Janson, Mark Burgman, and Richard Wrangham. I would also like to thank Donald Gerhart for all his help and for suggesting the use of techniques from phenetics for analysis of association data, Gary Schnell for suggesting the Mantel test, and Scott Ferson for computer assistance. This work was supported by a NSF Doctoral Dissertation Improvement Award and a grant from the Boise Fund.

REFERENCES

Badrian, A. J., and N. L. Badrian. 1984. Group composition and social structure of *Pan paniscus* in the Lomako Forest. In R. L. Susman, ed., *The Pygmy Chimpanzee: Evolutionary Biology and Behavior,* pp. 325–346. New York: Plenum Press.

Badrian, N. L., A. Badrian, and R. L. Susman. 1981. Preliminary observations on the feeding behavior of *Pan paniscus* in the Lomako Forest of Central Zaire. *Primates* 22:173–181.

Bygott, D. J. 1979. Agonistic behavior and dominance among wild chimpanzees. In D. A. Hamburg and E. McCown, eds., *The Great Apes,* pp. 405–427. Menlo Park, Calif.: Benjamin/Cummings.

Fossey, D. 1974. Observations on the home range of one group of mountain gorillas *(Gorilla gorilla berengei)*. *Animal Behavior* 22:568–581.

Fossey, D., and A. H. Harcourt. 1977. Feeding ecology of free-ranging mountain gorilla *(Gorilla gorilla berengei).* In E. McCown, ed., *Primate Ecology: Studies of Feeding and Ranging Behaviour in Lemurs, Monkeys and Apes,* pp. 415–447. London: Academic Press.

Galdikas, B. M. F. 1979. Orangutan adaptation at Tanjung Puting Reserve: mating and ecology. In D. A. Hamburg and E. McCown, eds., *The Great Apes,* pp. 194–233. Menlo Park, Calif.: Benjamin/Cummings.

Goodall, J. 1968. The behaviour of free-living chimpanzees in the Gombe Stream Reserve. *Animal Behavior Monographs* 1:161–331.

Goodall, J., A. Bandora, E. Bergman, C. Busse, H. Matama, E. Mpong, A. Pierce, and D. Riss. 1979. Inter-community interactions in the chimpanzee population of Gombe National Park. In D. A. Hamburg and E. McCown, eds., *The Great Apes,* pp. 13–53. Menlo Park, Calif.: Benjamin/Cummings.

Halperin, S. D. 1979. Temporary association patterns in free-ranging chimpanzees: an assessment of individual grouping preferences. In D. A. Hamburg and E. McCown, eds., *The Great Apes,* pp. 491–499. Menlo Park, Calif.: Benjamin/Cummings.

Harcourt, A. H. 1979a. The social relations and the group structure of wild mountain gorilla. In D. A. Hamburg and E. McCown, eds., *The Great Apes,* pp. 186–192. Menlo Park, Calif.: Benjamin/Cummings.

_____. 1979b. Social relationships among adult female mountain gorillas. *Animal Behavior* 27:251–264.

_____. 1979c. Social relationships between adult male and female mountain gorillas in the wild. *Animal Behavior* 27:325–342.

Hinde, R. A. 1976. Interactions, relationships and social structure. *Man* 11:1–17.

_____. 1979. The nature of social structure. In D. A. Hamburg and E. McCown, eds., *The Great Apes,* pp. 294–315. Menlo Park, Calif.: Benjamin/Cummings.

Horr, D. A. 1975. The Borneo orang-utan; population structure and dynamics in relationship to ecology and reproductive strategy. In L. A. Lindburg, ed., *Primate Behavior: Development in Field and Laboratory Research,* pp. 307–323. New York: Academic Press.

Jungers, W. L., and R. L. Susman. 1984. Body size and skeletal allometry in African apes. In R. L. Susman, ed., *The Pygmy Chimpanzee: Evolutionary Biology and Behavior,* pp. 131–171. New York: Plenum Press.

Kano, T. 1980. Social behavior of wild pygmy chimpanzees *(Pan paniscus)* of Wamba: a preliminary report. *Journal of Human Evolution* 9:243–260.

_____. 1982. The social group of pygmy chimpanzees *(Pan paniscus)* of Wamba. *Primates* 23:171–188.

Kuroda, S. 1979. The social group of the pygmy chimpanzee. *Primates* 20:161–183.

_____. 1984. Interactions over food among pygmy chimpanzees. In R. L. Susman, ed., *The Pygmy Chimpanzee: Evolutionary Biology and Behavior,* pp. 301–324. New York: Plenum Press.

MacKinnon, J. 1974. The behaviour and ecology of wild orangutan, *Pongo pygmaeus. Animal Behavior* 22:3–74.

_____. 1979. Reproductive behavior in wild orangutan populations. In D. A. Hamburg and E. McCown, eds., *The Great Apes,* pp. 256–273. Menlo Park, Calif.: Benjamin/Cummings.

Nishida, T. 1979. The social structure of chimpanzees of the Mahale Mountains. In D. A. Hamburg and E. McCown, eds., *The Great Apes,* pp. 72–121. Menlo Park, Calif.: Benjamin/Cummings.

Pusey, A. 1979. Inter-community transfer of chimpanzees in the Gombe National Park. In D. A. Hamburg and E. McCown, eds., *The Great Apes,* pp. 465–479. Menlo Park, Calif.: Benjamin/Cummings.

Rijksen, H. 1978. A field study on Sumatran orang-utans *(Pongo pygmaeus abelii;* Lesson 1827). Wageningen, The Netherlands: H. Veenman and Zonen.

Rodman, P. S. 1973. Population composition and adaptive organization among orangutans of the Kutai Reserve. In J. H. Crook, ed., *Behaviour of Primates,* pp. 171–209. London: Academic Press.

———. 1979. Individual activity patterns and the solitary nature of orangutans. In D. A. Hamburg and E. McCown, eds., *The Great Apes,* pp. 234–255. Menlo Park, Calif.: Benjamin/Cummings.

Schaller, G. B. 1963. *The Mountain Gorilla.* Chicago: University of Chicago Press.

Schnell, G. D., D. J. Watt, and M. E. Douglas. 1985. Statistical comparison of proximity matrices: applications in animal behavior. *Animal Behavior* 33:239–253.

Sokal, R. R., and F. J. Rohlf. 1981. *Biometry.* 2d ed. New York: Freeman.

Sugiyama, Y. 1968. Social organization of chimpanzees in the Budongo Forest, Uganda. *Primates* 9:109–148.

Thompson-Handler, N., R. K. Malenky, and N. Badrian. 1984. Sexual behavior of *Pan paniscus* under natural conditions in the Lomako Forest, Equateur, Zaire. In R. L. Susman, ed., *The Pygmy Chimpanzee: Evolutionary Biology and Behavior,* pp. 347–368. New York: Plenum Press.

White, F. J. 1986. Behavioral ecology of the pygmy chimpanzee. Ph.D. thesis, State University of New York at Stony Brook.

———. 1988. Party composition and dynamics in *Pan paniscus. International Journal of Primatology* 9:179–193.

———. In review. Distribution and relative frequencies of interactions in *Pan paniscus.*

White, F. J., and M. A. Burgman. In review. Social organization of the pygmy chimpanzee, *Pan paniscus,* multivariate analysis of intra-community associations.

White, F. J., and R. W. Wrangham. 1988. Feeding competition and patch size in the chimpanzee species, *Pan paniscus* and *Pan troglodytes. Behaviour* 105(1/2):148–164.

Wrangham, R. W. 1975. Behavioural ecology of chimpanzees in Gombe National Park, Tanzania. Ph.D. thesis, Cambridge University.

———. 1979a. On the evolution of ape social systems. *Social Science Information* 18:335–368.

———. 1979b. Sex differences in chimpanzee dispersion. In D. A. Hamburg and E. McCown, eds., *The Great Apes,* pp. 481–489. Menlo Park, Calif.: Benjamin/Cummings.

3

THE CHIMPANZEE MIND

ARE ANIMALS INTELLIGENT? WOLFGANG KÖHLER'S APPROACH

Emil W. Menzel, Jr.

THE PRIMORDIAL PROBLEM

Even today it is not uncommon to hear debates like this one, which took place in the 17th century:

> [Pierre Gassendi:] You say 'brutes lack reason.' But while doubtless they are without human reason, they do have a reason of their own.
> [Rene Descartes:] . . . your queries about the brutes are not relevant here, since the mind when communing with itself can experience the fact that it thinks, but has no evidence of this kind as to whether or not the brutes think; it can only come to a conclusion afterwards about this matter by reasoning a posteriori from their actions. (Wilson, 1969, p. 281)

Why doesn't the same logic apply to human beings other than oneself, as well as to nonhuman animals? In other words, how do I know that the reader of this paper is not as mindless as a baboon, or why should the reader give me the benefit of the same doubt? If I were a Cartesian, I would of course find it necessary to look for something special about people, by means of which we humans can recognize (if not commune with) one another as directly and unequivocally as each of us, in isolation, can recognize and commune with our own selves. And if I were a Gassendi-ite I would, in turn, subject that "something special" to precisely the same sort of criticism.

Actually, I am a bit of a Gassendi-ite, so let me note a few difficulties that exist today for common-sense Cartesianism.

First, it is dogma rather than psychological fact that the only mind one can know is one's own and that introspection is the primordial method by which one comes to know it. Developmentally speaking, it seems far more plausible that, insofar as one ever does know one's own mind or become a skilled introspectionist, it is through social learning processes, which in turn rest on attachment-formation and empathy with others. Not surprisingly, it is easier to empathize with and understand our own kind as opposed to aliens. But culturally acquired attitudes often determine whom we perceive as aliens.

Second, there is a big difference between subjective analogies and objective analogies (Mackenzie, 1977); and modern cognitive psychology rests on the latter, not the former. The issue is, in other words, what another being would be expected to do if it is assumed to be operating in accordance with this or that model, and which model's predictions are best confirmed by empirical data. In principle, this issue is just as resolvable for a nonhuman as for a human. Nor is it irrelevant here that the dominant recent models for developing theories of human cognitive processes are at present based on the very nonhuman digital computer.

Third, to argue, as Descartes did, that there is some so-called special "essence" of this or that species or this or that internal process is latter-day Platonism rather than zoology (Mayr, 1982). I will grant that I have never had, and never expect to have, any trouble in distinguishing parrots or dolphins or apes from people, and that what animals and people do and say is just as informative here as their physical morphology. But to jump from this to Platonic essences is to create more problems than one solves. What, for example, is the essence of human language; and what is the essence of this essence; and so on ad infinitum? What might well be a straightforward, empirical problem in behaviorally based taxonomy soon becomes a muddle of abstractions. It is important to go back to the empirical phenomena in question and make a fresh start!

The final and most serious problem of all is summed up by this passage from one of Darwin's notebooks: "Origin of man now proved. Metaphysics must flourish. He who understands baboon would do more for metaphysics than Locke." As a matter of fact, this may sum up the preceding points too.

A BOW, IN PASSING, TO DARWIN

I am sure that Darwin would not mind rewording this passage to add chimpanzee to baboon, and Descartes to Locke. I am equally certain that he would be greatly impressed by the strides that have been made in the past decade or two toward the goal of understanding chimpanzees. He might, of course, remind us that primatologists shall probably never fully reach this goal, for unless they lose their zest for primatological observation (which seems a contradiction in terms), there is always more to see, and any actual living beings are literally an inexhaustible subject matter. He would also probably enjoin us to study and ponder the works of our philosophical predecessors, for regardless of what one thinks of their methodology or their conclusions, many of the questions that they posed remain fundamental ones, which must be addressed by anyone who aspires to be a philosopher of nature as well as a bird (or chimp) watcher. Indeed, if you wish to understand either animal intelligence or chimpanzees, you simply must at some point study and ponder the works of Wolfgang Köhler.

KÖHLER'S PROBLEM

The fundamental question with which Köhler dealt in his classical monograph (1925) was derived immediately from Thorndike's even more influential work (1898). In a nutshell it was: Are animals intelligent – or are their actions completely explicable in terms of "instinct" or "chance" or some mixture thereof? Thorndike had said that in the course of his studies he had never seen any animal do anything that even seemed to entail reason or true intelligence (shades of Descartes!). He further hypothesized that "Animals might have no images or memories at all, no ideas to associate. Perhaps the entire fact of association in animals is the presence of sense impressions with which are associated, by resultant pleasures, certain impulses, and that, therefore, and therefore only, a certain situation brings forth a certain act" (1898, pp. 108–109). The process here, Thorndike said, is analogous to natural selection, in the sense that it entails no hypotheses, foresight, intentionality, or purposiveness on the part of the animal, but only (1) blind variations in behavior; (2) the chance occurrence of some variants that pay off (or are highly disadvantageous) in a given situation; and (3) the selective retention (or extinction) of these variants. In more formalistic terms, Thorndike's theory was the progenitor of S-R (stimulus-response, reinforcement) theories of learning.

Köhler was by no means opposed to natural selection. To characterize him as nonobjective or as a mentalist, as some authors have done, is to misinterpret him. He did, however, disagree very sharply with Thorndike. His studies in physics with Max Planck, as well as his studies in perceptual psychology with Max Wertheimer, led him to believe strongly that the foregoing sort of mechanistic reasoning was simplistic. Coincidentally, he published a book on gestalt principles in physics almost concurrently with his writings on gestalt principles in chimpanzee behavior. Many physical processes, he argued, entail higher order organizations (i.e., gestalts) that cannot adequately be described, let alone explained, as the simple sum of a discrete set of so-called elements. These processes include not only electromagnetic fields and the like but also patterns of activation in the cortex, visual perceptual fields, and those forms of overt chimpanzee behavior that Köhler described as intelligent or insightful. Some students of human perception had gone so far as to say that we cannot even perceive motion: all that one can really see at any given instant is the position of an object, and when we say that "it is moving," we are actually remembering what that object is and where it once was, perceiving where it is now, and inferring that it has moved. Psychologically speaking, objects and motion are not "out there" in the environment, or even in the eye of the observer, but in the mind. This theory of perception, said Köhler and other gestaltists, is nonsense.

By today, probably everybody would agree this far – especially inasmuch as specialized single receptor cells have been discovered which seem responsive to "motion as such." Still further, even the most radical and molecular of behaviorists have never refrained from talking about organized patterns of motion such as "turning left in a maze," "picking up an object," or even "approaching an object but then hesitating and circling it," as if these were objective events rather than purely subjective inferences on the part of the human observer. Why stop here, however, especially on the basis of a priori beliefs as to what levels of organization in natural events human observers can and cannot see? Might it not be that those behaviors that strike everyday observers as intelligent are organized differently than the trial-and-error sorts of performances that Thorndike described, and that it was Thorndike's placing of animals in arbitrarily devised puzzle boxes and his looking only at arbitrarily recorded "elements" of their performances that rendered him unable to see what it is that animals are actually capable of doing? To see intelligence – and in some cases you can literally see it if you care to, so Köhler said – a different approach is required.

KÖHLER'S APPROACH

Köhler never tired of the slogan, "Back to the phenomena as such!" He also practiced what he preached. His first step, methodologically speaking, was to pick some concrete examples of animals rather than argue about animal intelligence in the abstract. By a concrete example I do not, coincidentally, mean "the chimpanzee," for that is a mythical beast. I mean Rana and Sultan and Grande and the other individuals that Köhler studied. As Ernst Mayr puts it, "He who does not understand the uniqueness of individuals is unable to understand the working of natural selection" (1982, p. 47). Next to Jane Goodall (1986), no student of animal behavior has driven this point home more forcefully than Köhler. The long-term study of individually identified animals is, of course, probably the most important single difference between modern and earlier field studies, and it is the basis for most discoveries regarding social structure and the role of kinship. Students of animal learning have always been aware of individual differences and of the importance of recognizing one's animals individually. But long-term familiarity with the same individuals is still quite rare in this domain, and non-primatologists probably more often view it as a liability than as an asset, and get a fresh batch of animals for each study unless they simply can't afford it.

Köhler's second step was to watch the individuals he had selected in whatever situations he happened to find them, until he acquired a very good qualitative feel for, if not a detailed everyday description of, whatever they chose to do. The last chapter of his monograph, in particular, is loaded with detailed naturalistic or "baseline" observations that are at least as relevant and persuasive as the data from his formal tests. Until modern ethology came along, his descriptive accounts were unsurpassed; and even today, when much of his vocabulary is old-fashioned, one can literally visualize what his animals were doing, at least if one has had a few years experience with chimpanzees.

His film of a few animals performing in some of his specialized tests of box stacking and stick using ("Experiments on Ape Intelligence, 1914–1917") can be and should be viewed in as simple-minded and naturalistic a spirit as possible, and in the context of Thorndike's assertions. Before getting more formalistic, Köhler would ask that we simply watch the film and see what the animals do. Can you see anything that seems to suggest "reasoning or true intelligence," or would you agree with Thorndike that nothing of the sort is discernible? Should you have any trouble in reaching a

Figure 1. Do they look smart to you?

decision here, an alternative strategy would be to watch the film along with a sizable audience of people, and simply note whenever they laugh or clap their hands. It is a good bet that the events that were transpiring at these moments have something to do with intelligence (or stupidity) as humans actually perceive it; the problem is to specify more precisely what distinguishes these moments from more humdrum ones. The same strategy could, of course, be used in viewing anyone else's films or conducting live observations of animals, including humans (Fig. 1).

As Köhler himself said in this same connection, if formalistic definitions of intelligence are necessary at all, they are better attempted at the end of one's research than at the beginning. I would add: The better one's naturalistic account and the more complete one's analysis, the more likely it is that the question of whether one's animals are intelligent will become unimportant or semantic. Chimpanzee intelligence is whatever chimpanzees do, especially insofar as these actions give them an advantage over their competitors. It is most important to remember here that natural selection implies specialization of adaptations just as surely as it implies mental continuity. What chimpanzees do might or might not be similar to what humans do; but that is changing the study to one of human intelligence. Is it not chauvinistic for those who make their specialized living by talking, writing and solving arbitrary puzzles to assume that only living beings that can match or beat them at their own game have any intelligence worth talking about?

There is at least one point on which all students of intelligence (including both Köhler and Thorndike) do agree: Intelligence implies coping with novelty. We seldom can be certain that the situations which animals encounter "on their own" are indeed novel for them, or sufficiently so to constitute a convincing test of their ability to cope. Therefore, some control of their experience and some sorts of experimental tests are most desirable, if not at some point necessary. Köhler's third step was, accordingly, to conduct some tests of this sort.

Whereas Thorndike had used a Procrustean approach, involving situations that (deliberately) bore as little resemblance as possible to the sorts of problems his animals might encounter naturally, Köhler's approach was almost the opposite. He introduced almost the least degree of change he could – just enough to be reasonably certain that the situation or the required behavioral organization was indeed a novel one for his animals, and never so much that all the necessary information for coming up with an intelligent coping response was not visually present from the outset. If all the information is not available, he reasoned, how else could an animal solve the problem, other than by trial-and-error? If it is available, however, we should find it relatively easy to detect whether animals pick up this information and "see through" the problem. In the extreme case, the difference between the behavior that Thorndike would predict and intelligent behavior should be as obvious as the difference between an animal that happens to bump into a banana, without seeing it, after wandering about a field for some time quite at random, and, on the other hand, an animal that sees the same banana from a distance of a hundred meters and runs straight to it.

In brief, "insight" is not necessarily much more inferential a concept than ordinary "direct-sight." Even more importantly, there is a concrete and objective physical model of behavior that furnishes one the appropriate Thorndikean criterion of "total lack of intelligence." It is, so Köhler argued, physical Brownian motion, or a random walk. A physical particle that is moving at random will, given enough time, get to any given point in a finite space, but it will do so by means of a series of unrelated, discrete steps – and not by a single unified move that is organized from the outset with reference to this destination and any intervening obstacles. (Does one inanimate object even "approach" another in the same sense that we say one animal approaches another animal? Aristotle would have answered in the affirmative, but no modern post-Newtonian physicist would.) Hence, said Köhler, moves that qualify as "organized from the outset" are, by definition, contrary to Thorndike, and are insightful or intelligent, especially if they occur abruptly and on the first occasion that the animal encounters the exact situation in question.

The fact that chimpanzees' moves also resemble the sorts of moves that we ourselves might make introduces a very different and more problematical sort of criterion of intelligence. Or, I should say, it re-introduces the Cartesian standard. Here, the standard of comparison is presumed to be of maximal rather than zero intelligence, and one has to prove that the "null hypothesis" of no significant difference is true rather than that it can be rejected. In my opinion, such chimpanzee-human comparisons probably weakened rather than strengthened Köhler's case against Thorndike. Had he stuck to disproving the hypothesis of zero intelligence, it is hard to imagine that anyone could ever have disputed him; but as it was, he laid himself open to being accused of being overly anthropomorphic. Needless to say, it is difficult if not impossible to "prove" that any two phenomena are fundamentally the same, especially if neither one of them is very thoroughly understood, and one's opponents can always identify further features after the data are in. That's the bind that current students of "language in chimpanzees" are in, and their problem is exacerbated by the fact that no one has ever proposed a very clear-cut criterion of "zero language ability."

Fourth, and last, Köhler obviously had to deal with the problem of "instinct" as well as "chance." And this explains why he introduced problems a few cuts more complicated than simply spotting a piece of food at a distance and running straight to it. To be more precise, he used detour problems, in which a roundabout or indirect (and presumably novel) path had to be taken around some obstacle to get to the goal. How does tool using get into the picture? There was nothing particularly special about tool using per se for Köhler, except that it entails getting around an obstacle by means of manipulatory rather than locomotor behaviors, and it extends the discussion to less obviously "spatial" sorts of problems.

WHAT DID KÖHLER PROVE?

Regardless of what one might think of Köhler's theory, there are few of his empirical observations on chimpanzees that have not been confirmed by other observers. Indeed, once other observers began studying and testing chimpanzees in group-living, outdoor situations that approximated or surpassed Köhler's (naturalistically speaking), rather than observing isolated individuals in Thorndikean-type tasks, the performances that Köhler described soon started to sound tame rather than incredible. Köhler himself noted many apparent limitations in his animals. He would have been as amazed as Thorndike to learn of the sorts of performances that have been demonstrated by later researchers, especially in the domains of communication, social "politics," and "planning ahead."

In contrast to Köhler's observations, Thorndike's have fared poorly. Many cat and monkey behaviors which Thorndike believed impossible have, by today, been demonstrated in these and other animals, including insects. Thorndike's claim – that his tests furnish a more valid and representative picture of animals' actual abilities than field research – has not fared very well either, to put it mildly.

Does this mean that Köhler won the debate and Thorndike lost? Fortunately or unfortunately, that would be an oversimplification. No one today would question that animals' behavior often differs significantly from Newtonian or Brownian motion, even on their first trial in a novel test situation. No one would dispute that a highly molecular description of behavior might become simplistic when it breaks natural events apart into small, discrete "stimulus and response elements" and attempts to explain the directionality of an animal's behavior and its organization into various, hierarchically ordered levels of organization by "simple" associative principles. Most people would agree that chimpanzees and other animals do many things, both in controlled tests as well as "anecdotally," that look quite intelligent. Laypersons would probably call these behaviors intelligent if they saw a human performing the same response patterns. Few if any animal psychologists still adhere strictly to an S-R reinforcement theory of behavior; most by now employ some cognitive concepts. In all of these respects current thinking largely vindicates Köhler's position rather than Thorndike's. Nevertheless, one very important qualification remains.

The major single problem in Köhler's account is that it puts too much emphasis on immediately present variables and on visual perception and thus is too ahistorical for the tastes of most psychologists and ethologists. To be more precise, his description of behavior as intelligent or insightful seems unobjectionable, but his explanation of particular performances, and especially his account of their genesis, is incomplete. Thus, many authors have accused him of neglecting the ontogenesis of the behaviors he described and of not recognizing the importance of past experience (e.g., Birch, 1945; Harlow and Harlow, 1949; Schiller, 1952). And other authors have accused him of neglecting to identify the precise motor "elements" underlying complex performances and failing to recognize their species-typical or innate nature (e.g., Hall, 1963; Schiller, 1952). My opinion of these criticisms has been stated in more detail elsewhere (Menzel et al., 1970; Menzel and Juno, 1982, 1984). Suffice it to say here that whether or not experience or motor patterns are necessary, and how these experiences and motor patterns are organized and utilized, are two different questions which should not be confused with one another.

Such confusion is apparent, for example, in the statement made by Schiller, and endorsed by Hall, that so-called insightful behavior is nothing more than the chance occurrence of a species-typical motor pattern in a situation where it is followed by reinforcement (upon which, presumably, the "laws" of Thorndikean conditioning take over). This statement is much the same as Thorndike's position, except that it takes for granted as "elementary" the very behavioral organizations that require explanation. It could, however, be expressed more formalistically: it implies a simple random walk that is constrained in accordance with equations that are derived from a knowledge of a given species' functional anatomy. Could such a model predict the organization and the directionality of individual behavior in novel test situations? Could it predict when animals will and will not profit from chance? I would not bet on it.

SUMMARY AND CONCLUSIONS

In everyday life, most humans, and many nonhuman animals as well, seem to operate on the premises that (a) the more that other creature looks like us, the more likely it can do anything that we can do; (b) however, there is no substitute for direct experience, because appearances can be deceptive and living beings are full of surprises. The first premise, plus the assumption that "we" are intelligent, helps to explain why chimpanzees so often strike us as intelligent. The second helps to explain why laypersons so often turn to experts for further advice, and why experts hedge.

Some philosophers, and many animal psychologists, have argued that everyday intuitions are worthless and that the only way to secure an accurate, representative, and unbiased picture of any animal's ability is to assume, until it is empirically proved otherwise, that the animal has no more intelligence than an inanimate object. Proof of intelligence, in turn, must come from specially devised laboratory tasks that are "totally novel" for the animal and that are rigorously designed to assess the "essence" of intelligence, or some particular facet thereof, such as linguistic competence or ability to reason deductively. Such a strategy displaces the emphasis from understanding chimpanzees (or whichever species one is studying) to formalistic debates about "What is intelligence?" As often as not it turns out to be just as anthropocentric as one's initial, everyday intuitions.

An alternative approach is, however, possible. One of its prime proponents was Wolfgang Köhler. This paper describes his work but interjects additional commentary or opinions of my own that seem pertinent to contemporary problems, without particularly worrying about how accurately they reflect Köhler's views as such. The goal of this approach is to exploit and improve on our observational skills and our naturalistic feel for a given animal's everyday behavior. Perhaps the single most important substantive problem is to learn how the animal's behavior is directed and organized with respect to other animals or the environment – especially, but not exclusively, in those circumstances in which some change or novelty is introduced. This problem should be carefully distinguished from the problems of identifying all of the sensory and motor elements that might enter into a given organization and from problems of analyzing the genesis of these elements. The issue is not whether experience is necessary, but how experiences are utilized.

To see that animals are intelligent, it is necessary only to consider how their behavior deviates from that which one would expect if they were responding strictly in accordance with the known laws of physical science. It is a moot point as to whether "instinct" or "conditioning" is unintelligent; but in any event, before one seriously proposes that these alternatives too must be rejected before an animal is called intelligent, their predictive laws must be formulated in a fashion that is sufficiently precise to warrant the test.

To require that an animal do precisely what humans do before calling it intelligent seems anthropocentric and anti-Darwinian, and it also erroneously assumes that we know precisely what humans do. The importance of nonhuman models, even for the understanding of human intelligence, should accordingly not be underestimated. How many constraints on pure Brownian motion, and precisely what kinds of constraints, would one have to introduce in order to produce a movie simulation of a chimpanzee that would fool scientists into thinking that they were looking at a genuine article? That would be an interesting project for some aspiring young Cartesian.

ACKNOWLEDGMENTS

Supported by National Institute of Child Health and Development Grant HD-0616 and National Institute of Health Animal Resources Branch Grant 00165 to the Yerkes Regional Primate Research Center. This article was written to commemorate the centennial of Wolfgang Köhler's birth.

REFERENCES

Birch, H. G. 1945. The relation of previous experience to insightful problem solving. *Journal of Comparative Psychology* 38:367–383.

Goodall, J. 1986. *The Chimpanzees of Gombe.* Cambridge, MA: Harvard University Press.

Hall, K. R. L. 1963. Tool using performances as indicators of behavioral adaptability. *Current Anthropology* 4:479–494.

Harlow, H. F., and M. K. Harlow. 1949. Learning how to think. *Scientific American* 180:36–39.

Köhler, W. 1925. *The Mentality of Apes.* (Translated by E. Winter.) New York: Harcourt Brace.

Köhler, W. *Experiments on Ape Intelligence 1914–1917.* 16mm, 9 min. 1975. Distributed by Audio Visual Services, The Pennsylvania State University, University Park, PA.

Mackenzie, B. D. 1977. Behaviourism and the limits of scientific method. *Atlantic Highlands.* New Jersey: Humanities Press.

Mayr, E. 1982. *The Growth of Biological Thought.* Cambridge, MA: Harvard University Press.

Menzel, E. W., and C. Juno. 1982. Marmosets *(Saguinus fuscicollis):* Are learning sets learned? *Science* 217:750–752.

_____. 1984. Are learning sets learned? Or: perhaps no nature-nurture question has any simple answer. *Animal Learning and Behavior* 12:113–115.

Menzel, E. W., R. K. Davenport, and C. M. Rogers. 1970. The development of tool using in wild-born and restriction-reared chimpanzees. *Folia Primatologica* 12:273–283.

Schiller, P. H. 1952. Innate constituents of complex responses in primates. *Psychological Review* 59:177–191.

Thorndike, E. L. 1898. Animal intelligence. An experimental study of the associative processes in animals. *Psychological Monographs* 2(8):1–109.

Wilson, M. D., ed. 1969. *The Essential Descartes.* New York: The New American Library, Inc.

CROSS-FOSTERED CHIMPANZEES:
I. TESTING VOCABULARY

R. Allen Gardner and Beatrix T. Gardner

The critical role of early experience in the behavior of organisms is well known and well documented. Many animals have to learn to identify with their own species, many birds have to learn their mating songs. Even such habits as migration or overwintering in the same place are profoundly influenced by species-typical rearing conditions. In *cross-fostering*, the young of one species are reared by foster parents of another species. So deep is the belief in the effect of rearing conditions on human behavior that even alleged but unverified cases of cross-fostering such as the wolf children of India (Singh and Zingg, 1942) and the monkey boy of Burundi (Lane and Pillard, 1978) attract serious scholarly attention. For obvious ethical reasons it is unlikely that we shall see any experimental account of a human child reared by nonhuman foster parents. But in the twentieth century, the Kelloggs with Gua and the Hayeses with Viki pioneered the logical alternative, a form of cross-fostering in which the subjects are chimpanzees and the foster parents are human beings.

Before Project Washoe, the human foster parents in these experiments spoke to their adopted chimpanzees as parents speak to hearing children. In contrast to the resemblance to young children in other aspects of development, the chimpanzees acquired hardly any speech. For decades, the failures of Gua and Viki to learn to speak were cited and recited to support the traditional doctrine of absolute, unbridgeable discontinuity between human and nonhuman. Other scientists, aware of the silent habits of chimpanzees, looked for a technique that would not require speech. This was the innovation of Project Washoe. For the first time, the foster family used a gestural rather than a vocal language.

In Project Washoe, speech was replaced with American Sign Language (ASL), the manual language of the deaf in North America. Washoe learned signs from her human companions and used these in a childish and rudimentary way that resembled the early acquisition of speech and sign by human children. Because there are human children who learn ASL as a first language, Washoe's stage-by-stage acquisition could be compared with the stage-by-stage acquisition of speech and sign by human children

(B. Gardner and Gardner, 1971, 1974, 1975, 1980, 1985; R. Gardner and Gardner, 1969, 1974, 1978; R. Gardner et al., 1989).

With the introduction of ASL, the line of research pioneered by the Kelloggs and the Hayeses moved forward dramatically. In 51 months, Washoe acquired at least 132 signs of ASL and used them for classes of referents rather than specific exemplars. Thus, DOG was used to refer to live dogs and pictures of dogs of many breeds, sizes, and colors, as well as the sound of barking by an unseen dog. OPEN was used to ask for the opening of doors to houses, rooms, cupboards, or lids of jars, boxes, bottles — and even (an invention of Washoe's) for turning on a water faucet. Washoe also understood many more signs than she herself used (B. Gardner and Gardner, 1975).

She signed to friends and to strangers. She signed to herself and to dogs, cats, toys, tools, even to the trees. She asked for goods and services, and she also asked questions about the world of objects and events around her. When Washoe had about eight signs in her expressive vocabulary, she began to combine them into meaningful phrases. YOU ME HIDE and YOU ME GO OUT HURRY were common. She called her doll, BABY MINE; the sound of a barking dog, LISTEN DOG; the refrigerator, OPEN EAT DRINK; and her potty-chair, DIRTY GOOD. Along with her skill with cups and spoons, and pencils and crayons, her signing developed stage-for-stage much like the speaking and signing of human children (B. Gardner and Gardner, 1971, 1974; R. Gardner and Gardner, 1988; Van Cantfort and Rimpau, 1982).

Project Washoe was followed by a second, more advanced venture in cross-fostering (B. Gardner and Gardner, 1980; R. Gardner and Gardner, 1978; Drumm et al., 1986; Rimpau et al., 1989; Van Cantfort et al., 1989). Washoe herself was captured wild in Africa and arrived in Reno when she was about ten months old. Moja, Pili, Tatu, and Dar were born in American laboratories and each arrived within a few days of birth. In general, the human participants in the second project had a higher level of expertise in ASL and chimpanzee psychology because some of them were veterans of Project Washoe. Many of the new recruits were deaf or the hearing offspring of deaf parents or had other extensive experience with ASL before they joined the staff. All had learned ASL and studied the procedures and results of Project Washoe beforehand. And, while Washoe had been the only chimpanzee in Reno, the foster chimpanzees of the second project had each other as frequent companions.

TEACHING

The development of verbal behavior, as we know it in the human case, is inextricably bound up in the rest of the conditions of a human childhood. It still remains to be seen whether anything comparable could develop under other conditions. It seems equally unlikely that the other aspects of human intellectual growth could flourish without the development of verbal behavior. Thus, the purpose of ASL in the young lives of Washoe, Moja, Pili, Tatu, and Dar was to satisfy the requirements of cross-fostering. Sign language in this laboratory was a means rather than an end in itself, a means by which the chimpanzees could express their intelligence in ways that would permit comparison with human beings.

The laboratory procedures assumed that all aspects of intellectual growth are intimately related. For young chimpanzees, no less than for human children, familiarity with simple tools (such as keys), devices (such as lights), and articles of clothing (such as shoes) is intimately involved in learning signs or words for keys, lights, shoes, opening, entering, lighting, and lacing. The laboratory was well stocked with such objects and activities, and the subjects had free access to them, or at least as much access as human children usually have. While no more free than human children to go outdoors without permission, they were free of mechanical restraints both indoors and out. They not only learned to eat human-style food, they learned to use cups and spoons and to clear the table and wash the dishes after a meal. They not only learned to use human toilets (in their own quarters and elsewhere), but they learned to wipe themselves and flush the toilet, and even to ask to go to the potty to postpone lessons and bedtimes.

At the height of the so-called "Chomskian revolution" in psycholinguistics, it was frequently claimed that human children acquire their first language with incredible speed by the innate unfolding of a uniquely human mental process, and more or less independently of adult input (Lenneberg, 1967, p. 137; McNeill, 1966). Needless to say, this claim was always in conflict with common experience. More recently, a large body of painstaking research has supported the more traditional view, that human parents teach their children. As Snow (1977) puts it:

> The first descriptions of mothers' speech to young children were undertaken in the late sixties in order to refute the prevailing view that language acquisition was largely innate and occurred almost independently of the language environment. The results of those mothers' speech studies may have contributed to the widespread abandonment of this hypothesis about language acquisition, but a general shift from syntactic to semantic-cognitive aspects of language acquisition would probably have caused it to lose its central place as a tenet of research in any case (p. 31). . . . all language learning children have access to this simplified speech register. No one has to learn to talk from a confused, error-ridden garble of opaque structure. Many of the characteristics of mothers' speech have been seen as ways of making grammatical structure transparent, and others have been seen as attention-getters and probes to the effectiveness of the communication (p. 38).

In teaching sign language to Washoe, Moja, Pili, Tatu, and Dar, we observed human parents with young children, and we imitated them. There was constant chatter about the everyday events and objects that might interest the young chimpanzees. Many of the comments were aimed at teaching vocabulary, for example, THAT CHAIR, SEE PRETTY BIRD, MY HAT. Many events and objects were introduced, just so we could sign about them. There were frequent questions to see what was getting across, and we tried hard to answer the youngsters' questions and to comply with their requests. By expanding on fragmentary utterances, we could use the fragments to teach and to probe. We also followed the parents of deaf children in using an especially simple and repetitious register of ASL and making signs on the youngsters' bodies to capture their attention (cf. Schlesinger and Meadow, 1972; Rimpau et al., 1989; B. Gardner and Gardner, this volume).

RECORDING

While teaching was spontaneous and informal, as in a human nursery, the methods of recording results were precisely defined and meticulously followed. Each sign had to meet detailed criteria of form and usage before it was listed as a reliable item of vocabulary. In terms of form, it had to correspond to a sign made by human adults, or to an immature variant. Decisions were guided by the judgement of fluent signers who were also familiar with signing in young children. In the case of usage, spontaneity was defined in terms of informative prompting. If the sign was produced by the chimpanzee subject without informative prompting, such as direct modeling or guidance that would induce any portion of the target sign, then it was judged to be spontaneous. To be appropriate, however, it had to be prompted by the verbal and situational context and the presence of a suitable addressee (B. Gardner and Gardner, 1971, 1975, 1980; R. Gardner and Gardner, 1978; B. Gardner et al., 1989).

Appropriate usage was judged on the basis of context notes, with the understanding that infant usage can be either narrower or broader than adult usage. Nevertheless, the chimpanzee usage had to have some major overlap with adult usage. Thus, in these early records, Washoe and Tatu used OUT for both leaving and entering their quarters (it was only later that they divided this referential domain between OUT and IN).

A TEST OF COMMUNICATION

When Washoe was 27 months old, she made a hole in the then flimsy inner wall of her house trailer. The hole was located high up in the wall at the foot of her bed. Before we repaired the hole, she managed to lose a toy in the hollow space between the inner and outer walls. When Allen Gardner arrived that evening, she attracted his attention to an area of the wall down below the hole at the level of her bed, signing OPEN OPEN many times over that area. It was not hard to understand what the trouble was and eventually to fish out the toy. When the toy was found, it was exciting to realize that a chimpanzee had used a human language to communicate information. It was not long before such situations became commonplace. For example, Washoe's playground was in the garden behind a single-story house. High in her favorite tree, Washoe was often the first to know who had arrived at the front of the house, and her companions on the ground learned to rely on her to tell them who was arriving and departing.

Washoe could tell her human companions things that they did not already know – things that she alone had witnessed. In our view, this is the essential ingredient of what is called displacement (Hockett, 1977, pp. 147–150). Some years before Project Washoe began, Hockett discussed the enormous evolutionary advantages of the change from primate communications that were limited to the here-and-now to an early hominid system that permitted communication about food sources and predators that only one member of the social group had seen (Hockett, 1960).

This is what was absent in the performances of Clever Hans. Clever Hans, it will be remembered, was a German horse that seemed to do arithmetic by tapping out numbers with his hoof. Not the circus trainers or the cavalry officers, not the veterinarians or the zoo directors, not even the philosophers and the linguists who studied the case could explain how Clever Hans did it. Nevertheless, an experimental psychologist, Oskar Pfungst (1911), unravelled the problem with the following test. Pfungst whispered one number into Clever Hans' left ear and Herr von Ost, the trainer, whispered a second number into the horse's right ear. When Clever Hans was the only one who knew the answer, he could not tap out the correct sums. He could not tell his human companions anything that they did not already know.

Since then, controls for "Clever Hans Errors" have been standard procedure in comparative psychology. To date most, if not all, research on human children has been carried out without any such controls. It is as if students of child development believed that, whereas horses and chimpanzees may be sensitive to subtle nonverbal communication, it is safe to assume that human children are totally unaffected.

METHOD

Early in Project Washoe we devised vocabulary tests to demonstrate that chimpanzees could use the signs of ASL to communicate information. The earliest versions are described in B. Gardner and Gardner (1971, pp. 158–161) and R. Gardner and Gardner (1973, 1974, pp. 11–15). The first objective of these tests was to demonstrate that the chimpanzee subjects could communicate information under conditions in which the only source of information available to a human observer was the signing of the chimpanzees. To accomplish this, nameable objects were photographed on 35mm slides. During testing, the slides were back-projected on a screen that could be seen by the chimpanzee subject, but could not be seen by the observer. The slides were projected in a random order that was changed from test to test so that the order could not be memorized either by the observer or by the subject.

The second objective of these tests was to demonstrate that independent observers agreed with each other. To accomplish this, there were two observers. The first observer (O_1) served as interlocutor in the testing room with the chimpanzee subject. The second observer (O_2) was stationed in a second room and observed the subject from behind one-way glass, but could not see the projection screen. The two observers gave independent readings; they could not see each other, and they could not compare observations until after a test was completed.

The third objective of these tests was to demonstrate that the chimpanzees used the signs to refer to natural language categories – that the sign DOG could refer to any dog, FLOWER to any flower, SHOE to any shoe, and so on (Rosch and Lloyd, 1978; Saltz et al., 1972; Saltz et al., 1977). This was accomplished by preparing a large library of slides to serve as exemplars. Some of the slides were used in Pretests that served to adapt subjects, observers, and experimenters to the testing procedure. The slides that were reserved for the tests were never shown during Pretests, so that the first time that a particular chimpanzee subject saw any one of the test slides was on a test trial, and no test slide was shown on more than one test trial. Consequently, there was no way that a subject could get a correct score by memorizing particular pairs of exemplars and signs. That is to say, scores on these tests depended upon the ability to name new exemplars of natural language categories.

The vocabulary items that appeared in Moja's single test and in the second test of each of the other three subjects are listed in Table 1. All of the vocabulary items tested in this way had to be names of picturable objects (see R. Gardner and Gardner, 1984, for details). Differences among the subjects in Table 1 reflect differences in their vocabularies as well as a strategy of overlapping tests that would sample the range of picturable objects in the vocabularies without making the tests excessively long. For each test we chose four exemplars of each vocabulary item to illustrate the range of objects that a subject could name with the same sign. Different breeds represented CAT and DOG, different species represented BIRD and BUG, different makes and models represented CAR. The number of vocabulary items and the resulting number of trials (items x exemplars) appear in Table 2.

The correct sign for each vocabulary item was designated in advance of the tests. That sign and that sign only was scored as correct for that item. Although there were aspects of the pictures for which superordinate terms such as FOOD, or descriptive terms such as BLACK, might be scored, neither the presence or absence nor the correct or incorrect use of such terms was considered in the scoring of these tests.

Most of the replies consisted of a single sign which was the name of an object. Sometimes, the single noun in the reply was contained in a descriptive phrase, as when Tatu signed RED BERRY for a picture of cherries, or when Dar signed THAT BIRD for a picture of a duck. These replies contained only one object name and that was the sign that was scored as correct or incorrect. Occasionally, there was more than one object name in a reply, as when Washoe signed FLOWER TREE LEAF FLOWER for a picture of a bunch of daisies. In such cases, the observers designated a single sign for scoring (usually the first) without looking at the picture themselves. For each trial and each observer, then, one sign and one sign only in each report was used to score agreement between O_1 and O_2 and agreement between the reports of the observers and the name of the exemplar.

RESULTS

Table 2 shows how the major objectives of the tests were accomplished. The agreement between O_1 and O_2 was high for all seven tests; except for Moja, the agreement ranged between 86% and 95%. Note that this is the agreement for both correct and incorrect signs. The agreement between the signs reported by O_1 and O_2 and the correct names of the categories is also high; except for Moja, correct scores ranged between 71% and 88%.

The line labelled "Expected" in Table 2 needs some explanation. When we first

Table 1. Vocabulary items in the tests of four chimpanzees.

ITEMS	CHIMPANZEES				ITEMS	CHIMPANZEES			
	W	**M**	**T**	**D**		**W**	**M**	**T**	**D**
ANIMATES					**FOODS**				
baby	+		+	+	apple		+	+	+
bird	+	+	+	+	banana		+	+	+
bug	+	+	+	+	berry		+	+	+
cat	+	+	+	+	carrot			+	+
cow	+	+	+	+	cereal		+		
dog	+	+	+	+	cheese	+		+	+
horse		+			corn		+	+	+
					fruit	+			
PLANTS					grapes		+		
flower	+	+	+	+	gum	+			
leaf	+	+			ice cream		+	+	+
tree	+	+	+	+	meat	+		+	+
					nut	+	+	+	+
CLOTHING					onion		+		
clothes	+				orange		+		
hat	+	+	+	+	pea / bean		+		
pants	+				peach		+	+	+
shoe	+	+	+	+	sandwich			+	+
					tomato	+			
GROOMING									
brush	+	+	+		**OTHER**				
comb		+	+		ball			+	+
hankie		+			book	+	+		
lipstick		+			car	+		+	+
oil	+		+	+	hammer	+			
toothbrush	+	+	+	+	key	+	+	+	
wiper	+				knife		+		
					peekaboo			+	
SENSORY					pipe	+			
listens	+	+			smoke	+	+		
looks	+	+	+						
smells	+								
DRINKS									
coffee		+	+						
drink	+								
milk		+							
soda pop		+	+	+					

Notes: W = Washoe; M = Moja; T = Tatu; D = Dar.

described this testing procedure (B. Gardner and Gardner, 1971), we estimated the expected chance performance as 1/N where N is the number of vocabulary items on a test, and all items are represented by the same number of exemplars. This estimate was based on the assumption that only the chimpanzees were guessing and that their guessing strategies could only be randomly related to the random sequence of presentation. But this estimate may be too low because it does not take into account the possibility that the observers were guessing. In random sampling without replacement, the probabilities of later events in a sequence depend on earlier events. The observers could have used their knowledge of the items that had appeared earlier to predict the items that would appear later. Thus, players who can remember the cards that have already been played can win significant amounts at games such as black jack. Diaconis (1978) and Read (1962) deal with a similar problem in demonstrations of extrasensory perception. When highly motivated subjects in ESP experiments can see each target card after each prediction, their later predictions tend to improve.

Table 2. Scores on the vocabulary tests of four chimpanzees.

| | Chimpanzee subject | | | | | | |
| | Washoe | | Moja | Tatu | | Dar | |
Test	1	2	1	1	2	1	2
Vocabulary items (n)	16	32	35	25	34	21	27
Trials (n)	64	128	140	100	136	84	108
Inter-observer agreement (%)	95	86	70[a]	89	91	90	94
Correct chimpanzee responses as scored by:							
Observer 1 (%)	86	72	54[b]	84	80	79	83
Observer 2 (%)	88	71	54[b]	85	79	80	81
Expected * (%)	15	4	4	6	4	6	5

a. Based on 135 trials; Observer 2 missed 5 trials.
b. Based on 132 trials; 8 unscorable trials.
* Assuming that the observer was guessing on the basis of perfect memory for all previous trials that that observer had seen. (For detailed explanation, see text.)

To estimate the effect of informed guessing by the observers on chance expectancy in the tests reported here, we used a formula which assumes that both observers (1) saw each slide after each trial, (2) had perfect memory for the number of exemplars of each vocabulary item that had appeared before the beginning of each trial, and (3) guessed the correct sign on the basis of the number of exemplars of each vocabulary item that remained to be presented. With this formula we calculated, for each of the seven tests, the expected scores in Table 2 (Patterson et al., 1986). In all cases, this estimate is a small fraction of the obtained scores. Since O_1 and O_2 reported extra-list intrusions (signs that were not on the target lists), they were using a less efficient strategy than that assumed in the mathematical analysis. Hence, small as they are, the values in the expected line of Table 2 overestimate chance expectancy.

The expected score for Washoe's first test is appreciably higher than the expected scores for the other six tests for two reasons. First, this test was shorter than the other tests, and predictability depends on the number of vocabulary items – the fewer the items, the greater the predictability. Second, and more significantly for this discussion, predictability increases as we approach the end of the test. The last trial is completely predictable since there is only one vocabulary item that could have any remaining exemplars. The next to the last trial may be completely predictable, but there are at most two vocabulary items that could still appear, and so on. In all cases, except for Washoe's first test, both O_1 and O_2 were assigned to test sessions in such a way that no individual served as an observer for more than half of the trials of any single test. The device is similar to the way gambling casinos can defeat card-counting customers by reshuffling the deck. The smaller number of items and the assignment of the same two observers to both sessions of Washoe's first test account for the higher, but still quite small, expected score on that test.

CONCEPTS

To make sure that the signs referred to conceptual categories, all of the test trials were first trials; that is, each slide was shown to the subject for the first time on the one and only test trial in which it was presented to that subject. All of the specific stimulus values varied, as they do in natural language categories; that is to say, most human beings would agree that the exemplars in each set belong together. Apparently Washoe, Moja, Tatu, and Dar agreed with this assignment of exemplars to conceptual categories.

When teaching a new sign, we usually began with a particular exemplar – a particular toy for BALL, a particular shoe for SHOE. At first, especially with very young subjects, there would be very few balls and very few shoes. The same situation is common in human nursery life. Early in Project Washoe we worried that the signs might become too closely associated with their initial referents. It turned out that this was no more a problem for Washoe or any of our other subjects than it is for children. The chimpanzees easily transferred the signs they had learned for a few balls, shoes,

Figure 1. Moja (three years old) at brunch with the Gardners. She is signing APPLE.

flowers, or cats to the full range of the categories wherever they found them and however represented (Fig. 1), as if they divided the world into the same conceptual categories that human beings use.

It is reasonable to suppose that nonhuman animals use natural language concepts outside of the laboratory. A wild monkey that finds a ripe mango in a tree must learn something general about ripe mangoes, because it is certain that that monkey will never get to pick that particular mango again. A young lion that brings down an impala must learn something general about hunting impalas, because it is certain that that lion will never get to hunt that particular impala again. The same must be true of young hawks hunting field mice. It seems unlikely that any creature with a natural world as complex as that of a wild pigeon could earn its living without using some natural language concepts.

So much experimentation has been limited to precisely repeated stimulus objects or to objects that vary only in simple dimensions, such as color and size, that it would be easy to form the impression that the conceptual abilities of nonhuman beings are severely limited. But there have been notable exceptions. Hayes and Hayes (1953) working with chimpanzee Viki, Hicks (1956) and Sands et al. (1982) working with monkeys, and Herrnstein et al. (1976) working with pigeons, for example, have all demonstrated that nonhuman beings can use natural language concepts when they are presented with suitably varied stimulus material.

ERRORS

On these vocabulary tests, the chimpanzee subjects were free to use any sign in their vocabularies. Unlike the experimenter-defined errors of the forced-choice tests that have been traditional in laboratories of comparative psychology, the errors made in these free-choice tests can be analyzed for patterns. Analysis of the errors (cf. R. Gardner and Gardner, 1984, pp. 393–398) showed that most of the errors fell into one of two patterns: conceptual errors and form errors. Thus, DOG was a common error for a picture of a cat, SODA POP for a picture of ice cream, and so on, showing that conceptual groups such as animals and foods were a major source of confusion. Similarly, signs made on the nose such as BUG and FLOWER were confused with each other, as were signs made on the hand such as SHOE and SODA POP, showing that the cheremic structure of ASL (the shape of a sign) was a source of errors the way the phonemic structure of English is the source of errors in the verbal behavior of human beings.

CONCLUDING REMARKS

Although concern with grammar has occupied so much of the efforts of developmental psycholinguists, in our view it would be a mistake for psychobiologists to neglect method and theory in the study of reference. For if the development of human verbal behavior requires any significant expenditure of biological resources, then it must confer significant selective advantage on its possessors. In order to confer any selective advantage, however, a biological trait must operate on the world in some way; it must be instrumental in obtaining benefit or avoiding harm. If clarifying one's ideas confers selective advantage, it must be because in some way clarified ideas provide superior means for operating in the biological world. As for establishing social relations, a system of displays and cries is sufficient to maintain group cohesiveness in most animals. The selective advantage of a wider variety of signals would seem to be the communication of more information. But unless verbal behavior refers to objects and events in the external world, it cannot communicate information, and it cannot have any selective advantage. From this point of view, reference is the biological function of verbal behavior, and the role of grammar or structure in verbal behavior must be to enlarge the scope and increase the precision of reference.

The essence of Darwinism is the search for the fundamental laws of biology that act and interact to yield the rich diversity of species. Darwin himself predicted that where continuities in blood and bone might be easy to accept, continuities of mind and feeling would be another matter. As expected, our sign language studies of chimpanzees have stirred up intense controversy. There has been a long and lively debate on the question of whether or not these cross-fostered chimpanzees acquired something called "syntax." Indeed, there has also been a long and lively debate as to what "syntax" may consist of in terms of the observable behavior of human children – or human adults for that matter (e.g., Harris, 1984). But most commentators have agreed that the sign language studies of chimpanzees have demonstrated a significant degree of intellectual continuity between the cross-fostered chimpanzees and human children (e.g., Brown, 1970; Dingwall, 1979; Hewes,1973; Lieberman, 1984; Stokoe, 1983). Even among those who remain faithful to the doctrine of an unbridgeable gap between language and the rest of human behavior, and between human behavior and animal behavior, most have conceded that there is now more evidence of continuity than they would have expected.

ACKNOWLEDGMENTS

Research discussed in this article was supported by grants from the National Institute of Mental Health, the National Science Foundation, the National Geographic Foundation, the Grant Foundation, the Spencer Foundation, and the UNR Foundation.

REFERENCES

Brown, R. 1970. The first sentences of child and chimpanzee. In R. Brown, ed., *Selected Psycholinguistic Papers*, pp. 208–281. New York: Macmillan.

Diaconis, P. 1978. Statistical problems in ESP research. *Science* 201:131–136.

Dingwall, W. O. 1979. The evolution of human communication systems. *Studies in Neurolinguistics* 4:1–95.

Drumm, P., B. T. Gardner, and R. A. Gardner. 1986. Vocal and gestural responses of cross-fostered chimpanzees. *American Journal of Psychology* 99:1–29.

Gardner, B. T., and R. A. Gardner. 1971. Two-way communication with an infant chimpanzee. In A. Schrier and F. Stollnitz, eds., *Behavior Of Nonhuman Primates* 4:117–184. New York: Academic Press.

_____. 1974. Comparing the early utterances of child and chimpanzee. In A. Pick, ed., *Minnesota Symposium On Child Psychology* 8:3–23. Minneapolis: University of Minnesota Press.

_____. 1975. Evidence for sentence constituents in the early utterances of child and chimpanzee. *Journal of Experimental Psychology: General* 104:244–267.

_____. 1980. Two comparative psychologists look at language acquisition. In K. E. Nelson, ed., *Children's Language* 2:331–369. New York: Gardner Press.

_____. 1985. Signs of intelligence in cross-fostered chimpanzees. *Philosophical Transactions of the Royal Society,* B 308:159–176.

Gardner, B. T., R. A. Gardner, and S. G. Nichols. 1989. The shapes and uses of signs in a cross-fostering laboratory. In R. A. Gardner, B. T. Gardner, and T. E. Van Cantfort, eds., *Teaching Sign Language to Chimpanzees.* Albany, NY: SUNY Press.

Gardner, R. A., and B. T. Gardner. 1969. Teaching sign language to a chimpanzee. *Science* 165:664–672.

_____. 1973. *Teaching Sign Language to the Chimpanzee, Washoe.* (16mm sound film). State College, PA: Psychological Cinema Register.

_____. 1974. Teaching sign language to the chimpanzee, Washoe. *Bulletin D'Audio Phonologie* 4(5):145–173.

_____. 1978. Comparative psychology and language acquisition. *Annals of the New York Academy of Sciences* 309:37–76.

_____. 1984. A vocabulary test for chimpanzees. *Journal of Comparative Psychology* 98:381–404.

_____. 1988. Feedforward vs. feedbackward: an ethological alternative to the law of effect. *Behaviorial and Brain Sciences* 11:429–449.

Gardner, R. A., B. T. Gardner, and T. E. Van Cantfort, eds. 1989. *Teaching Sign Language to Chimpanzees.* Albany, NY: SUNY Press.

Harris, R. 1984. Must monkeys mean? In R. Harre and V. Reynolds, eds., *The Meaning of Primate Signals,* pp. 116–137. Cambridge: Cambridge University Press.

Hayes, K. J., and C. Hayes. 1953. Picture perception in a home-raised chimpanzee. *Journal of Comparative and Physiological Psychology* 46:470–474.

Herrnstein, R. J., D. H. Loveland, and C. Cable. 1976. Natural concepts in pigeons. *Journal of Experimental Psychology: Animal Behavior Processes* 2:285–302.

Hewes, G. W. 1973. Pongid capacity for language acquisition: An evaluation of recent studies. *Symposia of the Fourth International Congress of Primatology* 1:124–143.

Hicks, L. H. 1956. An analysis of number-concept formation in the rhesus monkey. *Journal of Comparative and Physiological Psychology* 49:212–218.

Hockett, C. F. 1960. Logical considerations in the study of animal communication. In W. E. Lanyon and W. N. Tavolga, eds., *Animal Sounds and Communication,* pp. 392–430. Washington, D. C.: American Institute of Biological Sciences.

_____. 1977. *The View From Language.* Athens: University of Georgia Press.

Lane, H., and R. Pillard. 1978. *The Wild Boy of Burundi.* New York: Random House.

Lenneberg, E. 1967. *Biological Foundations of Language.* New York: John Wiley and Sons.

Lieberman, P. 1984. *The Biology and Evolution of Language.* Cambridge, MA: Harvard University Press.

McNeill, D. 1966. The creation of language by children. In J. Lyons and R. Wales, eds., *Psycholinguistics Papers,* pp. 99–114. Edinburgh: Edinburgh University Press.

Patterson, J. C., B. T. Gardner, and R. A. Gardner. 1986. Chance expectancy with trial-by-trial feedback and random sampling without replacement. *American Mathematical Monthly* 93:520–530.

Pfungst, O. 1911. *Clever Hans* (C. L. Rahn, Trans.). New York: Henry Holt.

Read, R. C. 1962. Card-guessing with information – a problem in probability. *American Mathematical Monthly* 69:506–511.

Rimpau, J. B., R. A. Gardner, and B. T. Gardner. 1989. Expression of person, place, and instrument in ASL utterances of children and chimpanzees. In R. A. Gardner, B. T. Gardner, and T. E. Van Cantfort, eds., *Teaching Sign Language to Chimpanzees.* Albany, NY: SUNY Press.

Rosch, E., and B. B. Lloyd, eds., 1978. *Cognition and Categorization.* Hillsdale, NJ: Lawrence Erlbaum.

Saltz, E., E. Soller, and I. Sigel. 1972. The development of natural language concepts. *Child Development* 43:1191–1202.

Saltz, E., D. Dixon, S. Klein, and G. Becker. 1977. Studies of natural language concepts. III. Concept overdiscrimination in comprehension between two and four years of age. *Child Development* 48:1682–1685.

Sands, S. F., C. E. Lincoln, and A. A. Wright. 1982. Pictorial similarity judgements and the organization of visual memory in the rhesus monkey. *Journal of Experimental Psychology: General* 111:369–389.

Schlesinger, H. S., and K. P. Meadow. 1972. *Deafness and Mental Health: A Developmental Approach.* Berkeley: University of California Press.

Singh, J. A. L., and R. M. Zingg. 1942. *Wolf Children and Feral Man.* Hamden, CT: Shoe String Press (Reprinted in 1966 by Harper and Row).

Snow, C. 1977. Mother's speech research: from input to interaction. In C. Snow and C. Ferguson, eds., *Talking to Children,* pp. 31–49. Cambridge: Cambridge University Press.

Stokoe, W. C. 1983. Apes who sign and critics who don't. In H. T. Wilder and J. de Luce, eds., *Language in Primates.* Bloomington: Indiana University Press.

Van Cantfort, T. E., and J. B. Rimpau. 1982. Sign language studies with children and chimpanzees. *Sign Language Studies* 34:15–72.

Van Cantfort, T. E., B. T. Gardner, and R. A. Gardner. 1989. Developmental trends in replies to Wh-Questions by children and chimpanzees. In R.A. Gardner, B.T. Gardner, and T. E. Van Cantfort, eds., *Teaching Sign Language to Chimpanzees.* Albany, NY: SUNY Press.

CROSS-FOSTERED CHIMPANZEES:
II. MODULATION OF MEANING

Beatrix T. Gardner and R. Allen Gardner

MODULATION OF CITATION FORMS

Those whose study of American Sign Language (ASL) is limited to standard dictionaries and manuals can easily get the impression that signs must be made "just so" – that there is only one good or correct way to form each sign. In fact, just as in the case of spoken words, there are individual, regional, and other variations in signs. Moreover, dictionaries of ASL, like dictionaries of spoken languages, show signs in citation form – the form that is seen when an informant is asked, "What is the sign for X?" In normal conversation, fluent signers will modulate their signs in a variety of ways. The sign GIVE, for example, may start near the signer's body and move out toward the addressee to indicate, "I give you," or the movement may be reversed, to indicate, "You give me." Modulation makes signs more versatile and more expressive; a single lexical item can become several different signs.

Among the languages of the world, English is unusual in its heavy reliance on word order for the modulation of meaning. Most human languages rely more heavily on inflectional devices. ASL is one of the inflecting languages of the world (Klima and Bellugi, 1979, p. 299; p. 314). While books that teach vocabulary describe signs in citation form, descriptions of inflected forms can be found in books that teach conversational usage. Fant's (1972) introduction to ASL consists of dialogues; these are lessons on "how to put signs together the way deaf people do" and include many examples of modulated signs. The sign LIKE appears with two hands and with one hand; the two-handed version shows that you like something very much (p. 35). The sign EAT, when made with both hands moving alternately and repeatedly, indicates a banquet or a big meal (p. 46). Normally, the sign ALL RIGHT is made with just one movement, but in excusing a minor social error, such as someone forgetting your name, two small quick movements are used to emphasize that it is a casual, unimportant circumstance (p. 7).

As Fant (1972) points out, an important set of modulations for verbs such as ASK, GIVE, and MEET makes use of the sight line:

> This is an imaginary line between signer and observer, i.e. "speaker" and "listener." Whenever a sign such as SEE moves along the sight line

234

toward the observer, the pronouns "I" and "You" are implied, thus they need not be signed. That is to say, instead of signing I SEE YOU, you need to sign only SEE. Since the sign moves from "I" towards "you," the pronouns are built into the movement (p. 2). . . . The sign ASK-QUESTION is extremely versatile. When you sign it toward the observer, it means "I ask you." When the signer signs it toward himself, coming from the observer, it means "You ask me." When the signer signs it toward himself coming from an angle to the sight line, it means "(a third person or persons) ask me." To convey the idea, "I asked several people," move the sign in an arc as you repeat it, as if the people you asked were standing there in front of you (p. 75).

Thus, changes in form produce semantic, pragmatic, and grammatical effects. Indeed, a given change in form, such as reiteration, can be used for all three. Reiteration makes HOUSE into TOWN and TREE into FOREST – a semantic effect; reiteration also makes HOPE emphatic – a pragmatic effect; and reiteration changes the nouns WEEK and MONTH into the adverbs WEEKLY and MONTHLY – a grammatical effect. Recent work on ASL by psycholinguists has concentrated on grammatical effects, particularly on the line of sight inflection described by Fant, in which reference to participants is incorporated into signs for action (Wilbur, 1980, p. 19).

In studies of ASL, adult signers are treated as informants. They are asked to convey, in gestures, English words or sentences that embody the distinctions of interest: distinctions of degree, as in Fant's example of "like" and "like very much"; distinctions in person reference, as in "I give you," "I give them," "I give to each of them," "They each give me," and so on. Video records of these productions are examined to see whether each inflectional process changes the appearance of classes of signs in a characteristic way (Klima and Bellugi, 1979, pp. 299–315).

In studies conducted with children, more naturalistic methods are used. Investigators examine film and other records of conversations between children and familiar caretakers, and search for signs that vary in appearance from citation form. Two variations often mentioned as characteristic of "baby sign" are reiteration of signs (Hoffmeister et al., 1975) and making signs on the body of the adult addressee (Schlesinger and Meadow, 1972). Schlesinger and Meadow noted that parents frequently make signs on infants (as we did with the cross-fostered infant chimpanzees; see R. Gardner and Gardner, this volume). They suggested that this is a device for getting a young child's attention, and also a warm, approving, or playful way of signing (pp. 67–68).

The adult use of inflectional devices is the product of a long developmental process (Cokeley and Gawlik, 1974). Ellenberger and Steyaert (1978) studied how participants and places are incorporated into action signs by a child learning ASL as his first language. They noted that:

One might expect that such a representational system, because of its somewhat pictorial nature, could be easily grasped by a child, and would thus be acquired at an early age (p. 264). . . . They are, in fact, relatively late acquisitions, perhaps because such representations may require a fairly advanced mastery of cognitive skills involving spatial relationships (p. 268).

When Ellenberger and Steyaert analyzed a longitudinal series of films, they found that these inflections appeared in a childish, immature form when their subject, the deaf child of deaf parents, was between four and five years old. Before that, action signs such as BITE and BREAK occurred rarely and only in citation form. Beginning at 54 months, the deaf subject indicated the participants in an action, sometimes by placing the action sign on the body of the person and sometimes by moving the action sign in the direction of the person as adults do (Ellenberger and Steyaert, 1978, p. 265).

FIELD RECORDS

In our cross-fostering laboratory (R. Gardner and Gardner, this volume), a human member of the foster family was with Washoe, Moja, Pili, Tatu, and Dar from the time that these chimpanzees awakened in the morning until they fell asleep at night. One of the responsibilities of the human adults was to describe, in the daily log, what went on while they were with the subjects: meals, naps, chores, games that were played, and, of course, sign language conversations. The daily field records contain detailed descriptions of perceptual-motor, social, and communicative behavior, as well as notes on food intake, toileting, etc.

Descriptions of the shapes of signs were based on the system developed by Stokoe (1960) and used by Stokoe, Casterline, and Croneberg in compiling the Dictionary of American Sign Language (1965). The Stokoe et al. dictionary provides a written notation by means of three sets of symbols that refer to distinctions within three aspects of a gesture. The first aspect is the place on the body or in space where the sign is made (e.g., the cheek, the chest, in front of the signer). The second aspect is the configuration of the hand (e.g., whether the hand is fisted or open, which fingers are extended, and how the hand is oriented toward the place). The third is the type and direction of movement (e.g., contact, or grasp, or rub up or down or in a circle).

The Place, the Configuration, or the Movement, or often two or all three of these basic aspects of form, distinguish one sign from another. Thus, NUT and TOOTH-BRUSH are identical in Place but different in Configuration and Movement, while DOG and HAT differ in Place but are identical in Configuration and Movement.

The field notes contain detailed descriptions of the form of signs, including deviations from citation form. From these reports, we can identify types of modifications that have parallel effects on many different signs. For example, there are reports that COME, DOG, SORRY, and other signs that are normally made with one hand were duplicated with both hands. That is, both hands made the same sign at the same time. At least 20 different types of form modifications appear in the field descriptions. Each type can be characterized by an aspect of sign form (e.g., Place or Movement) and by the way in which this aspect differs from citation form.

Many of the form modifications that were reported for the cross-fostered chimpanzees (e.g., making a sign on the adult, or with the hands of the adult, or making it very large) produce conspicuous changes in the appearance of the sign. Yet all the modifications (except for blended signs in phrases) preserve the basic Place, Configuration, and Movement that are characteristic of the sign and enable the observer to identify it. In this respect, the modifications in the form of signs are very different from errors in diction (Gardner and Gardner, 1984, pp. 393–397).

A commonly reported modification involves Place: the young chimpanzees placed signs on the body of the adult addressee instead of on their own body. Thus, SWALLOW, for which the citation Place is the throat, and TICKLE, for which the citation form is the back of the hand, were placed on the corresponding body part of the addressee. The following entry in the daily log for 58-month-old Dar is typical (observer's initials and date of entry in parentheses):

> GRG was hooting and making other sounds to prevent Dar from falling asleep. Dar put his fist to GRG's lips and made kissing sounds. GRG asked WHAT WANT? and Dar replied QUIET, placing the sign on GRG's lips (GRG 5/19/81).

There were other types of Place modification. In one of these, the chimpanzees placed signs on or near the referent, as when OPEN was placed on various doors, containers, or on the addressee's mouth; and when HURT was placed on various cuts and bruises, or on the signer's head, after a fall. In a different type of Place modification, the sign is shifted beyond the normal signing space, either further away from the body or lifted unusually high, as in the following entry for Moja at 59 months:

> Moja made the sign COOKIE with her hands below the level of the windowsill. When KW (on the other side of the window) looked at Moja with a questioning expression, Moja raised her hands up to eye level and repeated the sign (JR 9/23/77).

Other types of modification involve Configuration. Instead of forming the configuration of the sign with their own hands, the chimpanzees molded the hands of the addressee, as in the following for Washoe at 28 months:

> After her nap, Washoe signed OUT. I was hoping for Washoe to potty herself and did not comply. Then Washoe took my hands and put them together to make OUT, and then signed OUT with her own hands, to show me how. Again today, Washoe signed UP and then took my hand and signed UP with it (NR 11/15/67).

The most commonly reported modification in Movement was to make the movement of the sign more vigorous or emphatic, as in the following for Moja at 26 months:

> Moja signed DOG on RB and me, and looked at our faces, waiting for us to "woof." After several rounds, I made a "meeow" instead. Moja

237

signed DOG again, I repeated "meeow" again, and Moja slapped my leg harder. This went on. Finally, I woofed and Moja leapt on me and hugged me (TT 1/4/75).

There were additional ways in which the young chimpanzees modified Movement, for example, by enlarging the movement and also by prolonging the sign.

Two commonly reported modifications involve the sign as a whole rather than its Place or its Configuration or its Movement. Signs were reiterated one or more times, and there were also two-handed versions of signs normally made with one hand, as in the following for Moja at 45 months:

Moja just does one no-no [prohibited activity] after another as soon as we get out in yard. She even jumps the fence. Finally, I retrieve her, tell Moja she must play in garage. Moja signed SORRY SORRY. I then sign YOU SURE SORRY? NOW GOOD? and she replies SORRY SORRY. The first three SORRY are in the standard, one-hand form, but she uses both hands for last SORRY (SW 9/6/76).

Several of the modifications that were described involve whole utterances rather than individual signs. Both signs of a phrase are involved in blended signs. The following is for Moja at 56 months (ICE CREAM is made by rubbing the index edge of the fist down over the lips, NO by shaking the head):

Moja stares longingly at Dairy Queen as we drive by. Then for a minute or more signs NO ICE CREAM many times by shaking head while holding fist to mouth, index edge up (RAG 7/9/77).

The modification in which gaze is prolonged is known as the "questioning look" in ASL (Covington, 1973; Van Cantfort and Rimpau, 1982, pp. 36–38). Like the rising pitch in a spoken question, this type of gaze distinguishes questions from statements. The following is for Tatu at 12 months:

Tatu takes a picture book and looks through it. I join her. Seems to be especially interested in pictures of flowers and of chimps. Tatu points to flowers, [and signs] THERE?, with raised eyebrows and prolonged eye contact. I sign FLOWER, then Tatu turns back to another page (KW 12/23/76).

The field notes show that each of the different types of modification was used by all the subjects. The modifications are also productive, in that each type of modification was used with different signs. Thus, as early as month 11, Pili was placing HEAR, QUIET, TICKLE, SLEEP, and three other signs on the body of the adult as well as on his own body. Most of the other types of modification also appear early, before the end of the first year. In that first year, however, only a few types are used regularly and productively with several different signs. The field records illustrate developmental trends in the number and types of modifications that are used by the cross-fostered chimpanzees.

VIDEOTAPE RECORDS

Tape-recorded conversations between the chimpanzees and human members of their foster family are especially suitable for studying modulation. There are eight videotapes of Dar, in the same setting, interacting with the same long-term member of his foster family, Tony McCorkle. In each of the sessions, Dar and his companion are sitting on a sofa, engaged in quiet activities such as food sharing, grooming, or tickling. The videotaped sessions last from ten minutes to 25 minutes.

For his dissertation research, James Rimpau has transcribed these eight tapes of ASL conversations that took place when Dar was between 40 and 49 months old (Rimpau, 1985; Rimpau et al., 1989). To demonstrate that his transcriptions were reliable, he enlisted two other long-term participants in the cross-fostering research to transcribe one session each. By comparing his own transcript with that of the second observer on an item-by-item basis, agreement was calculated for signs identified by both observers. Inter-observer agreement on the identity of the signs was 81% in one session and 84% in the other. Other measures of inter-observer agreement were also obtained. Of particular interest here is the high agreement for two types of form modifications: 98% for Place modifications (measured in one sample only), and 85% and 88% for reiteration of signs within an utterance.

In all, Dar made 1,531 signs in this record, and 20% of these were modified in form. There were 152 signs that were not in the citation Place but instead were placed either on the addressee or on the referent, and 145 signs that were reiterated. Because Place modifications appeared often, the circumstances in which Dar used them could be studied. Other modifications that we have discussed, such as molding signs with the hands of the addressee, emphatic movement, and questioning gaze, were recorded also, but there were only one or two instances of each in this particular sample of 1,531 signs.

As illustrated in the previous section, field notes usually included context descriptions: what the chimpanzee was looking at, what the adult was doing and signing, how the two were positioned, and so on. For a systematic study of the contexts in which Place modifications are used, Rimpau (1985) compared the modified and the citation forms of signs in the videotaped sample of conversations between Dar and Tony. These samples showed that when Dar modified the Place of signs for actions, he was including reference to persons and locations in the form of the sign. The sign BRUSH, for example, was modified into the phrase YOU-BRUSH and the phrase BRUSH-THERE. Three lines of evidence support this finding.

First of all, Place modifications appeared predominantly for signs that were verbs (e.g., CRY, OPEN, GROOM), or signs that could be used either as verbs or as nouns (e.g., BRUSH, COMB, EAT). While Dar used 71 different vocabulary items in the videotaped sample, four items – TICKLE, GROOM, BRUSH, and GUM – were modified so often that they accounted for 85 or slightly over half the 152 Place modifications. TICKLE, GROOM, and BRUSH are signs that are used for actions. In the sample, Dar made these signs in citation form on his own hand (TICKLE, GROOM) or forearm (BRUSH). Dar also made them on the corresponding part of Tony's body, and sometimes on his own neck or other places on his own body, that were different from the Place in citation form.

Second, the immediately preceding utterance by Tony influenced the form of these three action signs. Many of Tony's immediately preceding utterances were questions. When Dar signed TICKLE, BRUSH, or GROOM on Tony, he was replying to questions where some form of "YOU action" would be appropriate, questions such as WHAT ME DO? WHAT WANT? or WHO TICKLE? Furthermore, Dar replied to WHERE BRUSH? and WHERE GROOM? by signing somewhere on his own body (other than the citation Place).

Finally, when modification was absent, that is, when Dar used the citation form of TICKLE, BRUSH, and GROOM, he added signs for names and pronouns to the sign for action. In one case, Dar used BRUSH in citation form as a reply to WHAT THIS? in reference to a hairbrush. In the remaining cases, TICKLE, BRUSH, and GROOM were used as actions, and Dar then included a name sign or a pronoun, or also added the same sign in Place-addressee form to indicate the agent of action in his utterance, as in BRUSH GOOD YOU DAR and BOY TICKLE (citation form) TICKLE (on Tony's hand).

This analysis shows that Dar was using Place inflections. The way in which Dar incorporated participants and places into signs for action is very similar to the incorporations used by the deaf child studied by Ellenberger and Steyaert (1978). Although young chimpanzees and children place signs on the body of the person instead of just moving the signs in the direction of the person, the system that the youngsters of both species use is clearly related to the adult line-of-sight inflection (Fant, 1972).

EXPERIMENTAL MANIPULATIONS

Modulated signs also appeared in experiments in which contexts were systematically manipulated. In Drumm et al. (1986), for example, the human interlocutors announced coming events to Dar and Tatu under experimentally controlled conditions. The coming events were either positive, negative, or neutral. The chimpanzees were significantly more likely to reiterate the signs in their replies to the signed announcements of positive events. Thus, in response to the announcement, TIME ICE CREAM NOW, Tatu replied, ICE CREAM, ICE CREAM, ICE CREAM, ICE CREAM, ICE CREAM, ICE CREAM. When Keenan (1977) found examples of this kind of reiteration in the speech of human children, she interpreted them as a pragmatic device indicating agreement and emphasis.

CONCLUDING COMMENT

In R. Gardner and Gardner (this volume) we report the results of a test of vocabulary which show that cross-fostered chimpanzees use the citation form of ASL signs to communicate information. In this report we have discussed evidence that modifications in the form of the signs produced by these chimpanzees also communicate information. In making signs in modified form, the chimpanzees adopted and used the material we provided in new ways: they made the signs we taught them their own. When modified by these young chimpanzees, the signs became more visible, more expressive, and more effective in communication.

ACKNOWLEDGMENTS

Research discussed in this article was supported by grants from the National Institute of Mental Health, the National Science Foundation, the National Geographic Foundation, the Grant Foundation, the Spencer Foundation, and the UNR Foundation.

REFERENCES

Cokely, D. R., and R. Gawlik. 1974. Childrenese as Pidgin. *Sign Language Studies* 5:72–81.

Covington, V. C. 1973. Juncture in American Sign Language. *Sign Language Studies* 2:29–38.

Drumm, P., B. T. Gardner, and R. A. Gardner. 1986. Vocal and gestural responses of cross-fostered chimpanzees. *American Journal of Psychology* 99:1–29.

Ellenberger, R., and M. Steyaert. 1978. A child's representation of action in American Sign Language. In P. Siple, ed., *Understanding Language Through Sign Language Research,* pp. 261–269. New York: Academic Press.

Fant, L. J., Jr. 1972. *Ameslan: An Introduction to American Sign Language.* Northridge, CA: Joyce Motion Picture Co.

Gardner, R. A., and B. T. Gardner. 1984. A vocabulary test for chimpanzees. *Journal of Comparative Psychology* 98:381–404.

Hoffmeister, R. J., D. F. Moores, and R. L. Ellenberger. 1975. Some procedural guidelines for the study of the acquisition of sign languages. *Sign Language Studies* 7:121–137.

Keenan, E. Ochs. 1977. Making it last: repetition in children's discourse. In S. Ervin-Tripp and C. Mitchell-Kernan, eds., *Child Discourse,* pp. 125–138. New York: Academic Press.

Klima, E. S., and U. Bellugi 1979. *The Signs of Language.* Cambridge, MA: Harvard University Press.

Rimpau, J. B. 1985. Communication skills of cross-fostered chimpanzees: modulation of meaning. Paper presented at annual meeting of the American Association for the Advancement of Science, Los Angeles.

Rimpau, J. B., R. A. Gardner, and B. T. Gardner. 1989. Expression of person, place, and instrument in ASL utterances of children and chimpanzees. In R. A. Gardner, B. T. Gardner, and T. E. Van Cantfort, eds., *Teaching Sign Language to Chimpanzees.* Albany, NY: SUNY Press.

Schlesinger, H. S., and K. P. Meadow. 1972. *Deafness and Mental Health: A Developmental Approach.* Berkeley: University of California Press.

Stokoe, W. C. 1960. Sign language structure: an outline of the visual communications systems of the American deaf. *Studies in Linguistics,* Occasional Papers 8, University of Buffalo.

Stokoe, W. C., D. Casterline, and C. G. Croneberg. 1965. *A Dictionary of American Sign Language.* Washington, D.C.: Gallaudet College Press.

Van Cantfort, T. E., and J. B. Rimpau. 1982. Sign language studies with children and chimpanzees. *Sign Language Studies* 34:15–72.

Wilbur, R. 1980. The linguistic description of American Sign Language. In H. Lane and F. Grosjean, eds., *Recent Perspectives On American Sign Language,* pp. 7–31. Hillsdale, NJ: Lawrence Erlbaum.

SIGNING INTERACTIONS BETWEEN MOTHER AND INFANT CHIMPANZEES

Deborah H. Fouts

There are five members in our chimpanzee research family. Four of the five chimpanzees involved in our research acquired their signs from their human companions while being cross-fostered by R. Allen and Beatrix T. Gardner. The Gardners' cross-fostering project began over 20 years ago on June 21, 1966 (1971; 1978; 1984).

The project now has a different character to it as compared to the Gardners' original cross-fostering project. The chimpanzees are grown now, or nearly so, and they require different treatment than they did as infants. However, in our laboratory we still agree with the Gardners that an essential element in any project such as this is the environmental and social enrichment of the chimpanzees. As the Gardners reasoned when Project Washoe was begun, if Washoe was going to learn signs she would have to have interesting people to talk to and interesting things to talk about.

Today these five chimpanzees are no longer infants, and the role of the scientist has changed accordingly. Whereas the Gardners were in the nurturing roles of cross-foster parents, our role is something that is perhaps best described as the "Domestique" method of caring for the chimpanzees. This is because we spend the major part of our time taking care of the daily chores and individual needs of the chimpanzees. We act as their maid, their cook, and their trusted family butler. We clean their rooms, cook all three of their meals, prepare their snacks, make sure that they have good magazines available for browsing and that the young as well as the adult chimpanzees have interesting toys with which to play. We also play the role of the understanding and reassuring trusted friend when one of the chimpanzee family members becomes upset with another member. Indeed our primary role is to socially enrich and support these five chimpanzees. A parental role is not really appropriate when one considers that Washoe is older than most of the student assistants involved in the project.

We have our scientific role to play as well. And it is not too far from the way trusted family servants find things out about the family whom they serve. The servants quietly and unobtrusively observe the family members interacting, and sometimes they even watch when the family members don't know that they are being watched.

The two main domestiques who serve this family of five chimpanzees are Roger Fouts, who has been a friend and caregiver to Washoe for 19 years, and I. We might be considered the butler and cook, but I don't want to carry this analogy too far, no matter how well it fits. In addition, we have devoted graduate and undergraduate students who serve as well. We try to encourage freshmen to apply to our program so that the chimpanzees don't have to adjust to many new people. But with the nature of university life and the length of this project, new people are expected and planned for. Two of the human caregivers have been with us for the past four years, others for three years or less. Each school year we have between 10 and 15 students who are trained. However, only two or three of these are encouraged to stay.

There are strict prerequisites and requirements as well as a training regimen for prospective members of this project. Before being admitted to the training program, each student must have taken at least one quarter of a university-level sign language course. Training in the lab begins with a three-month introductory course which includes directed readings of books and articles that are pertinent to this field, and the viewing of captive and wild chimpanzees through films and videotapes. The student is then taught to recognize the individual chimpanzees and specific chimpanzee behaviors. This requires the students to learn the 40-page taxonomy of behaviors and to memorize the abbreviated terms of the taxonomy of behaviors, and they are tested on their knowledge and understanding of the taxonomy. The next step is one of viewing taped interactions of the chimpanzees. The viewing is done with an experienced observer (defined as a person who has had at least two years experience with American Sign Language and with these five chimpanzees), who explains each behavior numerous times with specific examples. As the student observes the taped interactions, he/she learns to record the chimps' behaviors by writing observations in longhand and then must transcribe each of these into abbreviations. We have found the videotapes to be an excellent teaching tool, as they allow for replay and discussion. The trainees must know the criteria for each sign to be judged a sign, and they must learn the PCMs (Stokoe et al., 1965) for signs – that is, the place where a sign is made, the hand configuration used to make the sign, and the movement involved in making the sign.

In addition to the signs, every other gesture is defined as a nonverbal behavior; these nonverbal behaviors are described, defined, and given an abbreviated form in the taxonomy. There are 338 abbreviated terms. Twenty refer to the actors in the interaction and 15 are related to contexts; there are 49 facial expressions and vocalizations and 117 verbs.

The reason that this is so important is that our criteria for each nonverbal behavior comes from this taxonomy. So just as there is a specific sign criteria based on the three elements – place, configuration, and movement of hand – there is also a specific criteria for each nonverbal behavior. By combining the sign criteria with the nonverbal behavior, the contextual correctness of sign usage is determined. For example, the PCM of a sign is analyzed with regard to the facial expression and any vocalization or other nonverbal behaviors that are used in conjunction with it. The vast majority of nonverbal behaviors have come from Jane Goodall's classification of wild chimp behaviors (1968). However, some seem to be unique to our chimps (e.g., the bronx cheer that Moja uses during grooming).

The most important members of this household we call a lab are, of course, the five chimpanzees. We have three females (Washoe, 23 years; Moja, 16 years; Tatu, 13 years) and one male (Dar, 12 years) who were cross-fostered by the Gardners. The fifth, another male, Loulis, now 10 years old, is Washoe's adopted son, and to date he has acquired 70 signs from the chimps in his social group.

SIGN ACQUISITION

The first study described in this paper has to do with how Loulis acquired his signs from Washoe and the other three chimpanzees (Fouts et al., 1982).

From March 1979, when Loulis was ten months old, through June 1984, when Loulis was six years old, the humans used vocal speech to communicate around Loulis except for seven signs: WHO, WHAT, WHERE, WHICH, NAME, SIGN, and WANT. During that period we continued to use an extensive repertoire of nonverbal gestures (both human and chimpanzee) as well as vocal speech. This was done to insure that the signs Loulis acquired would be from the chimpanzees rather than from the humans. This restriction did not really present a communication problem for the humans and Washoe, because she, like most chimpanzees who have been exposed to human speech, comprehends it fairly well. If anyone did err, and signed in front of Loulis, the sign and the instance were recorded. Over the five-year period there were less than 40 occurrences.

By June of 1984, at the age of six years, Loulis had acquired 55 signs, none of which were the seven signs listed above, and the humans began to sign again with all of the chimps.

When we began Project Loulis, we asked the questions: Was sign acquisition only a function of a chimpanzee's exposure to humans? Were humans forcing a behavior on the chimps that they wouldn't otherwise acquire? Or were the chimps active participants in the acquisition of signs? Were the Gardners biological alchemists? Or were they taking advantage of a gestural preparedness, a capacity already in the organism?

Our project with Loulis was designed to discover whether a chimpanzee could acquire signs of ASL without human intervention. We reasoned that the most likely place for cultural transmission across generations to occur would be in the mother-infant relationship. In 1978 Washoe became pregnant, her infant was born and, due to a tragic set of events brought on by unsafe environmental conditions, only lived for two months. Each day for three days after his death, Washoe would ask Roger, BABY? Washoe used both repetition of the BABY sign for emphasis as well as the eye gaze and gestural modulators used to indicate a question (she raised her eyebrows and held the last BABY sign). Roger replied BABY DEAD, BABY FINISHED, and Washoe would move away and drop her arms from the sign BABY which she had held throughout the interchange. Washoe withdrew from social interactions and didn't eat. In human terms, she showed the clinical symptoms of depression. (A similar behavior in wild chimpanzees has been reported by Jane Goodall.) Ten days after the infant's death, Roger was able to bring Loulis from Yerkes Regional Primate Center at Emory University to join Washoe and our project. When he first introduced Loulis to Washoe, Roger entered Washoe's cage area alone and signed BABY FOR YOU. Washoe hooted with excitement and began signing BABY BABY BABY repeatedly. Roger then left and returned with ten-month-old Loulis. At the sight of him, Washoe's arousal level lowered, she stopped her excited signing and looked mildly interested as she once more signed BABY. Yes, this was a baby but certainly not her baby. Loulis stayed clear of Washoe as well; she was large and a stranger. Loulis' standoff lasted all day and into the night, even though Washoe attempted picking him up during the day, initiated a tickle/play session with him, and tried to put him in her nest that evening. Loulis rebuffed her attempts and that evening climbed down and slept alone and away from Washoe. At 4:00 a.m. Washoe awoke, bipedally swaggered and stomped, backhand thumped the cage, and generally created a terrible racket (the chimps were housed in a corrugated metal barn). This noise woke Loulis, and when he awoke, frightened, Washoe signed COME HUG. Loulis' response was to jump into the closest hairy arms, which happened to be Washoe's. Her response was to take him to her nest, and the relationship has been strongly bonded ever since.

Loulis began imitating his first sign eight days after he was with Washoe (Fouts et al., 1982). This sign was one that Washoe used for George Kimball. The sign was originally a quiet sign, a name sign, GEORGE: the place for the sign was the back of the head, the hand configuration was a *g* hand (thumb and forefinger extended), and the movement was a *g* hand stroked down the back of the head as though to imitate the long hair that was fashionable in the early 1970s. Washoe frequently made this sign with a flat or *b* hand (palm open, all fingers extended). George was in charge of many breakfast meals, and his attention was frequently requested. A quiet sign is not very useful when someone's back is turned and Washoe adapted his name to a noisy slapping on the top of the head. This sign proved to be so effective in getting George's attention when he was not looking that Washoe began to use it to refer to people who did not have name signs and finally to anyone who was not looking at her. This has since been referred to in our lab as the HEY YOU! or PERSON sign and is used when Washoe or another chimpanzee is soliciting the attention of someone whose name sign is unknown, or who is not paying attention to the signer. Loulis first used this sign in the breakfast context. Later, he not only used it for humans but generalized its use to chimpanzees as well. He had overextended the sign much as a human child will do. For example, a human child learns the word *daddy* and for a time calls all men *daddy*. The generalized attention-getting sign (HEY YOU!) has been acquired by Moja, Tatu, and Dar from Washoe and Loulis and is used in chimpanzee-to-chimpanzee interactions as well as in chimpanzee-to-human interactions.

Given his rapid sign acquisition, Loulis had obviously missed many important months of sign exposure. Loulis began using two-sign combinations at 15 months, such as HURRY-GIMME and PERSON-COME.

We have observed three different types of tutoring by Washoe through the course of Loulis' sign acquisition. The following observation is of shaping and was recorded during the first eight days that Washoe and Loulis were together. It seemed to focus more on the comprehension of a sign than on its production. For their first three days together, Washoe would orient toward Loulis, sign COME to him, approach him, and then grasp his arm and retrieve him. For the next five days the sequence remained the same except that the last component of retrieval dropped out. Then, after the first eight days, Washoe stopped approaching Loulis and would simply orient toward Loulis and sign COME, and he would respond to her command.

Molding as a method of tutoring has been observed in several situations. One example occurred when Washoe was excitedly waiting for some special food and was food barking while signing FOOD in excited repetition. Loulis was sitting next to her, watching her. Washoe stopped signing, took Loulis' hand, molded it into the food configuration and touched it to his mouth.

Modeling, a third tutoring method, was observed when Washoe placed a toy chair in front of Loulis and demonstrated the CHAIR-SIT sign to him five times; we have never observed Loulis to use the CHAIR-SIT sign; he has acquired the FOOD sign.

Observational learning has played an increasing role in Loulis' acquisition of signs from Washoe and the other chimpanzees. Most of Loulis' signs appeared as delayed imitations of signs Washoe had modeled.

By using a nonintervention approach to our observations, we have been able to discover a consistency in how Loulis acquired some of his signs. For example, the signs TICKLE, DRINK, and HUG began as imitations of Washoe's signs. Loulis then went through a period where he used the correct PCMs in a variety of different contexts before he began to use them appropriately in context. If the context was consistent with the semantic referent of the sign, then the sign was judged as appropriate. We recorded the observations of interactions where he played with the sign, but we did not consider the sign to be in his vocabulary until he began to use it in an appropriate context and until it was observed in an appropriate correct context by three different observers.

Loulis' sign for TICKLE began as an imitation of one of Washoe's signs, later progressed to a sign with which he played, and eventually became a sign used in the correct context. Washoe often uses this sign with humans, in Loulis' presence, to solicit a scratch, a tickle, or a touch through the wire. Washoe first signs TICKLE and then presses her side against the caging so the human can tickle her. Loulis first began imitating Washoe by pressing his body to the caging, but he did not sign TICKLE. Usually the humans would tickle him if he did this. Later he began to imitate her TICKLE sign but did not press his body to the caging. Then he went through a period where he would play with the sign by signing TICKLE repeatedly to himself, but he still did not seek a tickle from the human. Finally, he began to sign TICKLE and to press himself against the caging for a tickle from humans.

Thirteen of Loulis' first 22 signs went through a transition of this sort before he used them in their appropriate contexts. In addition to achieving the correct PCM for each sign, the sign had to be associated with the appropriate nonverbal criteria. The criteria for each behavior are based on our taxonomy.

Loulis also acquired other behaviors from the chimpanzees in his group. For example, he learned to build a sleeping nest with blankets and occasionally with willow branches in the same manner as Washoe. He learned to use bowls and spoons as implements in the same manner as Washoe and the other cross-fostered chimpanzees with whom he resides.

In addition to Washoe, the other signing chimpanzees have had an influence on Loulis' acquisition of signs. When Loulis was 18 months old, Moja, a seven-year-old female chimpanzee cross-fosterling from the Gardners' second project, joined Washoe and Loulis.

By 36 months, when Loulis was using at least 28 different signs, two more signing chimpanzees were introduced to the group: Dar, a male 4 years 9 months old, and Tatu, a female 5 years 4 months old. The addition of Dar and Tatu gave Loulis even more to sign about. At that time he was the primary initiator of signs, and 90% were addressed to his mother. During the next eight months he initiated 1,292 signs. Fifty-four percent were to Washoe and 27% to Dar, his first and only same-sex peer. By the age of 5 years 3 months, Loulis had acquired 47 signs (Fouts et al., 1984).

ASSESSMENT OF SIGN ACQUISITION

As previously mentioned, in our research we use ethological observation techniques to record the chimpanzees' behavior and signing. We humans observe the chimpanzees either in a live observation situation or on previously recorded remote videotapes; other than this, the human role is primarily that of a domestique. In regard to this research, the humans involved in the project were in no way supposed to intervene with Loulis' acquisition of signs. Therefore, it was decided that in Loulis' case a human-structured testing situation would be inappropriate because it would certainly influence his signing behavior and consequently would provide an inaccurate picture of his natural acquisition of signs and their use. It was decided that a better test of Loulis' sign use and comprehension could be obtained from his interactions with the other chimps when humans were not involved in the signing or even present. We have been able to judge the correctness of Loulis' signing by determining whether or not a sign was used in a contextually appropriate manner and if Loulis' responses to the signs of the other chimpanzees were behaviorally appropriate.

Most behaviors we have observed in our lab have been found to be very similar to those described by Jane Goodall (1986). For example, the grin, the *whoo* face, backhand thump, and bipedal swagger are frequently observed in our lab. Other behaviors seem to be unique to this group of chimpanzees. By carefully recording how a sign is made and the nonverbal behaviors associated with it, we are able to determine the context, and the appropriateness of its use.

The following is an example of a play interaction that involved Tatu and Loulis. It is a play interaction because the participants exhibited behaviors categorized as indicative of play: for example, a play face and a play slap. Tatu also had previously met the reliability criteria for the use of the CHASE sign; that is, she had been observed, by three different observers, to use this sign in the correct context for 15 consecutive days. The CHASE sign is a context-specific sign. It is only appropriate in the play context and in an affinitive social context. For example, it is not appropriate in a discipline or eating situation. In this interaction Tatu signs CHASE and Loulis responds with play slap; he then quadrupedally approaches her, grasps her foot, and begins to pull her, and Tatu responds with a play face, orients to Loulis, and again signs CHASE. Tatu signs CHASE again, with an open mouth play face, and continues to be oriented toward Loulis. Tatu is then pulled along the floor by Loulis, and again Tatu signs CHASE, still with an open mouth play face. This low-arousal play interaction continued in much the same fashion for ten more minutes. It finally ended with Loulis moving away from Tatu and getting a drink from the spigot.

In addition to live observations, we have found remote videotaping, without humans present, to be an excellent way to begin to understand how and where the chimpanzees use their signs. During the summer of 1983 we began recording chimps' interactions using a remote video recorder. This procedure was repeated in the summer of 1984, 1985, and 1986. In essence, when we left the chimps alone and recorded any signs that occurred, we were testing the robustness of the phenomena. The remote videotape recording of their activities was an extension of the double-blind test, in that we were able to test whether or not they would sign to each other without humans present, and we were able to determine the appropriateness of their sign usage by the criteria of nonverbal behaviors, appropriate context, and the PCM of the signs.

The first use of this method was made when the chimps were confined to one room during annual maintenance of the caging and floors (Fouts, 1984).

The daily routine of the chimps was maintained: three meals and snacks per day, magazines with pictures, dress-up clothes, tree branches, and various other enrichments. We restricted human contact only during remote videotaping periods, at which times humans were out of sight and silent.

We used three cameras, with 75% of the total room recorded by the cameras. We recorded the chimpanzees' activities three times a day for 15 days. Each session lasted for 20 minutes. Each summer there were 45 twenty-minute sessions.

Each remote videotaping session was analyzed for two major communicative categories: nonverbal and sign. To determine if a gesture was indeed a sign, we viewed the entire interaction and then, using our context descriptions, categorized the interaction into a context. The context was defined by the behaviors that occurred within the interactions. For example, behaviors that help to identify a play situation are the foot extension (which is an invitation to a game of chase or play), the play face, the laugh, and the head bob (also found in greetings). If a sign was a contextually specific one, then we asked these questions: Was it used in the appropriate context? Did the sign meet the PCM criteria? Was the sign associated with appropriate nonverbal criteria?

These remote videotaping sessions without humans present served as blind tests for the appropriateness of Loulis' signing as well as for his sign comprehension.

In regard to more traditional tests in a structured situation, if an organism is presented with photos of a flower and signs FLOWER, that would be scored as correct because it was used in the correct context, that is, in the presence of a photo of a flower. Therefore, the traditional vocabulary testing technique is based on whether a sign is used in the correct context. With our procedure we were able to determine contextual correctness of signs, with proper controls for cueing, in a situation that was not structured by the experimenter but initiated by the chimpanzee.

What we found in this first analysis of the 45 twenty-minute remote videotaping sessions taped in 1983 was that during the random times sampled, the chimps rested, groomed, looked at the pictures in magazines, made nests with willow and apple branches, played with water balloons, put on dress-up clothes, and used toothbrushes to brush their teeth.

From the 15 hours we found that the interactions Loulis initiated accounted for 4 hours 13 minutes; of those, 43 minutes 20 seconds contained signs.

In the wild, chimps five years of age have been observed to spend more time with a same-sex peer and less with the mother (Goodall, 1971). One question we addressed was would Loulis display this behavior? And would he sign more to his peers than to his mother?

Loulis not only spent more time with his same-sex peer, Dar, he signed more to him. He played and socially interacted more with Dar than with any other chimp. However, when he got in a fight or became upset, he still sought reassurance from Washoe. Of the 451 signed and nonsigned interactions, Loulis addressed 40% to Dar. Of the 206 signs Loulis used, 55% were addressed to Dar. Loulis used 19 of his 47 signs.

The other chimps signed to Loulis and to each other as well. In one of the 20-minute sessions, there were 29 chimpanzee-to-chimpanzee signed interactions recorded. All of these chimpanzee signs were conversational without visible extrinsic rewards. Signed conversations were made up of one or more signed utterances and two or more turns.

All of the signing conversations have included some communicative behaviors in addition to signs (Fouts et al., 1984). For example, the following is a conversation that occurred during a play interaction between Loulis and Dar:

Turn 1 - Utterance 1: Loulis solicited a water balloon from Dar by signing HURRY HURRY while holding his hand toward the balloon.

Turn 2: Dar moved away from Loulis.

Turn 3 - Utterance 2: Loulis signed WANT to Dar.

Turn 4: Dar again moved away.

Turn 5 - Utterance 3: Loulis signed HURRY HURRY GIMME to Dar.

Turn 6: Dar terminated the interaction by putting the balloon into his mouth and turning away from Loulis, who also withdrew. This conversation had three signing utterances and six turns. In regard to the number of signs, it contained four separate signs, HURRY, WANT, HURRY, and GIMME. Both in Turn 1 and in Turn 5, HURRY was recorded as a repeated sign but not as two separate HURRY signs, since repetition of signs is a form of modulating semantic meaning (Van Cantfort and Rimpau, 1982). (Later, in reference to a piece of the same balloon, Loulis signed GUM GUM on Dar and then signed GUM on himself. Dar continued to chew on the balloon and Loulis did not get it.)

In our analysis of the remote videotapes, we found that the chimpanzees signed about things and events not present. A few examples are (1) Tatu signed TIME CLEAN to Washoe before a lunch meal should have been served, (2) Tatu signed BLANKET during the usual time for bedding down, and (3) Moja, while playing with Loulis' feet, signed SHOE.

We have also recently begun to use the remote videotapes to analyze how the chimps can affect the meaning of a sign by modulating it. For example, Loulis and all of the chimps use eye gaze, repetition, and establishment of loci (a change in the location or placement of the signs as a modulator). For example, during an interaction before lunch, Tatu was recorded on remote videotape as using both repetition for emphasis and the eye gaze and gestural modulators used to indicate a question. Tatu approached Washoe and signed TIME TIME TIME TIME EAT EAT. As she signed, Tatu looked steadily into Washoe's eyes and held the last EAT sign. We have also observed the use of question modulators by Loulis in his signing to the chimpanzees as well as in his signing to humans. The following is an example of an oft-repeated interaction with humans who are hosing down the cages during cleaning. With his eyebrows raised, Loulis approaches the person cleaning, looks straight into the person's eyes, and then signs DRINK, or THAT HURRY GIMME, or HURRY HOSE, holding the last sign of the utterance until the person says "OK" or "Go ahead" and allows Loulis to drink from the hose.

We are also analyzing the remote videotapes for occurrences of the chimpanzees signing to themselves. We refer to this as private signing, and we are now making a formal analysis of this type of signing. Some early examples of this signing follow: Dar would sign PEEK A BOO to himself while lying on the bench. While looking at pictures in a magazine, Loulis signed THAT FOOD. Loulis signed HAT while putting a wooden block on his head. Washoe signed BOOK after closing a magazine and tucking it under her arm. Tatu, while sitting alone on a bench, signed APPLE.

In addition to Loulis acquiring signs from the older chimpanzees, they too have acquired new signs from each other. For example, Washoe's sign for APPLE is based on a more generalized sign for FRUIT, *c* hand (fingers slightly curved toward palm) brushes cheek; Dar, Tatu, and Moja use the more specific APPLE made with an *x* hand, with the knuckle touched to the cheek; Washoe now uses both versions. Washoe uses COVER for blankets; Dar, Tatu, and Moja use the newer version, BLANKET, pulling up of covers; Washoe now uses both and will translate. One evening Roger asked Washoe if she were ready for her BLANKET, using the newer BLANKET sign, and Washoe responded in the affirmative using a repeated older COVER sign.

What these remote video recorded samples have shown us is that communication prospers in a socially enriched environment; Loulis and the other chimps signed to each other, they asked and answered questions, they commented on objects both in and out of their immediate environment. With proper controls for eliminating any chance of cueing, we have been able to determine the contextual appropriateness of Loulis' sign usage in an unstructured social situation.

In our laboratory we have demonstrated that chimpanzees can learn new signs from other chimpanzees. We have demonstrated that they sign to each other in conversations when no humans whatsoever are present. We have demonstrated that they sign to each other with such abundance that the signing interactions in only 15 hours of video can be analyzed for context, initiator, terminator, individual signs, number of signs, and number of conversational turns. We have also demonstrated that the signs used during interactions are so clear that we can obtain 93% agreement from observers fluent in American Sign Language. We have recently analyzed 56 hours of remote video recordings for instances of individual chimpanzees signing when alone. We found 367 utterances that fall into the ten categories used for private speech in human children (Furrow, 1984). These results indicate that, much like humans, chimpanzees talk to themselves in sign language.

REFERENCES

Fouts, D. H. 1984. Remote video taping of a juvenile chimpanzee's sign language interactions within his social group. Unpublished master's thesis, Central Washington University, Ellensburg, WA.

Fouts, R. S., D. H. Fouts, and D. Schoenfeld. 1984. Sign language conversational interactions between chimpanzees. *Sign Language Studies* 42:1-12.

Fouts, R. S., A. Hirsch, and D. H. Fouts. 1982. Cultural transmission of a human language in a chimpanzee mother/infant relationship. In H. E. Fitzgerald, J. A. Mullins, and P. Page, eds., *Psychobiological Perspectives: Child Nurturance Series,* vol. III, pp. 159-193. New York: Plenum Press.

Furrow, D. 1984. Social and private speech at two years. *Child Development* 55:355-362.

Gardner, B. T., and R. A. Gardner. 1971. Two-way communication with an infant chimpanzee. In A. Schrier and F. Stollnitz, eds., *Behavior of Non human Primates,* vol. 4, pp. 117-184. New York: Academic Press.

_____. 1978. Comparative psychology and language acquisition. *Annals of the New York Academy of Sciences* 309:37-767.

_____ 1984. A vocabulary test for chimpanzees *(Pan troglodytes). Journal of Comparative Psychology* 98:381-404.

Goodall, J. 1968. The behaviour of free-living chimpanzees in the Gombe Stream Reserve. *Animal Behavior Monographs* 1:163-311.

_____. 1971. *In the Shadow of Man.* Boston: Houghton Mifflin Company.

_____. 1986. *The Chimpanzees of Gombe.* Cambridge: Harvard Univ. Press.

Stokoe, W., D. Casterline, and C. Croneberg. 1965. *Dictionary of American Sign Language.* Washington D. C.: Gallaudet College Press.

Van Cantfort, T. E., and J. B. Rimpau. 1982. Sign language studies with children and chimpanzees. *Sign Language Studies* 34:15-72.

SPONTANEOUS PATTERN CONSTRUCTION IN A CHIMPANZEE

Tetsuro Matsuzawa

The present study demonstrates that a chimpanzee can learn language-like skills of combining "words" and also constructing the "words" from the elements. The chimpanzee spontaneously developed a favorite order in the "word" combination and construction, although no particular sequence was required by the research. The meaning of the order was analyzed in relation to the phylogenetic constraints of the chimpanzee's cognition. The implications for human language are also discussed.

Ai, a 10-year-old female chimpanzee, served as the subject in the following three experiments: number/color/object naming, individual recognition, and pattern reconstruction. She has learned a language-like skill aided by a computer system. When training started in April 1978, her age was estimated to be about 1.5 years.

A computer-controlled console terminal was used in these experiments. For the chimpanzee, the "words" are represented by symbols: geometric figures, arabic numerals, and letters of the alphabet (Fig. 1). Some "words" can be disassembled into the elemental figures ("graphemes"). Each symbol and each element was drawn on a key (2 x 2.5 cm) and could appear in various positions in a maximal 15 x 7 key matrix on the console attached to one wall of the experimental room (190 x 220 x 180 cm). This console interfaced with a computer (PDP11/V03 or NEC-PC9801) that controlled the experiment and recorded key choice. The experimenter sitting outside interacted with the chimpanzee through a display window located above the console.

"WORD-ORDER" IN NUMBER/COLOR/OBJECT NAMING

Before the present experiment, Ai had been trained to name 14 objects (Asano et al., 1982) and 11 colors (Matsuzawa, 1985a) by choosing among the set of symbols shown in Figure 1. The chimpanzee also mastered numerical naming from one to six by pressing the keys of arabic numerals (Matsuzawa, 1985b; Matsuzawa et al., 1986).

How could Ai's acquired skills of number/color/object naming be combined together? To answer this question, the chimpanzee was required to name the number, color, and object of 125 types of samples.

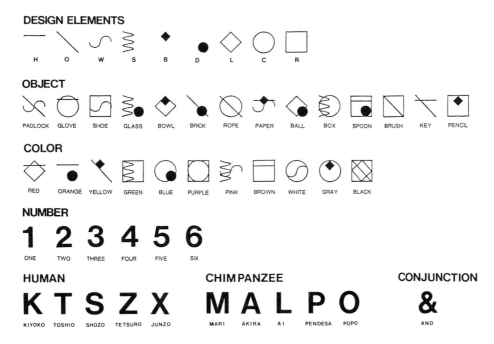

Figure 1. Symbols and design elements (graphemes) used in these studies. They were usually white figures on a black background.

Fifteen keys, each of which represented one of five numbers (1, 2, 3, 4, 5), five colors (red, green, blue, yellow, black), and five objects (pencil, paper, brick, spoon, toothbrush), were operative at one time. The sequence in which the keys were depressed was free, but the chimpanzee was required to select the three keys correctly describing each of the three attributes of 5 x 5 x 5 = 125 types of sample items. For example, when five red pencils were shown in the display window as a sample item, it was necessary for Ai to press keys of "5," "red," and "pencil" in any order (Fig. 2).

A session ended when 100 rewards had been delivered. Each reward required two consecutive correct trials. When the chimpanzee pressed an incorrect key, the "error" sound followed and the trial ended. When she pressed a correct key, she could proceed to the next choice. The pressed key and the keys belonging to the same attribute were darkened and inoperative, so that a repeated response such as "red/red" or "red/green" was blocked. It must be noted that the chance level of obtaining a reward is extremely low, that is, one out of 15,625 times. Nevertheless, in the final block of three sessions of the task, the chimpanzee received 604 problems during 127 minutes and made only three mistakes (99.5% accuracy).

Although no particular "word order" was required, the chimpanzee favored two particular sequences almost exclusively among six possible alternatives: color/object/number and object/color/number (left panel of Fig. 3). In both sequences, numerical naming was always last.

Figure 2. A view of the chimpanzee's console terminal in number/color/object naming. The experimenter had shown two blue spoons through the display window. Ai had just answered the color name (blue) and the object name (spoon) and was about to press the key for the number (two). On the front screens, the symbols chosen by the chimpanzee ("blue" and "spoon") were facsimiled.

The favored sequences continued to be used for naming 125 new sample items with "new color" (pink, brown, purple, white, and grey) and 50 new items with "new object" (padlock, glove, glass, ball, and key). Although these new samples had never been used in three-term naming, or even in simple numerical naming, the chimpanzee correctly described the three attributes of the samples by using the favored "word order." In the test of "limited number," the number of sample items was not varied but restricted to one of the five alternatives within a session. Decreasing the difficulty of numerical naming by using the same number of items in a given session had little effect on the favored sequence (right panel of Fig. 3 and details reported in Matsuzawa, 1985b).

Finally, the persistence of the favored sequences was proved in the "retest" repeating the original task after 16 months pause. The chimpanzee received various kinds of cognitive tasks, including numerical naming from one to seven but no training for object and/or color naming during the period. The results showed that the chimpanzee retained the "word order" in spite of the possible interference during the long delay interval (Fig. 3).

What is the determinant of the "word order"? The "readiness" with which the attributes are named might be the main determinant. Number names were learned

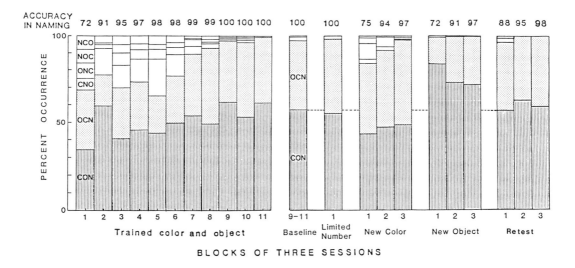

Figure 3. Percentage of six possible sequences in the three-term naming. N, number; C, color; O, object. For example, NCO refers to trials in which the sequence "number/color/object" was used correctly. The percentage of trials on which all three attributes of the sample items were named correctly is shown at the top of each column. The left panel shows the change of use of the six possible sequences as a function of three-term-naming training. The right panel shows the data of the tests investigating the determinant of the particular sequences. See the detailed explanation in the text. (Based on Matsuzawa, 1985b).

more slowly than were the names of objects and colors. Accuracy in numerical naming was always lower than that in object or color naming in the three-term naming task throughout the sessions. Numbers were named in the last position in the "word order" probably because of the relative difficulty in naming (Matsuzawa, 1985b). The difficulty, or the "readiness," of naming attributes of the perceptual world might depend on the ecological demands of a species. In the ecological niche of the chimpanzees, the visual discrimination of shape and color (size, depth, texture, and so on) might be important. The discrimination of number, however, might have lesser survival value for the chimpanzees.

Besides the speculation for the cause of organizing a specific sequence, two points should be noted. First, the chimpanzee spontaneously developed the favorite "word order," although no particular sequence was required. Second, the "word order," once established, was used to name a new set of items and persisted for a long time.

INDIVIDUAL RECOGNITION:
Naming Humans and Chimpanzees

Ai, at the age of six, learned to distinguish perfectly each letter of the alphabet in a matching-to-sample procedure, with 26 letters as choice alternatives (see details in Matsuzawa, in preparation).

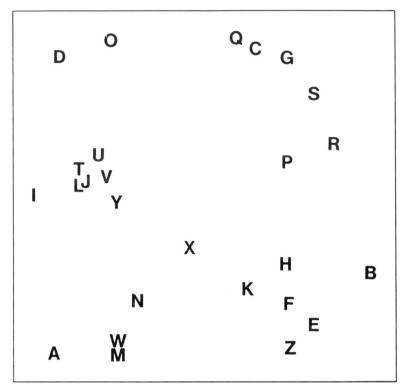

Figure 4. Two-dimensional scaling of the similarity of letters perceived by a chimpanzee. The distance between each pair of letters correlates to the perceptual dissimilarity of each pair. (Based on Matsuzawa, in prep.)

Each letter of the alphabet was drawn on a key in a 5 x 6 key matrix on the console. The letter format was the standard "Helvetica medium" (Letraset No. 721, 48 point in size). The experimenter exposed, one at a time, letters of the alphabet drawn on white cardboard (B6 size). The chimpanzee was required to press the matching key among 26 alternatives. The position of the letters on the console was changed from session to session to prevent the chimpanzee from memorizing positional cues. A daily session consisted of five blocks of 26 trials. The correct choice was rewarded.

Although the chimpanzee had never seen the letters of the alphabet previously, her accuracy was 41.5% in the first session (chance level is below 4%). The repeated training improved her performance to 98.0% accuracy in the 23rd session.

A sample-choice "confusion matrix" was used for two-dimensional scaling of the similarity of the letters. The result is shown in Figure 4. The distance between each pair of letters on the matrix corresponds to the perceptual dissimilarity of each pair. Where two letters are far apart, the chimpanzee seldom confused them. Where two letters are close together, the chimpanzee frequently confused them. The configuration is quite similar to those obtained in humans (Podgorny and Garner, 1979) and in pigeons (Blough, 1982). The product moment correlation of the chimpanzee data and the human data for the same letter format was .61, comparable to the correlation of .59 between different data on humans (Podgorny and Garner, 1979).

Once the letter discrimination was established, the chimpanzee was taught a symbolic matching-to-sample task in which a photograph of ten familiar individuals, five humans and five chimpanzees, had to be matched to an arbitrarily assigned letter (Fig. 1). The letters K, S, T, X, Z were used to represent five humans, and the letters A, L, M, O, P to represent five chimpanzees. The average confusion among the assigned letters is roughly the same in both humans and chimpanzees. Ai has been acquainted with these individuals, including herself, for at least 1.5 years so that each is very familiar to her.

While exposing the picture (a front view of the face of each individual), the experimenter lighted the ten keys on the console containing ten corresponding letters. The chimpanzee was required to press the matching key. Ai quickly learned to associate each letter with each picture within two sessions. Then, 16 different sets of photographs of the same individuals (including different views of face, upper half body, and whole body) were prepared for the transfer test. Table 1 shows an analysis made on the chimpanzee's 169 errors (in 1,221 trials in the transfer test). Confusion among humans was more frequent than confusion among chimpanzees. Confusion between humans and chimpanzees was very rare ($\chi^2 = 116.7$, p < .001 in chi-square test). Moreover, between-species confusion was restricted to cases in which photographs of humans were named as chimpanzees, never the reverse.

The accuracy of individual recognition reached a high level (more than 95% on the average) after repeated transfer tests using new sets of photographs. Then the chimpanzee was required to name two individuals in free order by pressing a conjunction key "and" when two photographs were presented at once.

Table 1. Frequency of types of confusion appearing in individual recognition, based on Ai's incorrect responses in the transfer tests for naming individual humans and chimpanzees. (Based on Matsuzawa, in prep.)

Incorrect naming	Photographs shown as samples	
	Humans	Chimpanzees
Incorrect human names	119 (70%)	0 (0%)
Incorrect chimp names	12 (7%)	38 (23%)

Table 2. "Word order" in naming two individuals. Column I shows the rank order resulting from the percentage of trials on which the individual was given first position (in other words, the priority defined as the position in the naming). Column II shows the rank order resulting from the accuracy of naming each individual. Column III shows the rank order resulting from the accuracy of identifying each letter in the beginning of letter-discrimination training. The data is based on the first 50 presentations of each letter.

Name	Symbol	Genus	Sex	I Priority in order (rank)		II Individual recognition (rank)		III Letter identification (rank)	
Ai	L	Pan	f	1	.630*	4	.951	9	.260
Akira	A	Pan	m	2	.627*	1	.976	2	.940
SHOZO	S	Homo	m	3	.583	10	.871	4	.720
Pendesa	P	Pan	f	4	.563	8	.889	7	.320
Popo	O	Pan	f	5	.520	1	.976	3	.820
Mari	M	Pan	f	6	.514	6	.890	5	.600
KIYOKO	K	Homo	f	7	.508	9	.875	8	.300
TOSHIO	T	Homo	m	8	.473	3	.963	6	.340
JUNZO	X	Homo	m	9	.310**	6	.890	9	.260
TETSURO	Z	Homo	m	10	.282**	5	.902	1	.980

Notes: * = $p < .05$, ** = $p < .01$ in binomial test.

Two photographs of different individuals were placed side by side in front of the chimpanzee. The 45 possible pairs of 10 individuals were presented randomly and equally often in a session. The left and right positions of the photographs were balanced. The chimpanzee was required to name two individuals in such a way as "A and B" where the order was free.

The data of four test sessions, 410 trials in total, were combined. The naming of each of 45 pairs was tested nine times on the average by using trial-unique photographs. Ai quickly developed a favorite sequence "A/B/and" among three possible alternatives to name the two individuals. The trials in which both individuals were named correctly were 350 out of 410 (85.4%). The average accuracy of individual recognition was 90.0% in humans, and 93.6% in chimpanzees.

Further analysis was done to examine the order of describing the two individuals. The percentage of trials in which the individual was given first position was calculated for each individual (Table 2). Although no particular sequence was required, Ai most often described herself first ($p < .05$ in binomial test) and "Tetsuro" (the experimenter) last ($p < .01$). The data might show the "ego-centricity" in an affective sense of self-centered choices or preferences. In general, chimpanzees rather than humans are given high priority in the naming.

The priority is not simply determined by the accuracy of naming each individual. The rank order of the accuracy of naming each individual is not congruent with the rank order of the priority ($r = .023$ in Kendall's rank correlation test, non-significant). Neither does the priority reflect the possible preference for the "names" (i.e., letters). Although the chimpanzee ultimately discriminated among the 26 letters at the level of 98% accuracy, she had shown differences of accuracy among the letters in the beginning of training. The rank order of the accuracy of discriminating each letter is not congruent with the rank order of the priority in two-individual naming ($r = .090$, non-significant, Table 2).

There were 25 pairs in which one human and one chimpanzee were presented as sample photographs. In these cases, a further analysis was done to examine the "species-centricity" in the sense of species-centered choices or preferences. The preference for naming chimpanzees before humans is shown in Table 3.

In a comparable test situation, human subjects were asked to judge the similarity among the pictures and concluded that the pictures of the chimpanzees were fairly confusable. For human subjects, the chimpanzees "all look the same" while the humans are different in many respects, such as hair style, moustache, wearing a pair of glasses, and so on. They also showed the tendency to name humans first before chimpanzees although the naming was much influenced by the left-right scanning sequence.

The results can be summarized as follows: The discrimination among pictures of chimpanzees was more accurate than discrimination among pictures of humans. When she was presented with a pair of pictures, one a human, the other a chimpanzee, Ai tended to name the chimpanzee first. Moreover, the pictures of Ai herself were especially likely to be named first. Ai spontaneously developed the "species-centric" and "ego-centric" order although no particular sequence was required. The results seem to reflect the phylogenetic constraints in individual recognition in the chimpanzee.

Table 3. The temporal order of naming two individuals. The preference for naming chimpanzees before humans was shown in the following two indices. The first was the number of correct trials in each of the two temporal orders. The second was the number of the pairs in which the temporal order was used more frequently. (In two out of 25 pairs, both were equally used.)

Temporal order of naming	Trials **		Pairs *	
Human and Chimpanzee	73	(37%)	5	(20%)
Chimpanzee and Human	122	(63%)	18	(72%)

Notes: * = $p < .05$, ** = $p < .01$ in binomial test.

Figure 5. A view of the chimpanzee's console terminal in "constructive matching-to-sample" task. A three-elements complex figure was presented as a sample. Ai had just chosen (H)orizontal line as a first element and was going to press the key for (C)ircle as a second element. Pressing a key produced a superimposed feedback facsimile of the elemental figure on the screen just under the sample.

PATTERN RECONSTRUCTION:
Constructive Matching-to-Sample from Graphemes

For the chimpanzee, the "words" are represented by symbols such as geometric figures, arabic numerals, and letters of the alphabet. Many researchers have shown that the great apes can combine the "words" into a "sentence" although the competence is still controversial (Gardner and Gardner, 1969; Premack, 1970; Fouts, 1973; Rumbaugh, 1977; Patterson, 1978; see criticism by Terrace, 1979). Ai also combined the "words" and spontaneously developed a consistent "word order" to describe the perceptual world.

Human languages have a "duplex" or two-level structure, that is, the sentences created from the words and the words created from the phonemes or the elements. Chimpanzees do not, so far as is known, construct "words" by assembling the explicit elements. In the writing system of human language, words consist of elements such as letters. For Ai, some of the "words" using geometric figures can be disassembled into nine fundamental elements called design elements (graphemes). Ai showed the ability to construct these "words" from the graphemes (Matsuzawa, 1986; Matsuzawa and Fujita, in preparation).

Ai, at the age of 7.5 years, was trained to do a task named "constructive matching-to-sample." The task is to create copies of samples from the elements. She had been trained to do "physical matching-to-sample" in which the subject is simply required to choose a "same" item matching a sample. Before the present study, Ai had experienced the physical matching-to-sample tasks using colors, geometric figures, arabic numerals, "Kanji" characters, household items, and pictures. In the present study, the chimpanzee was required to construct the complex geometrical figures from the elemental figures (Fig. 5).

The console terminal contained three sets of nine keys on which elemental figures were drawn: (H)orizontal line, (O)blique line, (W)ave, (S)nake, (B)lock, (D)ot, (L)ozenge, (C)ircle, and (R)ectangle (Fig. 1). Two special keys called a "request" key and an "end" key were also attached to the console.

The chimpanzee could start a trial whenever she wanted by pressing the "request" key. Pressing the key darkened it and lighted a set of ten keys in two rows containing the nine elemental figures and one blank key. At the same time, a sample figure automatically appeared on the screen of an IEE inline projector (or on the screen of a video terminal in the later stages). Pressing a key produced a feedback facsimile of the figure drawn on the key on another screen located just under the sample figure. Pressing the next key produced a superimposed facsimile on the feedback screen. The chimpanzee was required to construct a copy of the sample by assembling the elemental figures in any temporal order.

When the chimpanzee succeeded in constructing the copy, the "end" key was available to deliver a reward. When the chimpanzee did not choose the corresponding elements, the miscopied figure remained on the screen for one second with a feedback sound signalling the error. The entire process of the experiment was controlled by the computer while the experimenter observed the subject's behavior through a monitoring TV.

In the first stage of the task, the samples were 36 two-element figures exhausting the possible combinations of nine elemental figures. At the first trial of the first session, the sample was a complex figure disassembled into (W)ave and (S)nake. She repeatedly looked at both the sample figure and the elemental keys. It took 6.78 seconds to choose the first element (S)nake and another 12.40 seconds to choose the second element (W)ave. However, she failed at the second trial when a complex figure of (C)ircle and (B)lock appeared as a sample. Ai chose (C)ircle in 4.23 seconds and then chose (R)ectangle instead of (B)lock in 9.47 seconds. Her accuracy in the first block of 36 trials, in which each variant of the two-element figures appeared once, was 22%. It must be noted that the chance level of assembling the copy is less than 3%. Although the first element was easily recognized (80% correct in the first session), the second one was difficult (28% correct).

The data obtained on the first ten sessions (2,825 trials) were analyzed. The 36 variants of the two-element figures can be divided into two groups: the figures which had been used as "words" in Figure 1 (n = 22); and the figures which had never been

Tetsuro Matsuzawa

Table 4. The priority of reproducing each element in constructive matching-to-sample. The priority index was defined as the averaged position on which the element was reproduced. The nine elements are divided into four clusters and arranged in the order of confusedness within the clusters according to the perceptual similarity among the elemental figures in the chimpanzee.

| | Chimpanzee | | | | | | Human | |
| | 2-elements CMTS task (rank) | | 3-elements CMTS task (rank) | | 3-elements DCMTS task (rank) | | 3-elements DCMTS task (rank) | |
Elements								
Lozenge	5	1.42	1	1.42	1	1.59	3	1.80
Rectangle	3	1.32	2	1.61	2	1.70	1	1.40
Circle	4	1.37	4	1.83	4	1.84	2	1.67
Snake	1	1.07	3	1.64	6	1.99	6	2.12
Wave	2	1.30	6	2.09	3	1.71	7	2.19
Block	8	1.82	5	2.01	5	1.86	8	2.27
Dot	6	1.55	7	2.33	7	2.13	9	2.39
Horizontal	9	1.97	8	2.47	8	2.33	5	1.90
Oblique	6	1.55	9	2.56	9	2.59	4	1.88

Notes: CMTS = Constructive matching-to-sample.
        DCMTS = Delayed constructive matching-to-sample.

used in the previous experiments (n = 14). Although all the two-element figures were familiar to the subject, the accuracy of reproducing the named figures (37.4%) was superior to that of the unnamed figures (32.0%) in chi-square test (p < .01). The priority of reproducing each element is shown in Table 4. The priority index was defined as the averaged position on which the element was reproduced. The data revealed that the chimpanzee spontaneously developed a clear order to reproduce the elements although no particular sequence was required.

When the overall accuracy of reproducing two-element figures had reached the 99.5% level, the samples were abruptly changed to 84 variants of the three-element figures. Except for the four figures previously used for names, all other figures were new for the chimpanzee. Ai succeeded in constructing the figures at 62.5% accuracy in the first session of 168 trials while the chance level of construction was 1.2%. The priority index for each element was calculated based on the final six sessions (98.2%

correct out of 1,008 sessions, see Table 4). The order of priority differed slightly from that in the initial stage of two-element figures.

Finally, a delay was introduced between the sample presentation and the construction. When the chimpanzee pressed the "request" key, a sample was presented on a screen for only one second. The sample faded and then the keys for elemental figures were lighted. In this "delayed" constructive matching-to-sample task, no sample was available when the chimpanzee started to construct the copy.

In the first session, Ai constructed the copies without the samples (59.5% correct out of 84 trials). Her performance reached an accuracy of 94.9% in the final six sessions. Eight college students (age 18–19) received the same test with exactly the same apparatus and procedure. Accuracy among the students ranged from 72.6% to 95.2% (83.4% on the average) in 168 trials. Repeated experience slightly improved the performance of the students, but accuracy seldom exceeded 95%.

The priority of reproducing the elements was directly compared between the two species. Both species had a tendency to reproduce the sample first from the outer contours such as (R)ectangle, (C)ircle, and (L)ozenge. However, the priority of the other elements was different between the two species. The chimpanzee tended to reproduce straight lines last throughout the sessions.

According to a cluster analysis of the data obtained in physical matching-to-sample of the nine elemental figures, the figures are perceived by the chimpanzee as four clusters: straight lines such as (H)orizontal and (O)blique; solid shapes such as (B)lock and (D)ot; wavy lines such as (W)ave and (S)nake; and open shapes such as (L)ozenge, (C)ircle, and (R)ectangle. The confusedness within the clusters decreased along the above order.

The order of reproducing the elements might be partially determined by the confusedness within the clusters. Since the discrimination between (H)orizontal and (O)blique lines is more demanding in the chimpanzee than in the humans, the chimpanzee gave the two straight lines the last position of reproduction.

The spontaneous order shown in the construction illuminates some common perceptual/cognitive processes found in both human and chimpanzee subjects. On the other hand, the slight differences that appeared in the order of construction might reflect phylogenetic constraints on the ability to disassemble the complex figures into the components and to reassemble them.

DISCUSSION:
Pattern Construction and Its Implications for "Grammar"

The present study with Ai showed that language-like skills can be learned by chimpanzees on two levels. One is to construct a "sentence" (or a "phrase") from "words." Another is to construct a "word" from the design elements (graphemes). In both levels, the chimpanzee spontaneously established a consistent temporal order when no particular order was required.

The spontaneous organization described above cannot be explained in terms of "differential reinforcement" because no particular organization was required for reinforcement. It seems to reflect the internal perceptual/cognitive structure of the chimpanzee.

The tests for color perception, form perception, and visual acuity revealed no fundamental difference between Ai and human subjects (Matsuzawa, 1985a; Matsuzawa, in preparation). However, Ai showed a peculiar organization in the tasks demanding the construction of a sequence. She spontaneously organized her favorite temporal orders. Her "rules" were identified in her favorite orders: naming the number last in three-term naming, "ego-centric" and "species-centric" order in naming individuals, and constructing the outer contours first in pattern reconstruction.

In human languages, similar rules have been identified. Greenberg (1966, 1978), looking for language universals, found preferred orders. For example, he reported that the temporal order of the ordinary declarative sentence consisted of (S)ubject, (V)erb, and (O)bject. Although there are six possible orders for the (S)-(V)-(O) combination, the main order of the declarative sentence is SVO or SOV type in most human languages. A few languages belong to the VSO type. There appear to be no languages whose main order is OSV, OVS, or VOS. The universal rule for the declarative sentence can be summarized as a simple rule of "subject first, and then object."

In the human cognitive process, when we recognize an episode consisting of agent, action, and object, the agent might be spontaneously recognized before the object on which the agent acts. The orders spontaneously organized by Ai might reflect the cognition which is specific to chimpanzees. If so, the chimpanzee seems to utilize a rudimentary form of her own "grammar" to describe the perceptual world.

REFERENCES

Asano, T., T. Kojima, T. Matsuzawa, K. Kubota, and K. Murofushi. 1982. Object and color naming in chimpanzees *(Pan troglodytes)*. Proceedings of the Japan Academy 58(B): 118–122.

Blough, D. S. 1982. Pigeon perception of letters of the alphabet. *Science* 218:397–398.

Fouts, R. S. 1973. Acquisition and testing of gestural signs in four young chimpanzees. *Science* 180:978–980.

Gardner, R. A., and B. T. Gardner. 1969. Teaching sign language to a chimpanzee. *Science* 165:664–672.

Greenberg, J. H. 1966. *Language Universals.* Mouton: The Hague.

_____. 1978. *Universals of Human Language.* California: Stanford University Press.

Matsuzawa, T. 1985a. Color naming and classification in a chimpanzee *(Pan troglodytes)*. *Journal of Human Evolution* 14:283–291.

_____. 1985b. Use of numbers by a chimpanzee. *Nature* 315:57–59.

_____. 1986. Pattern construction by a chimpanzee. *Primate Report* 14:225–226.

_____. (In preparation). Form perception and visual acuity in a chimpanzee.

Matsuzawa, T., and K. Fujita. (In preparation). Constructing complex figures from the elements by a chimpanzee.

Matsuzawa, T., T. Asano, K. Kubota, and K. Murofushi. 1986. Acquisition and generalization of numerical labeling by a chimpanzee. In D. Taub and F. King, eds., *Current Perspectives in Primate Social Dynamics,* pp. 416–430. New York: Van Nostrand Reinhold.

Patterson, F. G. 1978. The gestures of a gorilla: language acquisition in another pongid. *Brain and Language* 5:72–97.

Podgorny, P., and W. R. Garner. 1979. Reaction time as a measure of inter- and intra-object visual similarity: letters of the alphabet. *Perception and Psychophysics* 26:37–49.

Premack, D. 1970. A functional analysis of language. *Journal of the Experimental Analysis of Behavior* 14:107–125.

Rumbaugh, D. M. 1977. *Learning by a Chimpanzee.* New York: Academic Press.

Terrace, H. S. 1979. Can apes create a sentence? *Science* 206:891–895.

SYMBOL ACQUISITION AND USE BY
PAN TROGLODYTES, PAN PANISCUS, HOMO SAPIENS

Sue Savage-Rumbaugh, Mary Ann Romski,
William D. Hopkins, and Rose A. Sevcik

Comparisons of cognitive skills between species of the great apes, though rare, are important in that they have the potential to reveal the raw material that nature had to work with during the emergence of early man. The simple basic mode of life does not differ drastically for any of the great apes, nor, in fact, are the foraging or survival strategies employed by apes obviously more sophisticated than those utilized by monkeys. Yet, if we wish to understand the evolution of the human brain, we must ask whether, and in what ways, cognitive skills may differ among extant apes.

PREVIOUS COMPARATIVE STUDIES OF GREAT APES

Rumbaugh (Rumbaugh, 1971; Rumbaugh and Pate, 1984) carried out the largest comparative study of great apes to date, presenting data on a common task (reversal of discrimination learning skills) from 15 orangutans, 15 chimpanzees, and 15 gorillas. He found large individual differences, but no significant species differences. Orangutans and gorillas tended, on the average, to show greater positive transfer during reversal tests. And while the best chimpanzee performed as well as the best gorilla or orangutan, the worst chimpanzee performed far below any gorilla or orangutan. However, there were large individual differences between the animals, particularly for the chimpanzees. The large variance, apparently the result of diverse rearing experiences, tended to obscure species variables.

Rumbaugh and Gill (1973) also tested great apes (five gorillas, five chimpanzees, and five orangutans) using a mediational object discrimination task, and again found no significant species differences. It is possible that the tests used in these studies were not sufficiently sensitive to the kinds of species differences which may exist among great apes. This view is supported by the fact that Rumbaugh and Gill also found no significant differences between the great apes and human subjects with mental retardation, even though some of the human subjects had at least minimal linguistic competence.

Language acquisition studies have been conducted with all great ape species and provide a possible metric for comparison of cognitive capacities. A casual reading of these studies suggests that the gorilla is more intelligent and conversationally skilled than other apes (Jolly, 1985). Unfortunately, the criteria used for measuring linguistic performance and vocabulary competency in the various studies differ widely. These differences make it impossible to determine whether the purported findings are a function of species' capacities or differential interpretations of data. A comprehensive research effort is needed in which apes of each species are reared in similar, if not identical, environments and are evaluated with the same data collection and test procedures.

Pan paniscus and *Pan troglodytes*

The two species of chimpanzee, *Pan paniscus* and *Pan troglodytes,* are closely related to one another, having separated only about 2.5 million years ago. While the question of cognitive differences between these species has long been of interest to scientists (Yerkes, 1929), there has been little comparative research which systematically addresses the issue. Studies which employ similar rearing and testing procedures with both species are nonexistent, and rarely has a given researcher had the opportunity to work with more than one species of great ape since the early studies of Robert Yerkes (Yerkes and Learned, 1925).

There are strong similarities among all living great ape species, and this is particularly apparent between the two species of *Pan*. These forms overlap morphologically to such an extent that anatomists did not recognize them as distinct species until 1929 (Schwarz, 1929). Even then, *Pan paniscus* was not fully described and accorded species status until 1933 (Coolidge, 1933). Indeed, unless one is trained to recognize the anatomical and behavioral differences which do exist between these forms, a member of one species is likely to be confused with the other, as has in fact happened repeatedly in zoological gardens.

There is no a priori reason to expect that one species of chimpanzee should perform better on cognitive tasks than the other. Both species lead similar nomadic existences, living in small groups which travel from place to place and build new nests each evening (Kano and Mulavwa, 1984). The unit groups of both species are flexible in that they divide into temporary parties to maximize feeding efficiency. It has been suggested, however, that the subgroups of *Pan paniscus* are somewhat more stable than those of *Pan troglodytes* and appear to be composed of males and females, as well as mothers and offspring (Badrian and Badrian, 1984; Kano and Mulavwa, 1984). Both species are primarily frugivorous, although hunting of small mammals has been documented in both species (Badrian and Malenky, 1984; Kano and Mulavwa, 1984). *Pan troglodytes* is found in a wider range of habitats, from moist forest to arid open savanna. *Pan paniscus* is found only in lowland forest and often in relatively swampy areas. Tool use of a wide variety has been documented extensively for *Pan troglodytes*, but has not been observed in *Pan paniscus*, although one account suggests that it does occur (Badrian et al., 1981). *Pan paniscus* employs a wider range of gestures and displays a greater diversity of social-sexual patterns (Kuroda, 1984; Savage-Rumbaugh et al., 1977; Thompson-Handler et al., 1984). Male dominance is much less pronounced in *Pan paniscus*, and males appear to play a greater role in infant rearing. Male-male cohesion is an important social factor among *Pan troglodytes*, while male-female bonds appear to be strongest in *Pan paniscus* (Kano, 1980, 1982; Patterson, 1979).

ISSUES IN COMPARATIVE WORK

It is often asserted that comparative evaluation of learning capacities is not possible because biological differences between species make it impossible to equate reasonably the difficulty of tasks across species. However, when comparisons are made between closely related organisms which share a common anatomy and manifest similar social behaviors, feeding strategies, and rearing patterns, these objections lose their validity. The extensive morphological and behavioral similarity between *Pan troglodytes* and *Pan paniscus* makes the technique of comparative rearing and evaluation a realistic goal.

How Should Cognitive Skills Be Evaluated?

It is doubtful that the cognitive capacities of apes can be adequately assessed through the use of standard primate learning tasks (i.e., discrimination learning, reversal learning, match-to-sample, oddity, etc.). These tasks are neither sufficiently complex nor diverse to reveal other than gross differences in competency, since even monkeys are capable of solving them. Language acquisition studies are, however, broad enough in scope and complexity to permit differences, should they exist, to manifest themselves. The acquisition and use of symbols places *no upper limit* on the skills which can be displayed, and the drive to communicate provides strong intellectual motivation. Moreover, in language acquisition studies, the subject is not constrained by a particular task that is set before him, but rather is able to display his competencies through a wide array of novel communicative devices. For these reasons, we regard the language acquisition paradigm as the method of choice for comparative studies of the cognitive capacities of *Pan paniscus* and *Pan troglodytes*.

THE LANGUAGE RESEARCH CENTER

Research on the topic of language acquisition in chimpanzees has been underway at the Language Research Center in Atlanta, Georgia, since 1970. The first ten years of the research program focused upon *Pan troglodytes*, the next six years concentrated upon *Pan paniscus,* and presently two animals, one of each species, are being reared together in a comparative study. Concurrently, since 1975, a companion project investigating the language acquisition skills of nonspeaking children and young adults with mental retardation has also been conducted.

This report focuses upon the behavioral and cognitive differences which have been observed between the two species, *Pan paniscus* and *Pan troglodytes,* when both are reared in an environment designed to foster language acquisition. Both species have learned the same graphic communication system, and both were reared and evaluated by a single research group using identical tests and criteria. This report also includes descriptions of the language acquisition and use skills of persons with mental retardation who are using the same keyboard communication system.

It is too early to conclude definitively that the differences observed between the two chimpanzee forms are due to species variables as opposed to environmental variables. It is the case that the rearing differences which have occurred inadvertently across time are far less extensive than those which exist between laboratories.

The Communication System

Much has been said about the different communicative systems employed in ape language projects (Gardner and Gardner, 1978; Jolly, 1985; Ristau and Robbins, 1982). Although it is generally observed that three different media have been used – plastic chips, graphic symbols, and manual signs – it is really only the last two media that have permitted the chimpanzee subject to go beyond the constraints of a predetermined problem and to communicate spontaneously and naturally.

All of these systems have also been employed with children with mental retardation who have difficulties producing speech (Carrier, 1974; Reich, 1978; Romski and Savage-Rumbaugh, 1986; Romski et al., 1984). While both manual sign systems and graphic symbol systems are effective for certain subgroups of children (Richardson, 1975; Romski et al., 1984), both systems have advantages and disadvantages. In the past, manual signs were employed with all children who had some movement of at least one upper extremity (Kiernan, 1977). Only children with more severe physical disabilities were assigned graphic symbol sets. More recently, graphic communication systems have been recommended for able-bodied children with severe cognitive impairments as well (Mirenda, 1985; Romski et al., 1984). Graphic systems are more easily interpreted by communicative partners (e.g., teachers and parents) because the printed English word typically appears above the symbol. In some computerized communication systems, synthetic speech is also available as feedback for the communicative partners.

The Language Research Center has elected to use a graphic symbol system (lexigrams) with both children and apes in view of the difficulties both groups of subjects evidence in producing adequately interpretable hand movements. Although some have argued that only a gestural system can be considered "natural," both we and others (Kiernan, 1977; Mirenda, 1985; Romski et al., 1984) have found that a symbol system becomes integrated into the life of the subject and becomes a "natural" means of communication as long as it is used by all members of the subject's community. The most important aspect of choosing a nonspeech communication system is to select one that subjects can acquire readily, can produce clearly, and can use effectively for communication. In our experience, subjects acquire graphic symbols more readily and produce them more clearly than manual signs. We are, in turn, able to provide more accurate responses to their communications, thereby facilitating the communicative process.

Figure 1. Kanzi communicates outdoors using a portable keyboard. When the subject touches a symbol, a speech synthesizer produces the English word. The unit is battery powered and can be carried about in the environment.

Figure 2. Mulika touches the symbol for banana on a nonelectronic keyboard. When the subject touches a symbol, no physical changes take place; however, touching a symbol is treated as the behavioral equivalent of speaking a word, and staff respond accordingly. These lightweight panels can be taken onto forest trails too narrow for the bulky, though portable, units shown in Figure 1.

The specific computer keyboard system, on which the visual-graphic symbols are displayed, has varied across time. The earliest keyboards, fixed in one location, had symbols that were illuminated and were projected above the keyboard each time a key was depressed. The only auditory feedback was the clicking noise produced by the contact within the key. The present keyboards can be freely moved about, and the keys are not depressed but are touch sensitive. Auditory feedback is provided by synthesized or digitized speech (Fig. 1). For the chimpanzees, a nonelectronic panel is also employed when traveling outdoors to areas that are heavily wooded, as shown in Figure 2. This panel is a photograph of the keyboard surface; when the lexigrams are touched, there is no physical change, either auditory or visual. Symbols are selected simply by pointing to the desired one.

Ape Subjects

The two *Pan troglodytes* subjects, Sherman and Austin, were born at the Yerkes Regional Primate Research Center in 1973 and 1974 respectively. Sherman spent the first 1 ½ years of his life with his mother before being placed in the nursery for one additional year. He was assigned to the language project at 2 ½ years of age. Austin was removed from his mother at two months of age due to a milk allergy and placed in the Yerkes' nursery where he remained until he was assigned to the language project at 1 ½ years of age. Upon being assigned to the language project, Sherman and Austin were housed with two other similarly aged chimpanzees (one male and one female) in the language project compound. They were uncaged for approximately eight hours per day, during which time they were engaged in language research work with human caretakers. A stable staff was maintained, with four people spending most of each day with the animals, and a total of 12 people in all participating in the study. (For additional details, see Savage-Rumbaugh, 1986a.)

Two *Pan paniscus* subjects, Kanzi and Mulika, were born at the Yerkes Regional Primate Research Center in 1980 and 1983 respectively. Kanzi was born at the Field Station compound and was kidnapped from his mother, Lorel, by another female, Matata, who was already rearing an infant of her own. Matata kept Kanzi and reared him as her son. Kanzi remained with Matata until he was 2 ½ years of age, at which time he was separated from her for four months while she was returned to the Field Station for breeding purposes. They were reunited upon her return. Mulika was born to Matata at the Language Research Center and was removed from her at four months of age due to a severe eye infection. Mulika was reunited with Matata six months after the infection healed. Upon being reunited, both Kanzi and Mulika had free access to their mother, though they typically chose to spend only part of each day in her indoor-outdoor compound.[1]

Kanzi's exposure to lexigrams began at six months of age as he watched Matata attempt to request different foods using a keyboard. Mulika's exposure to lexigrams began at two months of age by being shown slides of items paired with their appropriate lexigram. A stable staff of ten individuals has participated in the study with the *paniscus* subjects. Some of the staff members have worked with both groups of subjects. (For additional details, see Savage-Rumbaugh et al., 1986.)

Similarities of the Rearing Program for Ape Subjects

The basic characteristics of the language rearing program which have been constant for Sherman, Austin, Kanzi, and Mulika are delineated below:

1. The chimpanzees are in constant social contact with other chimpanzees and with human teachers all day, seven days per week. Every attempt is made to raise the animals in a manner that will produce an adult who likes and relates positively both to members of its own species and to members of the human species.[2]

2. Following the Gardners' original work, the language rearing program includes the chimpanzees in all aspects of daily life, from mundane cleaning to exciting outings. Many different activities are provided each day for the chimpanzees (e.g., foraging in the forest, painting and drawing, using tools, playing with other animals, social games, television programs designed and produced especially for them, video games, etc.). Every attempt is made to keep their environment as novel and as interesting as possible. The communicative use of lexigrams is modeled throughout all these activities.

3. A stable group of staff members works with the chimpanzees each day, in rotating shifts. It takes several years of experience to adequately train successful staff members. Constant daily interaction with the chimpanzees is needed for a relationship that is conducive to symbolic communication. The staff members (both graduate students and full-time employees) understand the goals of the project and their unique role in achieving those goals.

4. The laboratory operates 365 days a year, 24 hours a day, and the rearing environment of the chimpanzees is somewhat like that of an extended family.

5. The chimpanzees' health, welfare, and happiness is placed above all other considerations, personal and professional.

6. All instances of symbol usage, including the pragmatic function of the symbol usage (e.g., naming, requesting, stating, etc.), and whether or not the usage appears to be appropriate and interpretable given the context, are recorded.

7. Regular videotapes are made of all chimpanzee subjects. These tapes are used to document skills, to study the use of complex combinations of gestures, vocalizations, and lexigrams, and to assess the reliability of real-time observational coding (Savage-Rumbaugh, 1986a; Savage-Rumbaugh et al., 1986).

8. Lexigrams are used with all subjects throughout the day in all activities, and appropriate communicative usage is continually modeled for them. Spoken English accompanies most lexigram production by the modelers and is also used when there is no symbol available for a particular word. Signs and spontaneous gestures frequently accompany the modeler's speech and symbol usage. Synthesized or digitized speech is also produced from the computer-linked communication systems. Multi-media input by modelers is the norm, and lexigrams, gestures, and vocalizations, or any combination of them, are accepted in communications emanating from chimpanzees.

9. Individual training tasks are utilized as needed to promote vocabulary acquisition when modeling proves insufficient. (See Savage-Rumbaugh, 1986a, for a detailed description of the various training tasks employed with the *Pan troglodytes* subjects. To date, no training tasks have been needed with the *Pan paniscus* subjects.)

10. Regular assessment of naming, requesting, and comprehending skills is completed for all subjects using randomly presented alternatives and blind controls.

Differences of the Rearing Program for Ape Subjects

Because the subjects of the two species have been reared sequentially, a number of rearing differences exist between the two groups. Many of these differences have occurred because the laboratory was moved to a new location which provided additional outdoor space. The behavioral characteristics of the *Pan paniscus* subjects (such as their low level of aggressiveness) have made some activities possible that were not feasible with the *Pan troglodytes* subjects. For example, the continued contact between the mother and her two offspring in the case of the *paniscus* subjects occurred because language research had been underway with the mother since she was a juvenile, and she continues to remain exceptionally friendly and socially receptive to interaction with human beings (Savage-Rumbaugh, 1984).

The rearing differences which did exist between the two species are enumerated below:

1. The *paniscus* subjects experienced daily contact with their natural chimpanzee mother, while the *troglodytes* subjects had no contact with their natural mothers following removal from them.

2. The *paniscus* subjects were exposed to lexigrams at a younger age than the *troglodytes* subjects.

3. The *paniscus* subjects had an older chimpanzee model to observe as well as a human model. This model was Matata in Kanzi's case and Kanzi in Mulika's case.

4. The *paniscus* subjects had a much larger outdoor area in which to move about and spent many more hours outdoors each day during the summer months. Most of their food was located outdoors during warm weather.

5. Symbol training tasks were not employed with the *paniscus* subjects as they proved capable of acquiring symbols by observing their use before any training was initiated.

6. Portable keyboards have been used outdoors only with the *paniscus* subjects. Early attempts to do so with the *troglodytes* subjects proved unsuccessful.

7. A speech synthesizer was attached to the indoor keyboard used by the *paniscus* subjects after Kanzi demonstrated an ability to comprehend spoken English words.

SPECIES DIFFERENCES IN EASE OF SYMBOL ACQUISITION

Children typically begin to acquire language without any explicit instruction on the part of their parents. In fact, recent work indicates that word comprehension precedes word production throughout the single word stage and for some children is considerably more advanced than production (Benedict, 1979; Greenfield and Smith, 1979). Snyder et al. (1981) have suggested that the discovery of reference or the concept that words serve as replacements for objects and events probably occurs first in the comprehension of language. By the time a child begins to speak, he or she is already aware of the referential function of words and uses them in such a fashion from the start.

Figure 3. This figure illustrates the number of utterances, grouped according to the class of communicative function, that were produced across a typical day by Sherman *(Pan troglodytes)* and by Kanzi *(Pan paniscus)* at five years of age.

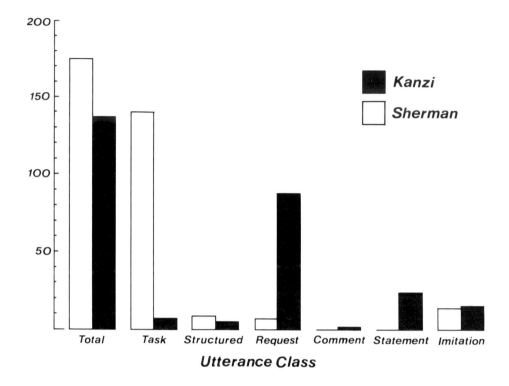

Utterance Classes:

Task = A task-related utterance, or one in which the occurrence of an utterance was called for by the constraints of the task. For example, food is hidden in a container and the chimpanzee had to request a particular tool to retrieve it.

Structured = The chimpanzee was asked a question, either directly or indirectly, and the utterance was a reply to that question. For example, the chimpanzee starts to go out a door, and the teacher says, "Tell me where you want to go."

Request = The chimpanzee asks for the experimenter to do something (play with him, take him outdoors, etc.).

Statement = The chimpanzee produces a statement about intent with no prior encouragement from the teacher. For example, the chimpanzee says "tickle" then rolls into the teacher's lap with a playface and begins to tickle the teacher.

Comment = The chimpanzee comments on an object or event without being asked to do so. For example, the chimpanzee is eating an apple, stops, comments "apple," then continues to eat the apple.

Imitation = The chimpanzee produces an utterance that was just produced by the teacher.

Unlike normally developing children, the *troglodytes* subjects showed no evidence of comprehending the referential function of lexigram symbols prior to the onset of their training. Even after simple productive skills were acquired (e.g., they could use specific symbols to request specific foods), receptive skills were still minimal. Even though they could ask for items by name, they could not retrieve those same items if someone else asked for them by name. (See Savage-Rumbaugh, 1986a, for a more complete account.)

The *troglodytes* subjects did not initially manifest a concordance between their symbol use and their behavior. For example, if they requested an item by name, and several different items were offered, they did not reliably select the item they said they wanted. Such symbol-behavior concordance appeared only after comprehension skills were taught.

By contrast, the *paniscus* subjects acquired symbols in the receptive mode first, much as normally developing children do. Consequently, when productive usage appeared, referential function was evident from the start. That is, a behavioral correspondence between what the *paniscus* subjects said and what they did was present from the outset and did not need to be taught.

Unlike the *troglodytes* subjects, the early productive usages by *paniscus* subjects were not restricted to requests, but included indicative comments and statements. Another important difference was that the early symbol usage of *paniscus* subjects often occurred in the absence of the referent (see Savage-Rumbaugh, 1986a; Savage-Rumbaugh et al., 1986; and Savage-Rumbaugh et al., 1985, for details). In general, the symbol usage of the *paniscus* subjects was characterized by a spontaneity and freedom from contextual constraint not seen in the *troglodytes* subjects. Figure 3 contrasts the types of utterances which typically occurred on any given day for Kanzi, a *paniscus* subject, and for Sherman, a *troglodytes* subject. Both subjects produced approximately the same number of utterances; however, Kanzi's utterances included a wider range of communicative functions. As can be seen in Figure 3, most of Sherman's utterances fell into the task-elicited category. This category is defined as those utterances which were elicited, in some manner, by the teacher (e.g., by asking Sherman to name an item). Very few of Kanzi's utterances fell into this category. By contrast, Kanzi produced spontaneous statements and comments which were not part of Sherman's repertoire.

The range of different vocabulary items used with any given topic was also much larger for Kanzi than for Sherman, even though their total vocabulary size was comparable, approximately 90 symbols at this time. As illustrated in Figure 4, whenever Kanzi addressed a particular topic, he did so with greater lexigram vocabulary flexibility than Sherman did. Thus, even though he and Sherman produced approximately the same number of lexigram utterances on any given day, Kanzi communicated about a much wider array of topics. When topics overlapped, as in the case of foods, actions, and locations, Kanzi used a wider variety of lexigrams within any given category than Sherman did. Overall, Kanzi's utterances covered twice as

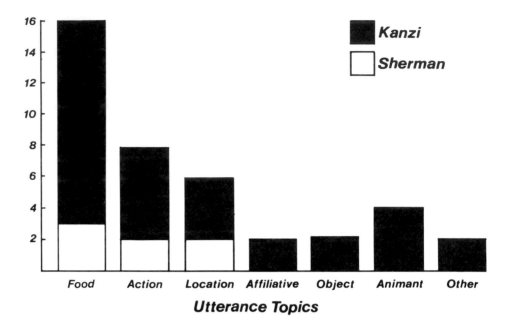

Figure 4. This figure illustrates the number of different lexigrams used with a given topic for Sherman and for Kanzi at age five during a typical day. Sherman's communications dealt with three topics (food, actions, and locations) with two to three lexigrams used per topic. Kanzi used a greater variety of lexigrams with each topic than did Sherman. (Austin typically used even fewer lexigrams than Sherman.) These data were collected when each subject was five years of age and had a vocabulary of approximately 90 lexigrams.

many topics as Sherman's utterances, and the number of symbols utilized for any given topic was two to four times the number utilized by Sherman. Austin's utterances (not illustrated) were even more restricted than Sherman's.

COMPREHENSION OF SPOKEN ENGLISH BY ANIMALS

The greater ease of acquisition, rapid grasping of referential function, and the greater variability of communications set Kanzi apart from Sherman and Austin. Why should symbol usage be easier for Kanzi? One possible explanation lies in Kanzi's capacity to understand spoken English. It has been reported that *troglodytes* subjects understand a limited amount of spoken English. Historically, however, the study of speech comprehension in apes, and other mammals, has been a rather neglected topic of investigation. This is unfortunate since speech comprehension is, for human beings, a more potent predictor of language knowledge than is language production. Many persons who are unable to speak, due to motoric disabilities such as cerebral palsy, are, nonetheless, capable of comprehending speech. When they are provided with an output device that permits them to communicate with artificially produced speech, they are immediately able to use it without going through a phase of "language learning" (Harris and Vanderheiden, 1980). Language, as a phenomenon, seems to develop appropriately in the cognitively normal, hearing individual, *regardless of whether or not he or she can speak.*

Many investigators have made the assumption that it is quite easy for apes (and for that matter dogs and even birds) to understand what is said to them. Even Darwin (1871) concluded that "that which distinguishes man from the lower animals is not the understanding of articulate sounds, for, as every one knows, dogs understand many words and sentences" (p. 85). This view holds that significant cognitive advances result from learning to use language, not from comprehending it. There is, however, almost no evidence to support these assumptions. Apparently, the view that dogs can readily understand words and sentences is so well accepted that almost no one has thought it worthwhile to determine whether or not there is any validity to this folk knowledge. Warden and Warner (1928) found that a trained show dog was *not* able to select an object from a group in response to spoken cues alone. While the dog in the Warden and Warner study did respond to ritualized multi-word commands, there was no evidence that it decoded novel sentences or distinguished between multi-word and multi-syllable commands.

Interest has also focused on whether or not dolphins and sea lions can learn appropriate responses to symbols and to novel combinations of symbols. Sea lions and dolphins can learn to respond to individual symbols (signs or whistles) and to carefully composed novel combinations of symbols (Herman et al., 1984; Schuster-man and Kreiger, 1984). However, as with Sherman and Austin, these skills have required repeated drilling, on individual utterances and the appropriate associated responses, before comprehension occurred. Typically, when training is complete, comprehension is demonstrated by a rather stereotyped motor response. It is the sequence of these stereotyped responses which can be ordered in a novel way to reflect a novel command. The symbols used with these subjects, however, have been either gestures or high-pitched whistles. Systematic evaluations of their ability to under-stand spoken English have not been conducted.

To date, studies of speech comprehension in apes have been limited and present a mixed picture. After rearing chimpanzees in their homes and attempting to teach them to speak, the Kelloggs (Kellogg and Kellogg, 1933) and the Hayeses (Hayes, 1951) concluded that the animals showed considerable deficits in comprehension as well as in production. When Viki and Gua seemed to comprehend spoken words, it was in context-specific situations or routine activities where the onset of speech readily served as a cue for the occurrence of a behavior that they already anticipated produc-ing in that context. In both cases, when explicit attempts were made to teach the chimps to respond reliably to various spoken words, they failed to do so. Viki was asked to point to her body parts (e.g., eye, ear, nose, etc.) on command, and Gua was asked to select one of several objects on command. Both apes received hundreds of practice trials on these tasks, but showed little improvement in speech comprehension.

More recently, Fouts, et al. (1976) and Patterson and Linden (1981) have reported that their ape subjects displayed some comprehension of spoken English. Yet, in both cases, the data supporting such conclusions are limited to a small number of phrases.

Documentation of Kanzi's Comprehension of Speech

Unlike Sherman and Austin, Kanzi seemed to comprehend many things that were said to him on the first occasion. In order to pinpoint the precise extent of his comprehension, two different comprehension measures were employed. The first measure evaluated whether or not Kanzi comprehended what was said to him in the context of normal use, when neither glances, gestures, nor the contextual information itself was sufficient to account for Kanzi's ensuing behavior. These data were ascertained by recording occurrences of Kanzi appropriately complying with vocal requests such as "Put on your shirt," "Hand me the knife," "Go get the towel." If pointing gestures or glances toward the encoded objects were needed in order to elicit the appropriate response, speech comprehension was not scored. Similarly, if the context was so routine that the behavior would probably have occurred whether or not the request was encoded symbolically, comprehension was not scored. It was often the case that entire sentences were responded to appropriately with no gestures or glances being emitted by the experimenter. Frequently, comprehension occurred for completely novel sentences; that is, the request which was appropriately complied with had not been made on a previous occasion. Only responses to utterances that were novel and spontaneous were recorded. Those utterances that were frequently used with Kanzi were ignored. Table 1 lists the multi-word requests that Kanzi appropriately responded to during a three-month period when Kanzi was age three years, ten months through four years of age. In all cases, Kanzi's behavior implied that he understood the utterances.

The second measure of comprehension consisted of formal tests in which the chimpanzee heard a single word and was asked to select its referent from three photographs. Pretraining for such tests was given to Sherman and Austin, but was not necessary for Kanzi or for Mulika. Testing included controls to prevent cueing (see Savage-Rumbaugh et al., 1985; Savage-Rumbaugh et al., 1986, for a complete description of these tests).

Kanzi (age 5 ½ years) and Mulika (age 2 ½ years) were easily able to listen to the English word and select its corresponding lexigram. Sherman and Austin, at ages eight and seven respectively, could not. Although they appeared to understand many words in a normal conversational context or in behavioral routines, these tasks illustrated that their comprehension was based on other than auditory cues. When limited to auditory cues alone, they failed. By contrast, Kanzi and Mulika continued to select easily any lexigram they knew the English name of even when their only cue was auditory input (Fig. 5).

It is significant that Kanzi and Mulika were able to associate graphic symbols with spoken words even though they were unable to pronounce the words. This finding would be analogous to a child without a voice learning to read at a very early age. Moreover, it implies that the "language" that Kanzi and Mulika learned is not "Yerkish," but spoken English. Concurrently, they learned that graphic symbols stand for certain spoken words. They, thus, acquired a double system of representation – the spoken word and the graphic symbol – at a very early age. Mulika, at 2 ½ years, was able to link photographs, lexigrams, and spoken words for 77 items (Brakke et al., 1986). This accomplishment would be an impressive feat even for a human infant and

Table 1. Utterances spontaneously comprehended by Kanzi without gestural or contextual cues.[1]

AT 46 MONTHS OF AGE

Let's play **BALL CHASE**.
Kanzi can you **BITE MULIKA**?
Let's play **BLANKET GRAB**.
Kanzi, **GO** get me a **ROCK**.
Let's **GO** to the **CHILDSIDE**.
GO to the **GROUPROOM**.
GO get the **STICK**.
Kanzi, can you **GROOM** the **DOG**?
Play **BALL CHASE** with **JEANNINE**.
JEANNINE is going to **BITE KANZI**.
JEANNINE will play **STICK CHASE**.
JEANNINE HIDE (hid) the **BALL**.
Can you **BITE JEANNINE**?
Let's play **BLANKET HIDE**.
Go **HIDE** the **BALL**.
KANZI, can you **HIDE**?
We can get **WATER** at the **TRAILER**.
Play **BALL CHASE** with **LINDA**.
TICKLE LINDA.
PLUG in the **VACUUM**.
Get the **RAISINS** and the **PEANUTS**.
Put the **ROCK** on the **VACUUM**.
ROSE and **MULIKA** are on the **CHILDSIDE**.
Let's play **STICK TICKLE**.
Can you **TICKLE** the **DOGS** with a **STICK**?
Can you **TICKLE MULIKA** with a **STICK**?
Let's **THROW** (toss) the **BALL** with the **STICK**.
TICKLE SHANE.
Let's play **VACUUM TICKLE**.
Where is the **WATER BALLOON**?
Let's play **WATER CHASE**.

AT 47 MONTHS OF AGE

KANZI, can you **HIDE** the **BALL**?
Please put the **APPLE** on the **FIRE**.
Let's play **BALL CHASE**.
Take the **CARROT** to the **CAMPFIRE**.
KANZI, play **STICK CHASE**.
Let's play **GRAB KANZI**.
KANZI, go **HIDE** the **EGG**.
JEANNINE wants to **HUG KANZI**.
KANZI, can you **TICKLE JEANNINE** with the
   FLASHLIGHT?
Would you **GO** get the **BLANKET**?
KANZI, please **GO** get the **FLASHLIGHT**.

KANZI, let's **GO** to "**MILK**" (the food site).
Why don't you **CHASE BILL** with the **STICK**?
The **MAGNET** is **HID(E)**ing under the
   UMBRELLA.
Could you put the **ONIONS** and the **TOMATOES**
   on the fire?
Could you **HIDE** the **RUBBERBAND**?
The **STRING** is **HID(E)**ing under the rock.
KANZI, please put the **POTATOES** on the **FIRE**.
The **STRING** is **HID(E)**ing in the
   REFRIGERATOR.
She is **HID(E)**ing by the **WATER**.
Would you **WASH** the **CARROT**?
Can you **WASH** the **MUSHROOMS**?
Let's make a **WATER BALLOON**.
Let's play in the **WATER** at the **TRAILER**.

AT 48 MONTHS OF AGE

Put the **APPLE** on the **FIRE**.
Get the **BALL** that's in the **BEDROOM**.
The **BALL** is **HID(E)**ing in the
   REFRIGERATOR.
Take the **GREENBEANS** to the **GROUPROOM**.
KANZI BITE me.
The **COKE** is in the **REFRIGERATOR**.
GO get the **BLANKET**.
Let's **GO** to the **COLONY ROOM** now.
GO use the **MAGNET** to get the **FOOD**.
Let's **GO** find **SUE'S CAR** on the **CHILDSIDE**.
Let's **GO** to **SUE'S GATE** and get **CARROT**(s)
   and a **SURPRISE**.
Can you **HIDE** the **LIGHTER**?
The **STRING** is **HID(E)**ing by the **WATER**.
Do you want **JEANNINE** to **CHASE PHIL**?
Do you want **JEANNINE** to **GO** get the **BALL**?
KANZI, **GO** get the **HAMMER**.
KANZI, **GO** get the **LIGHTER**.
KANZI, grab **MULIKA**.
PHIL and **KANZI** can go to **SUE'S OFFICE**.
PUSH PHIL.
Let's **GO** see **SHERMAN** and **AUSTIN**.
THROW the **BALL**.
KANZI, **TICKLE** me.
Take off **MULIKA'S** diaper.

1. All utterances were spoken. Words in uppercase were simultaneously indicated at the keyboard.

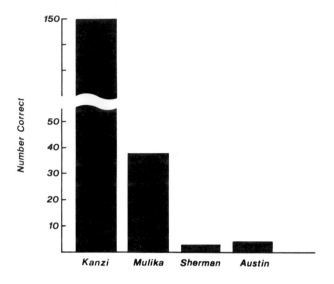

Figure 5. This figure illustrates the *Pan paniscus* and *Pan troglodytes* subjects' performance when they heard a spoken English word and were then asked to select the photograph that corresponded to that word.

suggests that the cognitive capacity of the infant *paniscus*, and perhaps that of other infant apes as well, has previously been grossly underestimated.

The *paniscus* subjects comprehended the speech of one person as readily as they did that of another. The tests of speech comprehension discussed above were administered by three different experimenters, each of whom was raised in a different part of the United States and thus manifested slightly different accents. The *paniscus* subjects did equally well regardless of which experimenter was testing them (Savage-Rumbaugh et al., 1986).

Understanding of Synthesized Speech

When an animal responds to human speech, it is often difficult to determine which component or combination of components of the complex auditory message is eliciting the response. Human speech includes affective information (conveyed by intonation and/or facial expression), phonetic information (conveyed by the acoustic pattern), and prosodic information (conveyed by speech rhythm). Mood, degree of stress, sex, and age of the speaker are also embedded within the speech signal. Although adequate data are lacking, it seems that dogs respond more reliably to the paralinguistic cues of speech than to the phonetic signal itself. Often, a command uttered by an individual other than the trainer, or a command uttered with a different rhythm and intonation, will not produce the correct response. Thus, it would seem pertinent to determine whether or not the *paniscus* subjects described in the present report are also responding to paralinguistic cues or phonetic cues.

The most straightforward method of eliminating paralinguistic cues is to present mechanical speech devoid of such cues. Tests of synthetic speech comprehension were given to one of the *paniscus* subjects, Kanzi, using a Votrax speech synthesizer. This synthesizer produces mechanical speech that lacks intonation, affect, and inflection. The Votrax synthesizer produces speech by assigning a sound signal to each letter of the alphabet. Words are formed simply by stringing together these individual sounds. Any word can be produced by the synthesizer simply by typing the appropriate sequence of letters into an Epson HX-20 computer which is attached to the synthesizer. It may be necessary to alter standard English orthography to produce an understandable word, but in any case all words are composed of the 26 sounds which correspond to the 26 letters of the English alphabet. Various recombinations of these sounds are recognized by English speakers as different words because of the unique sound pattern associated with each combination. That sound pattern shares sufficient characteristics with normal speech to permit word recognition by the English listener, or at least by the listener who is familiar with Votrax speech output. Figure 6 illustrates a spectrographic representation of the contrast between words produced

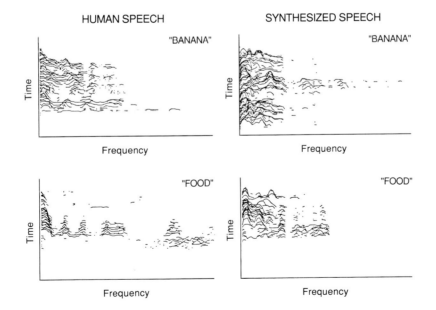

Figure 6. Spectrographic comparison of words produced by an adult human female and the Votrax speech synthesizer. The spectographs are rotated 90° so that duration of the sounds are displayed on the Y-axis, while frequencies are displayed on the X-axis. The cascaded regions depict amplitude trends for the sound samples.

naturally and words produced by the speech synthesizer. While the acoustic patterns of the words roughly resemble one another, no individual characteristic is the same for either word.

Kanzi's comprehension of synthesized speech was tested at 5 ½ years of age using the same procedures as were used to assess his comprehension of naturally produced speech, except that words produced by the speech synthesizer served as the input stimuli rather than the speech of a human companion. Kanzi received no advance training on this task, but for several years had heard the keyboard produce synthesized speech whenever he or others touched keys. Kanzi listened to the synthesized word, then pointed to the lexigram which corresponded to the word which he heard. Kanzi accurately identified 100 words produced by the Votrax speech synthesizer, compared to 150 words produced naturally. Approximately two-thirds of the words produced naturally were recognized when produced mechanically. By contrast, a four-year-old, normally developing child tested with the same set of 150 words and the same speech synthesizer was able to repeat only 33 (22%) of the synthesized words (Savage-Rumbaugh, 1986b). The four-year-old human subject who, unlike Kanzi, was not accustomed to synthesized speech would surely have performed the task more accurately with exposure to such mechanical speech. (This child had no difficulty repeating words presented naturally.) Still, the difficulty encountered by a normally developing human child when first presented with mechanical speech was not anticipated, and it emphasizes how different the acoustic pattern of the synthesizer is from natural speech. Kanzi's success on this task strongly implies that he was responding to the phonetic pattern of spoken words, not to specific inflection, intonation, or rhythmic cues, etc. This is not to say that Kanzi does not attend to such cues, only that his ability to associate particular spoken words with specific lexigrams is not exclusively dependent upon those aspects of the human speech signal.

Unfortunately, there are no data available for other animals on similar tasks. While synthetic speech sounds have been presented to other animals (Snowdon, 1979), these presentations have been limited to single syllable discriminations and are repeatedly associated with differential reward. The purpose of such tests was solely to determine whether or not the subjects could learn to form discriminations between single syllables. Such tasks are not comparable (either in terms of stimulus input or in training strategies) to the wide range of sequential auditory stimuli which were identified by Kanzi. Learning to discriminate between syllables such as *ba* and *pa* is quite a different task from associating 100 synthesized words with their referent without training.

Differences in Vocalizations

The vocal repertoires of the two chimpanzee species are distinct, though no complete contrastive analysis of the calls of both species has yet been published. It is beyond the scope of the present paper to focus upon the analysis of the species calls; however, special attention has been devoted to Kanzi's vocalizations because he appears to produce sounds that are not typical of other members of his species.

Moreover, many of these atypical vocalizations occur in response to vocal comments or to questions directed to him by his human companions.

To determine whether or not these vocalizations were in fact unusual sounds for a *paniscus* chimpanzee to produce, ten hours of audio and video recordings were obtained from a group of captive *Pan paniscus* housed at the Yerkes Field Station. The group consisted of one adult male, three adult females, and their offspring. One of the animals in the group was wild born. These recordings were contrasted with ten hours of recordings from Kanzi. Kanzi produced all of the vocalizations recorded for the group of animals. Sounds produced by all the *Pan paniscus* subjects, including Kanzi, are shown in Figure 7.

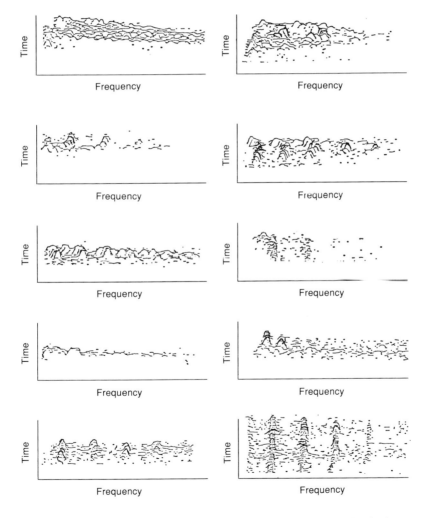

Figure 7. Spectral plots of 10 vocalizations that are produced by both Kanzi and the group-living *Pan paniscus* animals.

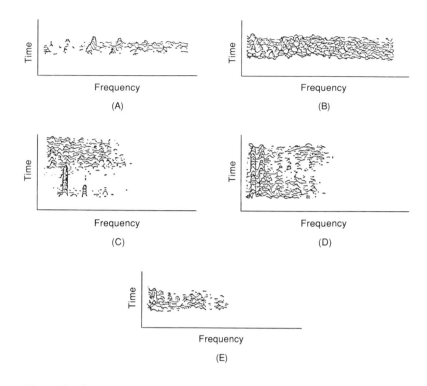

Figure 8. Spectral plots of five vocalizations that are unique to Kanzi's repertoire.

These recordings also revealed that Kanzi produced five additional vocalizations unlike any of the sounds produced by the group of *Pan paniscus* chimpanzees (Hopkins and Savage-Rumbaugh, 1986). These vocalizations are shown in Figure 8. Not only are these five additional vocalizations distinct, but some have patterns that do not resemble any of the calls produced by the group-living *paniscus*. For example, the distinct and abrupt frequency shift in call C, shown in Figure 8, is not seen in any of the calls produced by the other animals; instead, all of their frequency transitions are gradual.

It should be noted that Kanzi has been with his wild-born mother for several hours each day of his life except for one four-month period when his mother was returned to the social group for breeding purposes. His mother vocalizes often, and he characteristically vocalizes in response to her. Thus, his unique vocalizations are not readily attributable to the lack of an appropriate model as is the case with isolation-reared animals.

Vocal Response to Speech Input

Not only does Kanzi produce sounds that are uncharacteristic of the group-living *paniscus*, he frequently produces these unique sounds in response to speech directed toward him. By contrast, Sherman and Austin rarely tend to respond to speech input with vocal sounds of their own. To document this difference, data were analyzed from one videotaped work session with Sherman and Austin when they were five years of age. A similar session was also coded for Kanzi at five years of age.

An observer viewed the tapes and recorded all instances of human vocalizations which were directed toward a chimpanzee. The chimpanzee's response was then scored according to whether the vocalization was ignored or was responded to with a vocalization, with a gesture, or with a behavior. As can be seen in Figure 9, Sherman and Austin tended to respond with a specific behavior or with a gesture to speech directed toward them. Kanzi, by contrast, responded with either a behavior or a vocalization of his own. As Figure 9 also shows, Kanzi's spontaneous rate of vocalization was higher than either Sherman's or Austin's. This suggests that there are important species differences in the tendency to respond with vocal output to the speech of the caretaker.

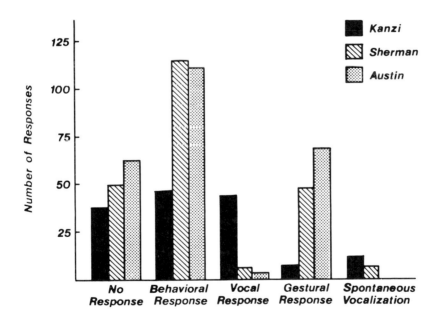

Figure 9. The types of responses chimpanzee subjects Kanzi, Sherman, and Austin made to human speech directed to them.

HUMAN SUBJECTS

Over the course of the last 11 years, another major focus of the research effort at the Language Research Center has been on the use of the nonhuman primate model applied to language intervention research with human subjects. With this approach, viable communication production modes and teaching strategies have been developed and employed with individuals with mental retardation who evidence concomitant severe oral communication impairments. To date, 22 youngsters and young adults – all of whom had been previously unsuccessful in learning to communicate with speech, manual signs, or other visual-graphic symbols (e.g., Blissymbols) – have acquired and used functional computer-based communication systems in the laboratory, in a residential living unit, and at home or at school.

Feasibility Study

In an initial feasibility study, begun in 1975, five institutionalized nonspeaking individuals with severe retardation acquired from 19 to 75 symbols in comprehension and production and employed them to convey a range of communicative functions. In addition, in conversational discourse, some individuals produced combinatorial utterances (e.g., ROYCE TOM EAT PUDDING) of up to seven symbols in complexity (see Romski et al., 1984, for a complete description). A probe, conducted 18 months after the completion of this study, revealed that these five individuals retained a mean of 83% of their comprehension vocabularies and a mean of 70% of their production vocabularies without systematic maintenance teaching on the learned symbols. Interviews with special education and living-unit staff corroborated this finding and described the ways in which the subjects continued to use their symbols in their everyday environment (Romski et al., 1985). This structured instructional approach, conducted in a laboratory setting, was derived from the nonhuman primate model. It incorporated technological and procedural adaptations and proved to be a viable teaching strategy for institutionalized individuals with mental retardation.

Study of Symbol Learning Process

The research with Sherman and Austin demonstrated that the symbol-learning process was a complex one. Explicit teaching paradigms were necessary to establish component skills in Sherman's and Austin's symbol repertoire. The next longitudinal language intervention study was also originally conducted in an experimental laboratory setting. It focused on the *process* of symbol learning by three additional individuals with severe or profound mental retardation who were nonspeaking and resided in a state mental retardation facility (Romski et al., 1986). Employing a request-based teaching paradigm, adapted from the one used with Sherman and Austin, the establishment of a conditional discrimination between a symbol and its referent was a critical feature in symbol acquisition and subsequent communicative use. Skills found necessary to establish referential symbol usage in Sherman and Austin were systematically evaluated (e.g., labeling, comprehension). For the human subjects, request skill did not initially generalize to labeling and to comprehension tasks, but additional request experience with the symbols resulted in consistent improvement in performance on both tasks. The emergence of subject-initiated symbol communications, the facilitation of spoken language comprehension and/or production, and changes in additional developmental domains (e.g., attention deployment) were also observed in the human subjects (Abrahamsen et al., in press).

These subjects' symbol vocabularies and system have since been transferred to an on-site residential locale. Successful generalization to additional communicative partners (i.e., direct-care staff) within the new setting was demonstrated (Romski and Sevcik, 1986).

Community Symbol Acquisition Study

Recent research with the *Pan paniscus* subjects, Kanzi and Mulika, provides us with a different approach to communicative symbol acquisition in work with humans. For Kanzi and Mulika, symbols are immersed in daily use with an emphasis on communication in semi-naturalistic settings. The most recent language intervention study with humans addresses a similar issue: microcomputer-based communication systems used in the homes and the classrooms of school-aged nonspeaking youngsters with moderate or severe mental retardation. Like the *paniscus* subjects, these subjects receive no specific training. Instead, symbols are used by parents and/or teachers for communication during everyday interactions.

The technology employed in this study is the Words+ Portable Voice II system consisting of a specially modified Epson HX-20 notebook computer and an adapted Votrax Personal Speech System voice synthesizer. Since the subjects are unable to utilize the Epson keyboard in a conventional manner, two options (a Unicorn Expanded Keyboard or a Welch-Allyn barcode reader) were obtained from other commercial sources to access the Words+ system. Activation of a symbol or lexigram is achieved by either touching the symbol on the display panel or by sweeping a barcode juxtaposed with the lexigram. In both cases, a synthesized equivalent of the spoken word for that symbol is produced.

This portable battery-operated communication system has been introduced into the homes of 7 subjects and the classrooms of 7 additional subjects ranging in age from 6 years 5 months to 20 years 3 months at the beginning of the study (Mean Age = 12 years 4 months).[3] All 14 subjects attended a local public school elementary or secondary special education program and resided at home with their families. The subjects evidenced moderate or severe retardation, were functionally nonspeaking, and were referred to the project by their special education teachers because they had made little progress learning to speak, sign, or use symbols.

Prior to the introduction of the communication system in the chosen setting, instruction was given to the communicative partners (e.g., parents and teachers) emphasizing the system's potential benefit as a functional communication system and the adult's role as a communicative partner. Technological modifications made by project personnel, involving both hardware and software components, made program access and execution easily attainable by the communicative partners. In this system,

Figure 10. Tim uses his portable communication system with a display panel to communicate with his teacher's aide in the school cafeteria.

it is not necessary for parents and teachers to learn the meanings of the symbols themselves, as the printed English word appears above each symbol, and with the activation of a symbol, or its accompanying barcode, a synthesized equivalent of the English word is produced. In addition, no data-recording demands are made on the communicative partners by the investigators. The many facets of intervention progress are monitored via three systematic measures: (1) communicative use probes, (2) vocabulary assessment probes, and (3) parent/teacher questionnaires. Nonparticipant observers engage in live, on-site coding of communicative behavior. They also obtain an audiotaped record of the interaction with each participant. Using both the coded and the audiotaped data, transcripts are then compiled of the communicative exchanges in the school or the home. In addition, monthly vocabulary assessments are conducted apart from the communicative context. Parent/teacher questionnaires provide weekly updates of communicative symbol use in the target settings.

All of the subjects successfully used the microcomputer to communicate on their first opportunity (Sevcik et al., 1986). Figure 10 illustrates one of the participants, Tim, using the display panel communication system in the school cafeteria with his teacher's aide. He is asking her for "milk."

Preliminary review of the three systematic assessment measures indicates that the subjects have learned both to comprehend and to produce the symbols on their display panels. Table 2 provides an excerpt from a communicative use transcript. It illustrates the use of the symbol system by both the child and his communicative partner, his teacher. As shown in Table 2, the student, Frank, integrates his symbol use with his

Table 2. Excerpt from a communicative use probe.[1]

In this exchange, Frank and his teacher are seated side-by-side during a breakfast preparation routine. Frank's communication system is positioned between them for access by both communicators. Life cereal, a bowl, a spoon, a glass, and Nestle's chocolate milk are all available on the table.

Frank: BOWL
Teacher: "You'd like a bowl. Very good, Frank. There you go."
Teacher: "What are you going to put in your BOWL?"
Frank: CEREAL
Teacher: "Oh wonderful, Frank, wonderful! How about some Life CEREAL? That's just great. Oh, it's new too."
Frank: Points to the container of chocolate milk sitting at a distance on the table and vocalizes, but his teacher is unclear as to what the referent of his pointing is.
Teacher: "What's that, Frank? What is that?"
Frank: CHOCOLATE MILK.
Teacher: "That's right, Frank. This is CHOCOLATE MILK. Good asking. Good asking."

---

1. All caps denote lexigram symbols + synthetic speech; quotes denote the spoken word.

already extant communicative patterns (i.e., vocalizations and gestures). Similarly, his adult communicative partner uses the symbols embedded within her verbal messages to communicate with him. The portable communication system, with its synthetic speech output, allows for a unique interface between visual-graphic symbols and spoken language, and increases the opportunity for effective communication in naturalistic settings.

Unique Human Competencies

Although there have been similarities in symbol acquisition between all three species, the human subjects have shown some competencies that are not extant in the chimpanzees. For example, over the course of study, increases in interpretable vocalizations have been documented for many of the human participants. Even more striking, some of the participants have demonstrated an increased propensity to produce intelligible spoken approximations of the words for the symbols they have learned (Romski et al., 1986; Romski et al., 1984). Observations of spontaneous rehearsal of such skills have recently been documented. Jon, for example, working independently at home, repeatedly activated symbols on his display panel, listened for the synthetic word equivalent, and then attempted to vocally produce the word himself.

Other subject variables, such as receptive language ability, have contributed to differential acquisition across species. That is, the greater the receptive capacity, the more readily symbols are acquired. The range of receptive language capacity in the human subjects parallels to some extent the cross-species range found in the apes. Those individuals with little or no receptive capacity for spoken English have shown symbol acquisition patterns similar to Sherman and Austin (Romski et al., 1986). Those with a significant amount of extant receptive capacity at the onset of training have demonstrated symbol acquisition patterns similar to those of Kanzi and Mulika.

Environmental variables, and their concomitant demands, have also contributed to differences in the ways in which the humans and the apes use their symbol systems. Since the human youngsters and young adults inhabit a much broader environment, they are able to use their skills with a wider variety of communicative partners and in a wider range of settings. Students, for example, have used their keyboards to communicate with unfamiliar store clerks in local shopping malls about items they wish to purchase.

DISCUSSION

The language research with the *Pan troglodytes* and *Pan paniscus* subjects provides important and unique input (at both the theoretical and technical level) to the study of symbolic processes in persons with mental retardation. The language intervention research project with humans, working from a nonhuman primate base, has specified viable communicative teaching strategies for use with individuals with mental retardation and associated severe oral language impairments. Further advances in language intervention research will come about through the continued interaction of these two research efforts at the Language Research Center.

The species differences between *Pan troglodytes* and *Pan paniscus* that have been observed to date are preliminary, and their final confirmation must await studies that equate rearing variables more completely than the present study does. The rearing differences which exist in the Language Research Center research program are minimal when contrasted with the rearing differences which have existed *between* other research studies. The rearing histories of Sarah, Moja, Washoe, Lucy, Alley, Viki, and Lana were very different, as were the communication systems they used. Nonetheless, with the exception of Viki, whose training focused on speech, reports indicate that all of these chimpanzees made remarkably similar progress. They all learned to "name" items and to produce symbol-symbol combinations. The evidence for speech comprehension in these different subjects varies, but in those cases where individual word comprehension was carefully measured and/or taught, appropriate responses were limited to less than ten spoken words.

It is also the case that a number of rearing variables differed between the Gardners' first and second sign language acquisition projects, just as rearing variables have existed in our work. The second group of animals was obtained by the Gardners at a much younger age than Washoe, and the teachers were more fluent signers than the teachers in the Washoe project. The younger animals in the second project also had the older animals as models. The second project was conducted on a ranch where the animals had a larger outdoor area in which to move about freely. Molding was stressed less as a training method during the second project than the first. These rearing differences between the Gardners' first and second signing projects closely parallel those between the Language Research Center *Pan troglodytes* and *Pan paniscus* projects. Since the Gardners held the species variable constant, it is instructive to compare the differences they reported between their first and second studies and those reported here. Their findings, when contrasted with those of the present study, should provide a preliminary base on which to begin to parcel out environmental and species variables.

The differences the Gardners (Gardner and Gardner, 1978) reported between their *Pan troglodytes* subjects were minimal in contrast to the differences observed between the two chimpanzee species in the present study. The main difference the Gardners reported was earlier symbol acquisition in the second group of subjects. Although training apparently was stressed less in the second group, both groups required training. The sign combinations of the second group of animals were very similar to Washoe's combinations. The second group of animals also employed signs of negation, while Washoe did not. The Gardners did not state whether signs of negation were trained or stressed more extensively during the second group's experiences. Since spoken English was not used with either group of subjects, comparison between their subjects on this important variable is not possible.

Since, fortuitously, the Gardners and we altered similar rearing variables between successive language acquisition teaching attempts, we can tentatively conclude that the findings of our present study are attributable, in the main, to the species variable and not to the rearing variable. Should this finding be verified by future studies, what would such species differences mean and how should they be interpreted?

Any attempt to interpret these differences must be tempered by the observation that *Pan paniscus* should not be viewed as more intelligent, in a general manner, than *Pan troglodytes*. Before such a conclusion is warranted, additional comparisons, encompassing a wider range of cognitive skills, must be made between the two species. In the wild, tool use appears to be more extensively developed in *Pan troglodytes* than in *Pan paniscus*. Moreover, *Pan troglodytes* inhabits a much wider range of environments, many of which appear to marginally support the existence of the species. The ability to use tools in flexible ways and the ability to adapt to a wide range of environments are often thought to be two qualities which signal evolutionary advance of cognitive skills. There is no evidence from the present work, or other studies, that *Pan paniscus* surpasses *Pan troglodytes* in these areas.

At the present time, there also appear to be no outstanding differences in the subsistence level or general lifestyle of the two species. If *Pan paniscus* does possess a greater capacity for symbolization and for linking auditory stimuli to representational processes, it is not clear that *Pan paniscus* is using this ability in the field to provide itself with a different lifestyle than *Pan troglodytes*. There does appear to be a distinct possibility that the social life of *Pan paniscus* is considerably more complex than that of *Pan troglodytes*. Male-female relationships are more elaborate and far more frequent in *Pan paniscus*. Infants appear to have contact with a much wider range of adults; the unit group is made up of larger, more stable subgroups; and preliminary observation of presumed attack behavior suggests a remarkable degree of coordination among a large number of individuals of both sexes.

Humphrey (1983) has recently advanced the theory that social interactions, as opposed to environmental variables, have driven the significant advances in intelligence among various primate species and, in particular, man. In this regard, it is interesting to note that Kanzi's most complex symbol communications occur in social interactions involving three or more individuals (Savage-Rumbaugh et al., 1986).

Taking an even broader view, Gardner (1983) argued that intelligence is not a unitary phenomenon but is made up of modules which are relatively independent of one another. In Gardner's view, verbal ability and nonverbal ability are separate modules; others include mathematical, musical, and spatial reasoning skills. Such a perspective would suggest that perhaps *Pan paniscus*, for reasons still unclear, evidences more highly developed skills of verbal intelligence. Other dimensions of intelligence, such as tool use and mathematical reasoning, may be more highly developed in *Pan troglodytes* (Rumbaugh et al., 1987). It is hoped that the co-rearing study, presently in progress, will help clarify many of the possible species differences reported here.

ACKNOWLEDGMENTS

The research described is supported by National Institutes of Health grant NICHD–06016 which supports the Language Research Center, cooperatively operated by Georgia State University and the Yerkes Regional Primate Research Center for Emory University. In addition, the research is supported in part by NIH grant RR–00165 from the Division of Research Resources to the Yerkes Center. The authors gratefully acknowledge additional support from the College of Arts and Sciences, Georgia State University. The Yerkes Center is fully accredited by the American Association for Accreditation of Laboratory Animal Care. This chapter is based on a paper presented at The Chicago Academy of Sciences Conference "Understanding Chimpanzees," Chicago, November 9, 1986.

ENDNOTES

1. Mulika died at 2 ½ years of age after accidentally becoming entangled in a bag that was placed in her mother's cage. No cause of death could be determined, and it was concluded that she died of fright.

2. To date, all of the chimpanzees reared in this way have developed normal sexual and social behaviors. None of them presently display deviant or abnormal social-sexual behavior patterns. The oldest subject, Lana, has successfully raised one infant and has recently given birth to a second offspring sired by either Sherman or Austin, both of whom competed for sexual access during her receptive phase, as is typical of *Pan troglodytes* males. All language project subjects continue to be housed and studied at the Language Research Center.

3. In the second year of the subjects' participation in the study, the communication system was available for the child's use in both the home and the school settings. The communication system was transported to and from school on the school bus with the child.

REFERENCES

Abrahamsen, A. A., M. A. Romski, and R. A. Sevcik. In press. Concomitants of success when persons with severe retardation acquire an augmentative nonspeech communication system: changes in attention, communication, and social interaction. *American Journal of Mental Retardation.*

Badrian, A., and N. Badrian. 1984. Social organization of *Pan paniscus* in the Lomako Forest, Zaire. In R. L. Susman, ed., *The Pygmy Chimpanzee: Evolutionary Biology and Behavior,* pp. 325–346. New York: Plenum Press.

Badrian, N., and R. Malenky. 1984. Feeding ecology of *Pan paniscus* in the Lomako Forest, Zaire. In R. L. Susman, ed., *The Pygmy Chimpanzee: Evolutionary Biology and Behavior,* pp. 275–300. New York: Plenum Press.

Badrian, N., A. Badrian, and R. L. Susman. 1981. Preliminary observations on the feeding behavior of *Pan paniscus* in the Lomako Forest of central Zaire. *Primates* 22:173–181.

Benedict, H. 1979. Early lexical development: comprehension and production. *Journal of Child Language* 6:183–200.

Brakke, K. E., E. S. Savage-Rumbaugh, K. McDonald, and W. D. Hopkins. 1986. A comparative analysis of symbol acquisition in two pygmy chimpanzees *(Pan paniscus). American Journal of Primatology* 10:391.

Carrier, J. K. 1974. Nonspeech noun usage training with severely and profoundly retarded children. *Journal of Speech and Hearing Research* 17:510–517.

Coolidge, H. J. 1933. *Pan paniscus:* Pygmy chimpanzee from south of the Congo River. *American Journal of Physical Anthropology* 18:1–57.

Darwin, C. 1871. *The Descent of Man.* London: John Murray. (p. citations from 1894 ed.)

Fouts, R. S., B. Chown, and L. Goodin. 1976. Transfer of signed responses in American Sign Language from vocal English stimuli to physical object stimuli by a chimpanzee *(Pan). Learning and Motivation* 7:458–475.

Gardner, B. T., and R. A. Gardner. 1978. Comparative psychology and language acquisition. In K. Salzinger and F. Dennaale, eds., *Psychology: the state of the art. Annals of the New York Academy of Sciences* 309:37–76.

Gardner, H. 1983. *Frames of Mind: The Theory of Multiple Intelligences.* New York: Academic Press.

Greenfield, P , and J. Smith. 1979. *The Structure of Communication in Early Development.* New York: Academic Press.

Harris, D., and G. C. Vanderheiden. 1980. Augmentative communication techniques. In R. L. Schiefelbusch, ed., *Nonspeech Language and Communication,* pp. 259–301. Baltimore, MD: University Park Press.

Hayes, C. 1951. *The Ape in Our House.* New York: Harper and Brothers.

Herman, L. M., D. G. Richards, and J. P. Wolz. 1984. Comprehension of sentences by bottlenosed dolphins. *Cognition* 16:129–219.

Hopkins, W. D., and E. S. Savage-Rumbaugh. 1986. Vocal communication in the pygmy chimpanzee *(Pan paniscus)* as a result of differential rearing experiences. *American Journal of Primatology* 10:407.

Humphrey, N. 1983. *Consciousness Regained.* Oxford: Oxford University Press.

Jolly, A. 1985. *The Evolution of Primate Behavior.* New York: Macmillan Publishing Company.

Kano, T. 1980. Social behavior of wild pygmy chimpanzees *(Pan paniscus)* of Wamba: a preliminary report. *Journal of Human Evolution* 9:243–260.

_____. 1982. The social group of pygmy chimpanzees *(Pan paniscus)* of Wamba. *Primates* 23:171–188.

Kano, T., and M. Mulavwa. 1984. Feeding ecology of the pygmy chimpanzees *(Pan paniscus)* of Wamba. In R. L. Susman, ed., *The Pygmy Chimpanzee: Evolutionary Biology and Behavior,* pp. 233–274. New York: Plenum Press.

Kellogg, W. N., and L. A. Kellogg. 1933. *The Ape and the Child.* New York: Whittlesey House.

Kiernan, C. C. 1977. Alternatives to speech: a review of research on manual and other forms of communication with the mentally handicapped and other non-communicating populations. *British Journal of Mental Subnormality* 23:6–28.

Kuroda, S. 1984. Interaction over food among pygmy chimpanzees. In R.L. Susman, ed., *The Pygmy Chimpanzee: Evolutionary Biology and Behavior.* New York: Plenum Press.

Mirenda, P. 1985. Designing pictorial communication systems for physically able-bodied students with severe handicaps. *Augmentative and Alternative Communication* 1:58–64.

Patterson, T. 1979. The behavior of a group of captive pygmy chimpanzees *(Pan paniscus). Primates* 20:341–354.

Patterson, F., and E. Linden. 1981. *The Education of Koko.* New York: Holt, Rinehart and Winston.

Reich, R. 1978. Gestural facilitation of expressive language in moderately/severely retarded preschoolers. *Mental Retardation* 16:113–117.

Richardson, T. 1975. Sign language for the severely mentally retarded and the profoundly mentally retarded. *Mental Retardation* 13:17.

Ristau, C. A., and D. Robbins. 1982. Language in the great apes: a critical review. In J. Rosenblatt, R. A. Hinde, C. Beer, and M. C. Busnel, eds., *Advances in the Study of Behavior,* vol. 12, pp. 141–255. New York: Academic Press.

Romski, M. A., and E. S. Savage-Rumbaugh. 1986. Implications of language intervention research: a nonhuman primate model. In E. S. Savage-Rumbaugh, *Ape Language: From Conditioned Response to Symbol,* pp. 355–374. New York: Columbia University Press.

Romski, M. A., and R. A. Sevcik. 1986. Augmentative communication use on a residential unit: implementation considerations. Paper presented at the annual meeting of the America Association on Mental Deficiency, Denver, CO.

Romski, M. A., R. A. Sevcik, and S. E. Joyner. 1984. Nonspeech communication systems: implications for language intervention with mentally retarded children. *Topics in Language Disorders* 5:66–81.

Romski, M. A., R. A. Sevcik, and J. L. Pate. 1988. Establishment of symbolic communication in persons with severe retardation. *Journal of Speech and Hearing Disorders* 53:94–107.

Romski, M. A., R. A. Sevcik, and D. M. Rumbaugh. 1985. Retention of symbolic communication in five severely retarded persons. *American Journal of Mental Deficiency* 89:313–316.

Romski, M. A., R. A. White, C. E. Millen, and D. M. Rumbaugh. 1984. Effect of computer-keyboard teaching on the symbolic communications of severely retarded persons: five case studies. *The Psychological Record* 34:39–54.

Rumbaugh, D. M. 1971. Evidence of qualitative differences in learning processes among primates. *Journal of Comparative and Physiological Psychology* 76:250–255.

Rumbaugh, D. M., and T. V. Gill. 1973. The learning skills of great apes. *Journal of Human Evolution* 2:171–179.

Rumbaugh, D. M., and J. L. Pate. 1984. The evolution of cognition in primates: a comparative perspective. In H. L. Roitblat, T. G. Bever, and H. S. Terrace, eds., *Animal Cognition,* pp. 512–523. Hillsdale, NJ: Lawrence Erlbaum.

Rumbaugh, D. M., E. S. Savage-Rumbaugh, and M. Hegel. 1987. Summation in the chimpanzee *(Pan troglodytes). Journal of Experimental Psychology: Animal Behavior Processes* 13(2):107–115.

Savage-Rumbaugh, E. S. 1984. *Pan paniscus* and *Pan troglodytes:* contrasts in preverbal communicative competence. In R. L. Susman, ed., *The Pygmy Chimpanzee: Evolutionary Biology and Behavior,* pp. 395–413. New York: Plenum Press.

_____. 1986a. *Ape Language: From Conditioned Response to Symbol.* New York: Columbia University Press.

_____. 1986b. Comprehension of spoken English and synthesized speech in a pygmy chimpanzee *(Pan paniscus)*. *American Journal of Primatology* 10:429.

Savage-Rumbaugh, E. S., K. McDonald, R. A. Sevcik, W. D. Hopkins, and E. Rubert. 1986. Spontaneous symbol acquisition and communicative use by pygmy chimpanzees *(Pan paniscus)*. *Journal of Experimental Psychology* 115:1–25.

Savage-Rumbaugh, E. S., R. A. Sevcik, D. M. Rumbaugh, and E. Rubert. 1985. The capacity of animals to acquire language: do species differences have anything to say to us? *Philosophical Transactions of the Royal Society of London* B 308:177–185.

Savage-Rumbaugh, E. S., B. J. Wilkerson, and R. Bakeman. 1977. Spontaneous gestural communication among conspecifics in the pygmy chimpanzee *(Pan paniscus)*. In G. H. Bourne, ed., *Progress in Ape Research*, pp. 97–116. New York: Academic Press.

Schusterman, R. J., and K. Kreiger. 1984. California sea lions are capable of semantic comprehension. *The Psychological Record* 34:3–23.

Schwarz, E. 1929. Le chimpanze de la rive gauche du Congo, *Pan satyrus paniscus*. *Bull. Cercle Zool. Cong.* 6:114–119.

Sevcik, R. A., M. A. Romski, and D. Washburn. 1986. Microcomputer communication system implementation in homes and schools. Poster session presented at the annual meeting of the American Speech-Language-Hearing Association, Detroit, MI.

Snowdon, C. T. 1979. Response of animals to speech and to species-specific sounds. *Brain, Behavior, and Evolution* 16:409–429.

Snyder, L. S., E. Bates, and I. Bretherton. 1981. Content and context in early lexical development. *Journal of Child Language* 8:565–582.

Thompson-Handler, N., R. K. Malenky, and N. Badrian. 1984. Sexual behavior of *Pan paniscus* under natural conditions in the Lomako Forest, Equateur, Zaire. In R. L. Susman, ed., *The Pygmy Chimpanzee: Evolutionary Biology and Behavior*, pp. 347–368. New York: Plenum Press.

Warden, C. J., and L. H. Warner. 1928. The sensory capacities and intelligence of dogs, with a report on the ability of the noted dog "Fellow" to respond to verbal stimuli. *Quarterly Review of Biology* 3:1–28.

Yerkes, R. M. 1929. *The Great Apes: A Study of Anthropoid Life*. New Haven: Yale University Press.

Yerkes, R. M., and B. W. Learned. 1925. *Chimpanzee Intelligence and Its Vocal Expression*. Baltimore, MD: Williams and Wilkins.

CURRENT AND FUTURE RESEARCH ON CHIMPANZEE INTELLECT

Duane M. Rumbaugh

INTRODUCTION

A perspective of research on the chimpanzee's intellect serves to underscore the importance of this species in our attempts to understand comparative primate cognition in general and the foundations of human cognition in particular. On the strength of research findings of recent years, we should feel comfortable in concluding that chimpanzees have the requisite capacities for mastering several dimensions of language. Notwithstanding the fact that language is "natural" only to our species, chimpanzees share with other apes the ability to learn both to comprehend and to produce symbols which can function as representations of referents that are not necessarily present. In that sense it may be said that the chimpanzee has the capacity for basic semantics.

On balance, the chimpanzee has served as a fine teacher in that our study of them has yielded (1) a more precise understanding of both language and requisites to language and (2) refined techniques for study of primate cognition. Present and projected research on other topics is in support of the contention that the chimpanzee can readily discern cause-effect relationships between behavior and the consequences of behavior. The chimpanzee is able to overcome problems of spatial separation of the loci of responses and the consequences thereof. Chimpanzees can learn complex tasks through observational processes. Recently they have been shown to have a capacity for primitive summation. Current work promises future data on the chimpanzees' ability to spell and to learn the relative values of arabic numbers. In sum, chimpanzees are relational learners oriented toward the detection/assertion of cause-effect relationships. They have the requisites for high-order relational processes that, as they become understood, assist our perspective of human behavior.

A PERSPECTIVE FROM COMPARATIVE PSYCHOLOGY

In some quarters comparative psychology is said to be focused on only one topic – studies that rank-order animals' intelligence. This assertion is at best a distortion of historical perspective. Comparative psychology is, rather, a systematic effort of scientists to understand the evolution of behavioral processes of adaptation (Rumbaugh, 1985). Some of these processes are primarily under the control of genetics and efficiently serve the survival and reproductive interests of animals that have relatively simple neural control systems and short life spans. Other behaviors are more open to the influence of experience and learning. These behaviors are associated with animals that have complex nervous systems and long life spans. Both kinds of behaviors are the ends of a continuum, but, in all probability, all behaviors are both made possible and constrained by the interactions of biological and environmental parameters.

The great apes (Pongidae) and humans *(Homo sapiens)* are noted for advanced brain development and long lives. The degree of their genetic relatedness is high. Although their appearances are distinct, all agree that there are similarities sufficient in their morphology to make it clear, even to the naive eye, that they are closely related. Scientific research has served to validate that perception. High degrees of genetic relationship make not only for physical similarities, they make for high degrees of psychological similarities. This paper addresses questions of intellect and cognition in great apes and humans – not to the end of rank-ordering, but rather to understand the dimensions of the similarities and differences. To many observers it is clear that there are great similarities in the intellectual processes of the young chimpanzee and the maturing child. These similarities can be so striking as to divert attention from the differences that, in fact, define the uniqueness of the species. On balance, however, it will be an understanding of both the differences and similarities of chimpanzee and human intellects that will be the contribution of a comparative psychological approach.

"HOW GREAT IS THE CHIMPANZEE'S INTELLECT?"

Happily, this question may be rejected out of hand. It is a question of no interest to this author because it cannot be answered. A more proper question is, "What kinds of psychological processes are supported by the chimpanzee's brain? How do these compare and contrast with the processes of other primate species? And with the human?"

Research of very recent years makes clear that we have profoundly underestimated the chimpanzee's intellectual operations. The data are now sufficiently strong to justify the conclusion that chimpanzees are capable of certain operations basic to language. That they do not have "all of language" does not mean that they have "none of it." The common chimpanzee's *(Pan troglodytes)* accomplishments with respect to language might be compared to those of a mythical bird that, while *normally flightless,* might manifest several of the basic parameters of winged flight as a consequence of refined rearing and learning opportunities engineered by humans. Notwithstanding that the bird possibly lacks both certain elements critical to full flight and the facility to integrate the elements of flight that it can master, it could engage in "flight behaviors" while not yet flying. While such a perspective might be generally fair to an effort designed to summarize the language achievements of the common chimpanzee, it would be less than fair if the referent were the bonobo *(Pan paniscus)* (Savage-Rumbaugh, 1986; Savage-Rumbaugh et al., 1986). It would be less than fair because the evidence is strong that the bonobo can learn its symbols spontaneously just as the human child learns its vocabulary, and it can understand specific English words produced by human speech, recorded speech, or digitized speech. Its comprehension is revealed by its ability to "translate" from English to lexigram word-symbols. It does not require special training to do so. And it is surely fair to assert that the bonobo Kanzi probably has learned more of our language than humans have learned of his species' communication systems.

Research of recent years has made clear that apes can come to use their symbols referentially for things not present (Savage-Rumbaugh, 1986); they can learn symbols from other members of their species (Savage-Rumbaugh, 1986; Fouts et al., 1984) and coordinate their own behavior with conspecifics through precise production and comprehension of symbols used in a dynamic social context (Savage-Rumbaugh, 1986); they can learn rules for ordering their words (Pate and Rumbaugh, 1983; von Glasersfeld, 1977); they make comments (Greenfield and Savage-Rumbaugh, 1984) and announce their intended actions (Savage-Rumbaugh et al., 1983); and their productions can be both spontaneous and devoid of imitation of others' symbol usage (Miles, 1983). Their capacity for syntax, though limited, is nonetheless extant and is exemplified in the bonobo's (i.e., Kanzi) self-generated, rule-based, structured use of arbitrary gestures in combination with learned symbols (Greenfield and Savage-Rumbaugh, 1986).

If we look for traces of human intellect in the chimpanzee, what will we find? The answer to this question is not simple. First, what we find will be a function of how we go about our search. It merits emphasizing that no scientific method can *prove* the absence of a talent or a skill. Thus, if through our search we find no evidence of a given skill or ability, all that can be properly concluded is that with the methods employed no positive evidence was obtained. What we learn, what we discover, is as much a function of how we go about our science as it is upon the facts of the real world. Second, as we look for traces of skills in chimpanzees, we should not expect to find the human expression of a given skill in simply a lesser version. Thus, as the search is continued for traces of language in apes, no one expects to find an ape that can literally speak, even in short, though grammatically correct, sentences. Rather, we should expect to find bits and pieces of language. And the science that emerges will be both supported and limited by two factors – the *behavior* of our subjects and our *interpretation* of those behaviors. Ultimate truth is not a realistic goal for science. Nevertheless, to us as humans, the consumers of science, the scientific effort accrues what appears to be an increasingly precise, hence intellectually satisfying, perspective of behavior.

New experimental procedures are making studies possible of animal cognition processes thought to be inaccessible just a few years ago (Roitblat et al., 1984). What are some of the recent probes into new dimensions of the chimpanzee intellect? Research with apes on questions of language has had center stage for many years, and evidence has become refined and compelling. Given that they have at least primitive language competence for "speaking, listening, and reading" (e.g., they have skills of production, comprehension of human speech in the case of *Pan paniscus*, and the ability to interpret nonspeech symbols), what of their abilities for other advanced processes? For instance, what is their competence with numbers and computations?

CAN CHIMPANZEES LEARN NUMBERS AND COMPUTATIONS?

If we ask, *Can apes use numbers in any manner that relates to human use of numbers?*, and *Are they capable of, say, addition?*, how might we start our search? Best we do that by examining relevant literature.

Contemporary Trends

To date, most number-related research has focused on training animals "to count" in some elemental fashion. Davis and Memmott's review (1982) concluded that even the most primitive manifestation of counting is a "last resort" for animals and requires substantial contextual support. And even when subjects discriminate and respond to a specific number of things or events, it does not appear analogous to human counting with cardinal numbers. Such performance has been characterized as being restricted to "number discrimination" (Fernandes and Church, 1982).

True counting requires the application of several requisite abilities, such as being able to discriminate differences between varying amounts, being able to cope with heterogeneous sets (varying in form, size, shape, texture, etc.) as well as homogeneous sets of materials, and the ability to establish one-to-one correspondence between paired items of different sets. Counting also entails the precise one-one "tagging" of individual items with a series of labels that exist in a stable order, the partitioning (i.e., the setting apart) of items already counted from those yet to be counted, and knowing that the number assigned to the last member of a set counted also represents the number of items which comprise that set or were otherwise partitioned through counting behaviors (i.e., the *cardinal principle*; Gelman and Gallistel, 1978).

Objective evidence of these several dimensions of counting has yet to be obtained in animal studies. Nonetheless, animals have demonstrated the ability to discriminate different numbers of things and events. Additionally, that some animals are also capable of ordinality is strongly indicated by Menzel's (1965) study. His chimpanzees readily learned to respond differentially to colored plaques as cues to obtain the larger of paired food incentives. They also readily learned the rank-order relationship among five opaque plaques (differentiated by brightness, thickness, and other dimensions) and the amount of food which the plaques covered.

Ferster (1964) trained two chimpanzees to "match to sample" eight binary numbers, ranging from 000 to 111 (0–7). After about 500,000 trials, they reached a 95% level of accuracy. Next, they learned to "write" the binary numbers, as labels, for displays of from one to seven items which varied in size, shape, arrangement, etc. of the items. Performance ranged between 95% and 99.8% correct. Yet Ferster did not conclude that the chimpanzees were truly counting in the sense that they were enumerating objects. He reported early in his work that the objects' arrangements or some other feature apart from number determined the basis for correct choice. For example, even a shift in the arrangement of three triangles produced suggestions of confusion in his subjects.

More, Less, and Middle

Thomas and Chase (1980) demonstrated that squirrel monkeys *(Saimiri sciureus)* could acquire a single response, based on a conditional cue, to a stimulus containing the "fewest" elements, the "most" elements, or an intermediate number. In line with these findings, Dooley and Gill (1977) taught a language-trained chimpanzee, Lana (Rumbaugh, 1977), to choose one of two quantities of linearly aligned bits of food of similar sizes. When the bits of food in the two lines were arranged in the ratios of 1:2, 2:3, 3:4, or 4:5, she chose the larger amount 95%, 90%, 90%, and 65% of the time respectively. Upon request she was also able to label one of two lines of washers, of varying sizes, as either "lesser" (88% correct) or "greater" (90% correct), a performance which reflected both relative number/length and the ability to differentiate and to label one in relation to the other. In all likelihood, it was some attribute of the linear arrangements rather than anything purely numeric that supported her performances.

Rohles and Devine (1966, 1967) added to the work of earlier investigators (Yerkes, 1934; Spence, 1939) on the "middleness" concept in chimpanzees. Their initial study (1966) showed that the experimental subject was able to select a middle item from arrays of 3, 5, 7, 9, 11, and 17 stimulus objects, laid out in a row, on greater than 90% of the trials. They also demonstrated that the concept of middleness was retained even when stimulus objects were arranged asymmetrically (i.e., with unequal spaces separating the objects).

More recently, Davis (1984) taught a raccoon to choose a transparent cube containing three items (e.g., grapes, raisins, small metal balls) from an array of cubes containing 1, 2, 3, 4, and 5 items; and Matsuzawa (1985) trained a chimpanzee *(Pan troglodytes)* to apply the appropriate arabic numeral to items of varied displays ranging in quantity from one to six.

Temporal Numerosity

There is other research that examines "counting" in reference to stimulus events presented serially, rather than simultaneously.

Davis and McIntire (1969) suggested that rats could "count to three" to predict safety from shock. When three unsignaled shocks were regularly superimposed upon a baseline of food-reinforced lever pressing, responding was suppressed. Immediately following delivery of the third shock, however, lever pressing recovered. They named this effect the "post-third shock-rate acceleration" and suggested that the rats were, indeed, "counting" the number of shocks. In addition, they demonstrated that this "conditioned acceleration" was not dependent on specific temporal cues within the 30-minute testing session, as the rats increased their responding following the third shock whether it occurred early or late in the session (12 min vs. 27 min).

Davis and Memmott (1983) subsequently tested the effect of a salient predictor of safety – a 60-second tone that preceded and terminated with the delivery of each of three shocks. The result was the total elimination of the post-third shock-rate accelera-tion. Their conclusion was that "counting" is not a natural behavior in animals and that it may be restricted to situations devoid of other more salient predictors.

Thus, Davis and his colleagues produced behavior that was sensitive to a specific number of events – under tightly controlled artificial circumstances. Notwithstanding, number-relevant discrimination appears to be a survival-related (i.e., prepared), proto-numerical skill that can support optimal foraging (Krebs and Davies, 1978). The postulation of such a skill presents a parsimonious explanation for the existence of "natural" number-related responses in a wide variety of circumstances, such as which cluster of berries on a bush should be approached and which branch bears the greater amount of buds. As Davis and his colleagues have shown, such a skill can be trained to high levels of accuracy, at least with small numeric values in specific contexts. Although there is no evidence that their animals were counting in the true sense, their responses were relevant to counting in that they were highly accurate and represented more refined discriminations than judgments of, say, "several" or "few."

Can Chimpanzees Summate?

At this point it is asked whether in animals there might be processes of summation not predicated on either counting or on prior training, processes that might establish a sensitivity to specific numbers of things or events. To be specific, given that humans can add numbers, might our close relative the chimpanzee be able to combine varying quantities to achieve a related conclusion? For humans, *addition* answers the sophis-ticated question, "What is the total number of things we have when we count them or when we combine numbers that represent totals of separate sets of things?" For the chimpanzee, the *summation* (e.g., combination) of varying quantities separated in space might answer a simpler question, "Which combination nets the greater amount?"

A series of five experiments (Rumbaugh et al., 1987) were designed to investigate the possibility of "summation" operations by common chimpanzees *(Pan troglo-dytes)*. Summation was said to have occurred if a subject reliably chose the pair of quantities whose overall sum was greater than the sum of other pairs presented (the experiment used a stated numerical range).

Our subjects were two adult male common chimpanzees *(Pan troglodytes)*, Sherman and Austin, with extensive prior research experience (Savage-Rumbaugh, 1986). Only bits of chocolate as incentives served to motivate them (e.g., they were not food deprived). The apparatus (Fig. 1) consisted of a flat wooden platform (0.46 m x 0.61 m) which served to support two identical stimulus trays.

Two pairs of food wells were located on the forward edge of two independently sliding trays. To start a trial, the experimenter pushed both trays forward as a single unit to within reach of the subject's fingers. The chimpanzee was permitted to choose and extract the chocolates from only one of the two panels. No facial or visual contact between subjects and experimenter was possible during trial preparation, the presenta-tion of trials, and the subjects' choices.

Unbroken pieces of semisweet chocolate bits served both as incentives and as the

Figure 1. Sketch of the test apparatus. The chimpanzee has selected the tray to his right, the unchosen tray has been withdrawn by the experimenter beyond the subject's reach, and all four wells have chocolates – a condition that occurred in all but the first experiment. The blind is shown in a raised position only for the purpose of this figure. In practice, during trial presentation it was never higher than the minimum necessary for the experimenter to see which well/tray had been selected.

Pat McNeely, Georgia State Univ.

stimuli to be discriminated. As soon as contact with one tray was made, the other tray was immediately drawn back far enough to be out of the subject's reach yet remain in his full view. There was no correction procedure for a choice of the lesser quantity – even if the subjects chose the lesser amount, they were still allowed to consume the chocolates. Thus, they were never "wrong" on any trial in the sense that they got no food.

The purpose of the first study was to ascertain whether our chimpanzees would initially choose by preference (e.g., without special training) the larger of two discrete quantities of consumables. Menzel's (1965) data suggested that they would be able to do so. From one to four chocolates were placed in only one member of each pair of food wells (e.g., one food well of the two per panel was empty). From the outset, and without specific training to do so, Sherman and Austin clearly preferred the larger quantity of chocolates (in greater than 95% of all trials).

That the subjects had shown a strong preference for the greater of two quantities enabled the next step, which was to determine whether they would show a similar preference between two *pairs* of quantities.

In subsequent studies, two pairs of quantities (0–4) were randomly generated for each trial. Examples of pairs of chocolates are 1 and 2 vs. 3 and 2; 2 and 3 vs. 3 and 3; 3 and 2 vs. 4 and 0; 2 and 1 vs. 4 and 4; etc. Other quantities that formed "meaningful comparison trials" were at first mixed in with all randomly generated pairs, then came to be the only type of trial presented in final tests. Meaningful comparison trials were those which afforded a more valid assay of summation through certain pair comparisons (e.g., 3 and 3 vs. 1 and 4; 0 and 4 vs. 3 and 2; 2 and 2 vs. 3 and 0; etc., occurred more frequently within each block of trials than provided for solely by chance). In meaningful trials, the pair with the greater *total* number of chocolates *did not* also have a greater number of chocolates in one of its wells. (Preference for such a pair might reflect simply an attraction to that quantity rather than to the combined total of the pair.) Identical pairs of quantities (e.g., 3 and 2 vs. 3 and 2; 4 and 1 vs. 4 and 1; etc.) were not viewed as "meaningful" comparisons because they did not constitute a real choice for our purposes. Any other comparisons that result in equal totals (e.g., 2 and 1 vs. 3 and 0; 0 and 4 vs. 2 and 2; etc.) were deleted because they left interpretation at issue (e.g., there is no choice for a greater total that might reflect summation). One, but not both, of the pairs could contain a zero value in one well (if both pairs had an empty well, the conditions of the first study would be reinstated for no purpose). Finally, a given quantity (e.g., 1, 2, 3, or 4) was not used in both pairs in the same trial.

Austin's and Sherman's choices of the larger sum rapidly improved across studies to better than 0.90 on meaningful comparison trials. Both subjects were generally more precise in the selection of the greater amount on trials where the proportionate differences were relatively large.

The last study of the series was to determine whether the summation processes of these chimpanzees would generalize to novel and more complex comparisons. After all, one of the best bases for inferring that a subject really "knows" something is to test for competence in new situations. For this purpose, we introduced the novel quantity of five items into the meaningful comparisons (0–5). Both chimpanzees maintained strong preferences for the greater sum. On *familiar* arrays (i.e., previously encountered trials with from zero to four chocolates per well), the subjects chose the larger sum on 0.84 trials on the first day; on *novel* trials (e.g., those with five chocolates in a well), they chose the larger sum on 0.86 trials of the first day! A comparison of accuracy levels across days revealed only a slight difference (0.07) favoring familiar arrays.

Whatever the processes were that supported high preference for the greater sum in previous studies, they generalized to novel displays without substantial decrement. This finding indicates that this preference was not because the subjects had learned to choose a specific pair from each display.

Overall, the results clearly support the assertion that whether or not they might have "true" counting skills, chimpanzees can readily perform basic summation operations. Moreover, these summation skills were rapidly acquired, were highly accurate, and were generalized without substantial decrement to previously unencountered novel pairs of food quantities.

It bears emphasizing that no correction procedures were used to encourage selection of the greater sum. Notwithstanding, preference of it steadily improved to high levels of accuracy with ratios that were as small as 7:8. The consequence of choices (e.g., obtaining the chocolates to eat) appears to have been sufficient for the combinatorial operation of elemental summation to come into play. Whatever the nature of that operation, it was one that the chimpanzees probably brought to the situation. It was neither intentionally shaped nor trained. And to many it is a surprise that a difference of even one chocolate bit on many trials could produce a reliable choice between trays to obtain the greater sum.

How Do Chimpanzees Summate?

An important question is, of course, "What is the nature of the process that supports choice for the greater amount?" Parsimonious explanations include the possibility that the summation process is simply an extension of the numerosity or volume discrimination abilities. Summation in our studies is, perhaps, essentially a perceptual process through which a pair of separate entities are joined or fused into a unified percept. The question as to which of two food wells (right pair vs. left pair) were to be fused must have been answered through the procedure of withdrawing the unchosen tray. After all, the two middle wells were no further apart than were those of each tray. Although the to-be-joined numbers of chocolates in this summation task were separated by 11 cm, the possibility exists that the left pair and the right pair were first "fused" perceptually so as to produce, in effect, only two quantities for comparison and choice – with one being "more" than the other (e.g., having a greater degree of numerosity). In support of the perceptual fusion hypothesis is the fact that Rohles and Devine (1966, 1967) demonstrated that chimpanzees could discount the unequal spacing of objects to select the one that was the "numeric" middle. Alternatively, the perceptual fusion of the contents per pair of wells might have provided the basis for a *volume* discrimination. Both animals had prior records of quite reliably "serving" the smaller of two portions of food to the other chimpanzee while taking the larger portion for their own consumption in a long series of food-sharing tasks (Savage-Rumbaugh, 1986).

It should be noted that, since surface area was not controlled, the animals' responses may have occurred as a function of surface area rather than number discrimination, per se. Nonetheless, even if surface area were the relevant variable, the "summation" response would still represent a combining of noncontiguous areas. (Neither the chocolate pieces nor the wells were contiguous.) As in other studies, the size and shape of the stimulus materials might be varied in order to control for surface area as a discriminative stimulus. However, in this study, such a manipulation would introduce a serious confounding because a variation in the magnitude of reinforcement (total chocolates obtained) would not be related to number or quantity. Such a manipulation must await future research that does not use edible materials as the number-related stimuli.

The preferred explanation incorporates the concept of *subitizing*. *Subitize* is a word coined by Cornelia C. Coulter for use by Kaufman et al. (1949). Its origin is the Latin verb *subitare*, or "to arrive suddenly." Kaufman et al. used the term in accounting for discontinuities they detected in data when their human subjects were asked to specify the number of dots briefly presented. With six or fewer dots, the subjects were very accurate; with more than six dots, their accuracy broke sharply. It was suggested that there were two defining operations – six or fewer dots were *subitized,* and more than six dots were *estimated.* Neither is the same as formal counting.

Subitization is possibly an unlearned phenomenon that allows for differentiations between small quantities (see Gelman and Gallistel, 1978; von Glasersfeld, 1982; and Cooper, 1984, for perspectives). Gelman and Gallistel argue, however, that it should not be viewed as ". . . a low-level, 'primitive' way of abstracting the numerosity of sets" (1978, p. 225). Just what its role might be in the emergence of true counting skills is subject to debate, and the specifics of its operations are currently unclear. Notwithstanding, infants do differentiate between small numbers of things (Starkey and Cooper, 1980), do so without formal training, and certainly do it without true counting skills. Subitization might be the process that supports such performance. Furthermore, seven-month-old human infants need no formal training to prefer looking at a numerical array that corresponds to the number of sequenced sounds that they concurrently hear (Starkey et al., 1983). The suggestion is that they come prepared to note similarities that are number based as they perceive their world through different modalities. Human infants of five months of age ". . . can abstract quantities when they are small numerosities (fewer than four or five)" (Strauss and Curtis, 1984, p. 149). They do this without formal counting competence. This ability does not improve markedly with age, however, until formal counting and language come into operation. And, even with language, children can encounter great difficulty in counting beyond three items. Descoeudres (1921) termed this problem the *un, deux, trois, beaucoup* phenomenon.

In summary, our preferred view is to suggest that Sherman and Austin chose the larger sum as a result of the following steps: first, they subitized the amounts of chocolates in each well for each pair of wells; second, they combined, through elemental relational summation, the subitized amounts for each pair; and, third, they then compared the summations as the basis of choice between the greater number of incentives. That they did not subitize the chocolates directly by pairs of food wells is suggested by the totals presented in the last study of the series. Even as the ratios there between pairs of wells became as small as 6:7 and 7:8, the chimpanzees' preferences for the greater sum held at 88% and 79% respectively. These numbers (6, 7, and 8) would tax the subitizing skills of many adult humans and well exceed those of children three to four years old, if not older. Clearly, Sherman and Austin were responding even to a paired quantity in excess of three as other than an undifferentiated *beaucoup.*

Just how they did so is open to conjecture. Gelman and Gallistel (1978) might give us a clue, however, as they discuss semi- and full-algebraic reasoning. This form of reasoning applies to manipulations not on specific numbers but upon numerical relations as variables. Our subjects were possibly concluding some semi-algebraic or relational operation between the two pairs of quantities they encountered. Whether it was to fuse the members of each pair into a single pair perceptually, thereby discounting the space that separated them, or whether to the contrary the spatial separation served to assist the summing of the estimated numerosities for each pair, which without physical separation would be too large for them to subitize, can be determined only by additional research.

Duane M. Rumbaugh

It must also be acknowledged that Sherman and Austin are very special animals. They have received extensive language training and other enriching experiences. Therefore, it would be of interest to replicate this study with chimpanzees with other histories and with subjects of other primate species for the building of a comparative perspective of summation. That perspective should include developing human children's choices before and during their learning how to count, and, more importantly, how to add.

WHAT OF THE CHIMPANZEE'S ABILITIES TO INFER CAUSE-EFFECT RELATIONSHIPS?

As early as 1962, Mason anticipated that chimpanzees, and other apes as well, might be distinguished among animals for their proclivity to perceive the linkage between their behaviors and the consequences thereof. They not only note the consequences of their behaviors, they seemingly "experiment" to see what alterations in their behaviors might generate altered consequences. An observation of recent years in our laboratory serves to illustrate (yet not prove) the point.

A heavy wind storm had felled a large tree next to the colony room which housed, at that time, Sherman and Austin. When the crew came to remove the tree, the animals were seemingly terrified by the sounds of power saws and trucks. (They could hear, but not see, the activities of the cutters.) To identify the source of the sounds to the chimpanzees, a color video camera was placed on the scene and a monitor was stationed near the chimps. Not only was the animals' apparent terror reduced, they became transfixed with the visual correlates of sounds. Sherman and Austin had long records of experience with television and had used it to solve problems of many different types (Savage-Rumbaugh, 1986). They also exhibited the ability to differentiate between real-time video portrayals of their hands reaching for banana segments otherwise out of view, and replays (hence useless) of early test sessions (Menzel et al., 1985).

Learning Control Tactics of Distal Events Through Observation

Our chimpanzees (Savage-Rumbaugh, 1986) have learned how to use joysticks to control the cursor on a television monitor. This activity was learned by observing an experimenter (Savage-Rumbaugh, 1986) do the task – not as a consequence of systematically having behaviors shaped via the reinforcement of approximations of competency. They learned in a few minutes by watching the experimenter use the joystick to control the cursor's movement so as to make contact with the target, which then resulted in the vending of prized incentives to eat. By doing thus, the chimpanzees have learned subsequently to track erratically moving targets, to select letters with which to "spell" words, and to select arabic numbers in training tasks designed to teach ordinal, if not precise, quantitative values. To date, no one has reported evidence to encourage the conclusion that a monkey could master use of the joystick, either observationally or by any other means, to engage in these complex tasks.

306

The Origin of Cause-Effect Perception and Learning

The importance of extending perceived, if not actual, control to animals and humans alike – to the end of enhancing their health, temperament, and longevity – is becoming generally recognized. Recently Rumbaugh and Sterritt (1986) formulated a theory of *control* and its evolutionary roots. It was posited that the initial emergence of sentience in hominoids was likely coupled with high states of anxiety and fear. The tenet was that to "know" and yet not to be able to "understand how to predict and/or control" natural events might well have left our precursors living in chronic fear and apprehension as they experienced and nonspecifically began to comprehend the power and tempo of natural events. The emergence of a proclivity to perceive and posit cause-effect relationships, valid or not, might have served to abate the extreme affective response to environmental events that were precisely perceived in detail yet understood very inadequately at best. Not only would such an evolutionary move serve to save energy otherwise wasted on high emotion, it would enhance general curiosity and inquiry into salient events – thus increasing the prospects that new and more critical behavior-consequence linkages might be learned. To the degree that such linkages enhanced survival, reproductive success would increase. If we are correct in this perspective, the path was then open for selection for increased brain development that would serve to enhance the processes of relational perceptions/ learning and inference. And when, from time to time, posited cause-effect relationships *were* valid, the prospects both for the control and productive exploitation of natural events and resources would have been profoundly advanced (Savage-Rumbaugh, 1984). *Invalid* cause-effect linkages would still have served the more limited consequence of abating the costs of high emotion, though serving the perhaps unfortunate end of spawning superstitions – a proclivity to which humans still remain painfully subject.

DO WE KNOW THE ESSENCE OF LEARNING?

At a recent national convention, this author had the joy of hearing, on the one hand, a brilliant paper on "Animal Learning Comes of Age" and, also, the consternation of hearing a division president question whether studies of animal learning had even been born! Although we know a tremendous amount about how behavior patterns are acquired and used in laboratory settings, psychology is still struggling to define the essence of the learning process. The trend is that learning is the acquisition of perceived relationships between stimuli and/or between stimuli and responses, with stimuli not being the elicitors of behavior but, rather, being *information* and predictors of consequences. This trend, now substantiated even with research on rats and pigeons, is totally compatible with what comparative psychologists have held to be true for decades – namely, that apes differ from monkeys and other animals in terms of what they learn and how they learn. They are proficient in learning through observing things done by "socially important others." But what they learn is *not* precisely the responses to execute so much as it is what *act* must be carried out for a prized goal or state of affairs to be achieved. Additionally, scientists at our research center are coming to suspect that apes learn observationally most efficiently if the behavior to be modeled is incidental to the natural stream of events of daily life. This is, at best, a working hypothesis at present, not one supported by hard data. Yet it is with new perspectives on old problems that science makes its way to closer approximations of truth.

This author continues to be of the opinion that it will be through the study of observational learning that many contemporary issues in learning research will be resolved, and new and powerful perspectives of behavior will be formulated.

WHAT ARE THE IMPLICATIONS OF CONTEMPORARY AND FUTURE RESEARCH ON THE CHIMPANZEE INTELLECT?

It appears that the inevitable consequences of present and future research will be as follows:

1. We will continue to garner a better understanding of the *evolution* of complex processes that have served humans well as they have come to occupy all major land masses on our planet and now reach out for new horizons in space research.

2. We will continue to gain a better understanding of the role of *environment* in the development of intellectual competence in all advanced primates – including humans.

3. We will come to appreciate more keenly the excellence of primates (through their behavior) as teachers to all of us, scientists and laymen alike, who are fascinated by the dynamics, the processes which generate complex behavior.

4. We will come to appreciate and to *conserve,* hence sustain, animal life in its diverse forms because of its recognized value as part of the natural world and as a biological legacy that ethically must not be denied to future generations.

5. Research will become more competent, more creative, more focused, and more responsibly conducted. Science will thereby more efficiently serve our desires both to learn more of our world and to learn information requisite to solving vexing problems of human and animal health.

In sum, research on the chimpanzee's intellect will, most assuredly, help make for a better world!

ACKNOWLEDGMENTS

This investigation was supported in part by NIH grant RR–00165 from the Division of Research Resources and by grant HD–06016 from the NICHD to the Yerkes Regional Primate Research Center. The Yerkes Center is fully accredited by the American Association for Accreditation of Laboratory Animal Care. Figure 1 was drawn by Mr. Pat McNeely, School of Art and Design, Georgia State University. The able assistance of Ms. Susan Hall and Ms. Judy Sizemore in the preparation of the manuscript is gratefully acknowledged. Parts of this paper were drawn from the paper by Rumbaugh et al. (1987) entitled "Summation in the Chimpanzee *(Pan troglodytes)."*

REFERENCES

Cooper, R. G., Jr. 1984. Early number development: discovering number space with addition and subtraction. In C. Sophian, ed., *Origins of Cognitive Skills.* Hillsdale, NJ: Erlbaum.

Davis, H. 1984. Discrimination of the number three by a raccoon *(Procyon lotor). Animal Learning and Behavior* 12:409–413.

Davis, H., and R. W. McIntire. 1969. Conditioned suppression under positive, negative, or no contingency between conditioned and unconditioned stimuli. *Journal of the Experimental Analysis of Behavior* 12:633–640.

Davis, H., and J. Memmott. 1982. Counting behavior in animals: a critical evaluation. *Psychological Bulletin* 92:547–571.

_____. 1983. Autocontingencies: rats count to three to predict safety from shock. *Animal Learning and Behavior* 11:95–100.

Descoeudres, A. 1921. Le developpement de l'enfant de deux a sept ans. Paris: Delachaux et Niestlé C. A.

Dooley, G. B., and T. V. Gill. 1977. Acquisition and use of mathematical skills by a linguistic chimpanzee. In D. M. Rumbaugh, ed., *Language Learning by a Chimpanzee: The Lana Project,* pp. 247–260. New York: Academic Press.

Fernandes, D. M., and R. M. Church. 1982. Discrimination of the number of sequential events by rats. *Animal Learning and Behavior* 10:171–176.

Ferster, C. B. 1964. Arithmetic behavior in chimpanzees. *Scientific American* 210:98–106.

Fouts, R. S., D. H. Fouts, and D. Schoenfeld. 1984. Sign language conversational interaction between chimpanzees. *Sign Language Studies* 42:1–12.

Gelman, R., and C. R. Gallistel. 1978. *The Child's Understanding of Number.* Cambridge, MA: Harvard University Press.

von Glasersfeld, E. 1977. The Yerkish language and its automatic parser. In D. M. Rumbaugh, ed., *Language Learning by a Chimpanzee: The Lana Project,* pp. 91–130. New York: Academic Press.

_____. 1982. Subitizing. *Archives de Psychologie* 50:191.

Greenfield, P. M., and E. S. Savage-Rumbaugh. 1984. Perceived variability and symbol use: a common language cognition interface in children and chimpanzees. *Journal of Comparative Psychology* 98(2):201–218.

_____. 1986. A chimpanzee's *(Pan paniscus)* use of syntax. Paper presented at the International Congress of Primatology. Bielefeldt, Germany.

Kaufman, E. L., M. W. Lord, T. W. Reese, and J. Volkman. 1949. The discrimination of visual number. *American Journal of Psychology,* 62:498–525.

Krebs, J. R., and N. B. Davies. 1978. *Behavioural Ecology: An Evolutionary Approach.* Oxford: Blackwell.

Mason, W. A. 1962. The primary role of primates in comparative psychology. Washington, DC: American Psychological Association Convention.

Matsuzawa, T. 1985. Use of numbers by a chimpanzee. *Nature* 315:57–59.

Menzel, E. W. 1965. Selection of food by size in the chimpanzee, and comparison with human judgments. *Science* 131:1527–1528.

Menzel, E. W., E. S. Savage-Rumbaugh, and J. Lawson. 1985. Chimpanzee *(Pan troglodytes)* spatial problem solving with the use of mirrors and televised equivalents of mirrors. *Journal of Comparative Psychology* 99(2):211–217.

Miles, H. L. 1983. Apes and language: the search for communicative competence. In J. De Luce and H. T. Wildon, eds., *Language in Primates: Perspectives and Implications.* New York: Springer-Verlag.

Pate, J. L., and D. M. Rumbaugh. 1983. The language-like behavior of Lana chimpanzee: is it merely discrimination and paired-associate learning? *Animal Learning and Behavior* 11(1):134–138.

Rohles, F. H., and J. V. Devine. 1966. Chimpanzee performance on a problem involving the concept of middleness. *Animal Behavior* 14:159–162.

_____. 1967. Further studies of the middleness concept with the chimpanzee. *Animal Behavior* 15:107–112.

Roitblat, H. L., T. G. Bever, and H. S. Terrace. 1984. *Animal Cognition.* New York: Columbia University Press.

Rumbaugh, D. M., ed. 1977. *Language Learning by a Chimpanzee: The Lana Project.* NY: Academic Press.

Rumbaugh, D. M. 1985. Comparative psychology: profiles in adaptation. In A. M. Rogers and C. J. Scheirer, eds., *The G. Stanley Hall Lecture Series,* vol. 5, pp. 7–53. Washington, DC: American Psychological Association.

Rumbaugh, D. M., and G. M. Sterritt. 1986. Intelligence: from genes to genius in the quest for control. In W. Bechtel, ed., *Science and Philosophy: Integrating Scientific Disciplines,* pp. 309–322. Boston: Martinus Nijhoff Publishers.

Rumbaugh, D. M., S. E. Savage-Rumbaugh, and M. Hegel. 1987. Summation in the chimpanzee *(Pan troglodytes). Journal of Experimental Psychology: Animal Behavior Processes* 13(2):107–115.

Savage-Rumbaugh, E. S. 1984. Ape language and implications for the evolution of human intelligence. Paper presented at the Association for Behavior Analysis Convention.

_____. 1986. *Ape Language: From Conditioned Response to Symbol.* New York: Columbia University Press.

Savage-Rumbaugh, E. S., K. McDonald, R. A. Sevcik, W. D. Hopkins, and E. Rubert. 1986. Spontaneous symbol acquisition and communicative use by pygmy chimpanzees *(Pan paniscus). Journal of Experimental Psychology: General* 115(3):211–235.

Savage-Rumbaugh, E. S., J. L. Pate, J. Lawson, S. T. Smith, and S. Rosenbaum. 1983. Can a chimpanzee make a statement? *Journal of Experimental Psychology: General* 112(4):457–492.

Spence, K. W. 1939. The solution of multiple choice problems by chimpanzees. Comparative Psychology Monographs 15:1–54.

Starkey, P., and R. G. Cooper. 1980. Perception of numbers by human infants. *Science* 210:1033-1035.

Starkey, P., E. S. Spelke, and R. Gelman. 1983. Detection of intermodal numerical correspondences by human infants. *Science* 222:179–181.

Strauss, M. S., and L. E. Curtis. 1984. Development of numerical concepts in infancy. In C. Sophian, ed., *Origins of Cognitive Skills.* Hillsdale, NJ: Erlbaum.

Thomas, R. K., and L. Chase. 1980. Relative numerousness judgments by squirrel monkeys. *Bulletin of the Psychonomic Society* 16:79–82.

Yerkes, R. M. 1934. Modes of behavioral adaptation in chimpanzee to multiple choice problems. *Comparative Psychology Monographs* 10:1–108.

4

CHIMPANZEE
CONSERVATION

POPULATION STATUS OF WILD CHIMPANZEES
(PAN TROGLODYTES)
AND THREATS TO SURVIVAL

Geza Teleki

INTRODUCTION

This report presents a continental overview of the status of wild chimpanzee populations and their prospects for survival. The genus *Pan* consists of two species that are endemic to equatorial Africa: *Pan troglodytes* (traditionally known as the common chimpanzee, although "common" is a misnomer now that the implied abundance no longer exists) and *Pan paniscus* (also known as the "pygmy" chimpanzee, now more often referred to as the bonobo [Hill, 1969a, 1969b; Horn, 1979]). The focus here is exclusively on *Pan troglodytes,* with the generic term "chimpanzee" used solely for that species and its member subspecies. An overview of the survival problems faced by *Pan paniscus,* which mirror those of *Pan troglodytes,* can be obtained from reports by MacKinnon (1976), Badrian and Badrian (1977), Susman et al. (1981), Kabongo (1984), Susman and Kabongo (1984), Kano (1984), and Malenky et al. (this volume).

Although classification of chimpanzee subspecies may vary in nomenclature (Hill, 1969a; Reynolds and Luscombe, 1971), taxonomists generally agree that *Pan troglodytes* can be further divided into three main populations that exhibit mutually exclusive geographical ranges: *P. t. verus* in western Africa, *P. t. troglodytes* in central Africa, and *P. t. schweinfurthii* in eastern Africa (Napier and Napier, 1967).[1]

Chimpanzees once spanned most of equatorial Africa, from southern Senegal to western Tanzania, encompassing all or parts of at least 25 countries (Yerkes and Yerkes, 1929; Reynolds, 1967; Hill, 1969a) (Fig. 1). The range of the western subspecies, *P. t. verus,* apparently extended from the Gambia River area to the west bank of the Niger River and included all or parts of at least 12 nations (Albrecht and Dunnett, 1971). The range of the central subspecies, *P. t. troglodytes,* apparently stretched from the east bank of the Niger River to the west bank of the Ubangi River and the west bank of the Zaire River, with a distribution covering all or parts of seven nations. And the range of the eastern subspecies, *P. t. schweinfurthii,* apparently extended from the east bank of the Ubangi River and along the north bank of the Zaire

River to Lake Victoria and Lake Tanganyika in the Rift Valley, including all or parts of at least six nations. Now, however, all subspecies exist in greatly reduced ranges (Fig. 2).

The historical distribution of chimpanzees was initially perceived as commensurate with the tropical forest belt that once stretched nearly unbroken across most of equatorial Africa, leading many early scholars to conclude that chimpanzees were forest-dwelling, tree-living, plant-eating apes (Hartmann, 1886; Hill, 1969a). Today it is evident that chimpanzees also occupy many other habitats and that not all tracts of rain forest contain chimpanzees (Kortlandt, 1965, 1972, 1976; Kano, 1972; Teleki, 1980; McGrew et al., 1981).

Chimpanzees exhibit omnivorous dietary habits and a wide range of locomotor skills (Kortlandt, 1972; Wrangham, 1975; Teleki, 1977; McGrew et al., 1981). At some sites, up to 330 food types (flora and fauna) are consumed during the year (Wrangham, 1977; Teleki, 1977, 1981). Chimpanzees utilize dry savanna-woodlands, mosaic grassland-forests, and humid canopy rain forests, and live from sea level to at least 3,000 meters elevation; however, the availability of permanent sources of surface water may be a key factor limiting the chimpanzees at the fringes of their geographical range (Kortlandt and van Zon, 1969; Kortlandt, 1972, 1983; McGrew et al., 1981). Communities inhabiting dry zones are more nomadic than those living in moist zones (McGrew et al., 1981). Travel across open grasslands can be quite extensive, with treks covering 10 kilometers or more at a time, but some tree cover is essential for foraging and nesting sites (Kortlandt, 1972, 1983; Teleki, 1977; Baldwin et al., 1982). The multidimensional mobility, the omnivorous dietary habits, and both the behavioral and social diversity of chimpanzees indicate that the species is characterized by much greater adaptive flexibility than was previously acknowledged (McGrew and Tutin, 1978; Galdikas and Teleki, 1981; McGrew, 1983; Kortlandt and Holzhaus, 1987).

Historically, survival options for chimpanzees in the natural environment were surely enhanced by an exceptional adaptive flexibility based on the successful interplay of reproductive biology and social organization. Now, however, these same reproductive and social characteristics greatly heighten susceptibility to the impact of extrinsic pressures that threaten all wild chimpanzee populations, and survival is not possible unless the requirements for both reproduction and a full social life are available.

The chimpanzee exhibits exceptionally long birth intervals (5.6 years on average) and a long maturation span (females start reproducing at about 12 years of age). Infants commonly nurse up to four or five years of age and are constantly carried about by their mothers for at least three to four years, and full independence does not occur until about eight years of age (Teleki et al., 1976; Clark, 1977; Goodall, 1986). Infants that are orphaned before the age of five years have a slim chance of surviving even in normal wilderness conditions (Goodall, 1983, 1986). With an expected reproductive span of only 25 years for the average female, the total number of expected live births is no more than five per lifetime (Teleki et al., 1976; Clark, 1977; Goodall, 1983;

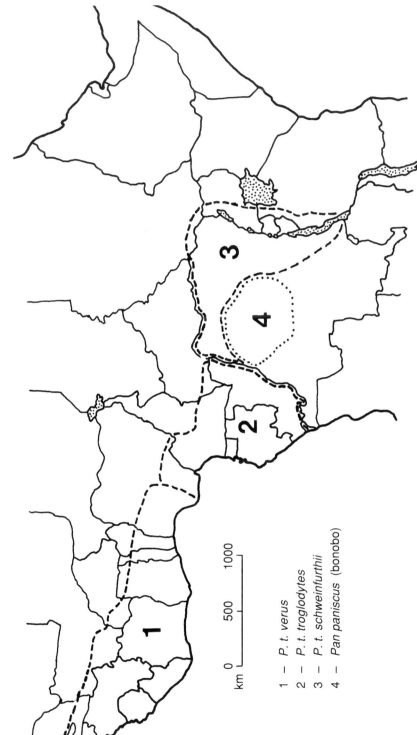

Figure 1. Historical range of *Pan troglodytes* and *Pan paniscus*.

1 – *P. t. verus*
2 – *P. t. troglodytes*
3 – *P. t. schweinfurthii*
4 – *Pan paniscus* (bonobo)

km 0   500   1000

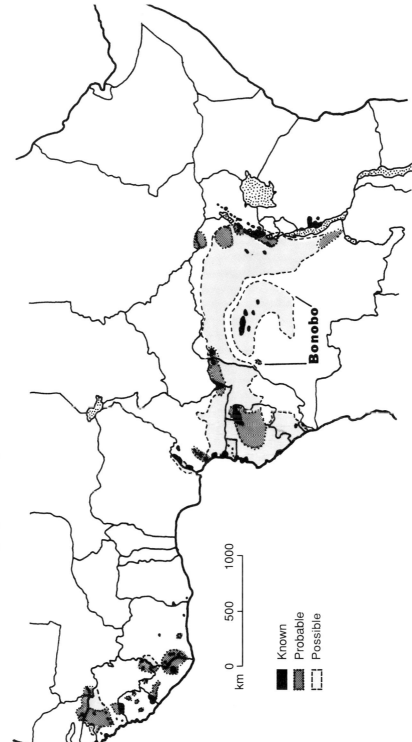

Figure 2. Current distribution of *Pan troglodytes* and *Pan paniscus.*

Hiraiwa-Hasegawa et al., 1984; Sugiyama, 1984). Not all mature females are reproductively capable, and single births are the norm, with twin births being both rare and
low in survival probability (Goodall, 1983, 1986). These reproductive limitations
lower the potential for recovery from any calamity. The risk increases as the remnant
units decrease in size.

Chimpanzee social organization is a highly complex structural unit dependent
upon continuity and stability (Sugiyama, 1973, 1979; Wrangham, 1979; Teleki, 1977;
McGrew and Tutin, 1978; Nishida, 1979; Ghiglieri, 1984; Goodall, 1986). Disruption
of lifetime relationships, strong family ties, and sexual bonds among chimpanzees has
both immediate and long-term detrimental consequences for social unit survival
(McGinnis, 1979; Goodall, 1986). Populations isolated in small units become susceptible to social disintegration, higher stress levels, and disease (Teleki et al., 1976;
Sugiyama, 1984; Ghiglieri, 1984; Goodall, 1986). The situation may be particularly
acute in areas long exploited for commercial trade, such as Sierra Leone, where
remnant social units occupying pockets of shrinking habitat have a marked deficiency
in reproductive females (Teleki, 1980; Teleki and Baldwin, 1980, 1981).

INFORMATION SOURCES AND METHODS

The information for this report is based on published survey reports, supplemented
by direct correspondence with field researchers, by vegetation maps and satellite
imagery, and by trade records from a variety of organizations.[2] The main field sources
are listed in Table 1, with other sources cited in the text and in the accompanying
bibliography.

In order to assess chimpanzee numbers and distributions in a region much larger
than the United States, two basic field parameters had to be measured: *population size*
and *habitat size*. Crude density, or the number of chimpanzees living in any pre-
scribed area, is a derivative of these two measures. One task, then, was to sketch a
template of distributions using available field data, while another task was to extrapolate from that template to a population estimate for the continent.

Original field studies. At the first and most exact level, the original density figures
provided by field researchers working in known habitat areas were tabulated and
mapped for each nation. Systematic survey work began during the early 1960s
(Kortlandt, 1965, 1983; Kortlandt and van Zon, 1969), and extensive survey data are
now available for nine countries inhabited by chimpanzees: Guinea (de Bournonville,
1967; Sugiyama and Soumah, 1988), Equatorial Guinea (Jones and Sabater Pi, 1971),
Tanzania (Kano, 1972), Uganda (Albrecht, 1976), Senegal (Baldwin et al., 1982),
Sierra Leone (Teleki, 1980), Gabon (Tutin and Fernandez, 1984), Mali (Moore,
1985a, 1985b), and Burundi (Trenchard, 1988). Less extensive survey coverage is
available for other countries, including Liberia, Ivory Coast, Ghana, Nigeria, Congo,
Central African Republic, and Zaire. Recent surveys yielding national estimates were
adopted without modification. New information supplanted early survey data, and
in some cases, such as Guinea, significantly altered previous figures (compare
de Bournonville, 1967, with Sugiyama and Soumah, 1988).

Table 1. Conservation conditions for chimpanzee populations in 25 nations.

COUNTRY	CONSERVATION CONDITIONS	MAIN SOURCES
ANGOLA	Limited to NW corner area (Cabinda) above Zaire River; distribution uncertain and survival threatened by hunting; extermination foreseeable; no protected habitats.	Bothma, 1975
BENIN	Extermination completed some decades ago.	Kortlandt, 1965; Kortlandt and van Zon, 1969
BURUNDI	Exterminated east of 29° 30' E and likely elsewhere by now; few isolated units may survive north of Lake Tanganyika near the Zaire border; threatened by hunting and farming; no protected habitat.	Kortlandt, 1965; Eibl et al., 1966; Verschuren, 1978; Trenchard, 1988
CAMEROUN	Exterminated in central area; isolated units along west coast, bulk of population along south border; threatened by hunting and logging; extermination along Nigeria border foreseeable; limited protection in some areas, including Campo (3,300 km^2), Duala-Edea (1,600 km^2), Dja (5,000 km^2).	Kortlandt, 1965; Gartlan and Struhsaker, 1972; Gartlan, 1975a, 1975b; Cousins, 1978
C. AFR. REP.	Exterminated above 4° N and east of 18° E; isolated units thinly dispersed in SW corner near Cameroun and Congo borders, all declining; threatened by hunting; no protected habitat, but one proposed at Dzanga Sangha (3,000 km^2).	Kortlandt, 1965; Kortlandt and van Zon, 1969; Carroll, 1986a, 1986b; Fay, 1987
CONGO	Isolated units reportedly present in north and southwest along Gabon border; all declining, threatened by hunting and logging; presence at Odzala (12,000 km^2) not confirmed; no other protected habitat.	Bassus, 1975; Tanno, 1987; Spinage, 1980
EQ. GUINEA	Thinly dispersed throughout central area with isolated concentrations around hill-top forests; rapidly shrinking range, all populations declining; threatened by hunting, logging, and farming; no protected habitats.	Jones and Sabater Pi, 1971; Sabater Pi and Groves, 1972; Sabater Pi and Jones, 1967; Sabater Pi, 1979
GABON	Widely but somewhat unevenly distributed in central and eastern areas; locally vulnerable to hunting and logging; no protected habitat, but some planned; rehabilitation tried.	Gandini, 1979; Tutin and Fernandez, 1983, 1984, 1985; Hladik, 1974; Harcourt and Stewart, 1980
GAMBIA	Extermination completed some decades ago; rehabilitation project ongoing.	Brewer, 1976, 1978; Carter, 1978, 1981

(continued)

Table 1. (continued)

COUNTRY	CONSERVATION CONDITIONS	MAIN SOURCES
GHANA	Exterminated throughout except along the Ivory Coast border; very scarce and declining; threatened by hunting and logging, even in protected areas of Bia (78 km^2) and Nini-Suhien (104 km^2); extermination foreseeable soon; rehabilitation project failed.	Booth, 1956; Jeffrey, 1970, 1974; Rucks, 1976; Asibey, 1978
GUINEA	Thinly dispersed in NW and SE areas at low densities; heavy exploitation prior to the 1970s for biomedical use; declining with exports via Sierra Leone; new survey of central areas urgently needed; no protected habitats.	de Bournonville, 1967; Kortlandt, 1965, 1986; Kortlandt and van Zon, 1969; Teleki, 1980; Sugiyama, 1984; Sugiyama and Soumah, 1988
GUINEA-BISSAU	Extermination probably completed in recent years, but few small units may survive in SE along Guinea border; no protected habitat.	Kortlandt, 1965; Kortlandt and van Zon, 1969
IVORY COAST	Thinly dispersed in isolated units, mainly in SW along Liberia border; threatened by hunting and logging; recent exports mostly of Guinea origin; extermination foreseeable unless protection improved at Mt. Nimba (50 km^2), Mt. Peko (340 km^2), Marahue (1,010 km^2), and Asagny (300 km^2); rehabilitation project planned.	Kortlandt, 1965, 1986; Bourliere et al., 1974; Boesch, 1978; Prince, 1985; Prince et al., 1986
LIBERIA	Thinly dispersed in NW and SE, mostly in concessioned forest blocks; all declining, threatened by heavy hunting and logging; exports high for decades; exploitation in the 1970s for biomedical use; export drain continues via Sierra Leone; nominal protection at Sapo park (1,300 km^2), but no other protected habitat; rehabilitation project ongoing.	Kortlandt, 1965; Jeffrey, 1977; Teleki and Baldwin, 1979b; Teleki, 1980; Anderson et al., 1983; Prince et al., 1986
MALI	Thinly dispersed in SW corner near Guinea and Senegal borders; marginal habitat is shrinking; no protected habitats, but one proposed at Bafing River (1,000 km^2).	Sayer, 1977; Moore, 1985a, 1985b
NIGERIA	Reportedly present in SE corner along the Cameroun border; rapid habitat shrinkage; extermination foreseeable; no protected habitats.	Kortlandt, 1965; Monath and Kemp, 1973; Hall, 1976
RWANDA	Extermination may have been completed in recent years, but isolated small units may exist in west along Zaire border; threatened by farming; rapid habitat shrinkage; presence at Volcanoes (120 km^2) not confirmed; no other protected habitat.	Curry-Lindahl, 1956; Eibl et al., 1966; Spinage, 1972

(continued) |

Table l. (continued)

COUNTRY	CONSERVATION CONDITIONS	MAIN SOURCES
SENEGAL	Thinly dispersed in southeast along Guinea and Mali borders; marginal habitat shrinking; small unit protected at Niokolo-Koba (9,130 km^2), but no other protected habitat; rehabilitation project failed.	Baldwin et al., 1982; McGrew et al., 1981; Brewer, 1976, 1978
SIERRA LEONE	Thinly dispersed in north and east along Guinea and Liberia borders, largely exterminated elsewhere; severely threatened by commercial exploitation; major exporter for several decades; nominally protected at new Outamba-Kilimi (980 km^2) park but no other protected habitats.	Phillipson, 1978; Teleki, 1980; Teleki and Baldwin, 1981; Whitesides, 1985; Wilkinson, 1974
SUDAN	Thinly dispersed in south along Zaire border; marginal habitat shrinking; extermination foreseeable; no protected habitats.	Kock, 1967, 1969; Kortlandt, 1965; Butler, 1966
TANZANIA	Thinly dispersed isolated units in east along Lake Tanganyika as far as 7° S; well protected at Gombe (32 km^2) and Mahale (1,500 km^2) but not elsewhere; rehabilitation successful at Rubondo Island.	Suzuki, 1969; Kano, 1972; Nishida, 1979; Teleki et al., 1976; Itani, 1979; Goodall, 1983; Borner, 1985
TOGO	Extermination completed some years ago.	Kortlandt, 1965
UGANDA	Widely dispersed isolated units in forest blocks along Rift Valley, parallel to the Zaire border; many units threatened by habitat degradation and some hunting; reasonably protected at Budongo (11 km^2), Toro (555 km^2), Kibale (560 km^2), and some other sites; no protection in most forest blocks.	Stott and Selsor, 1959; Reynolds and Reynolds, 1965; Sugiyama, 1968; Suzuki, 1971; Albrecht, 1976; Ghiglieri, 1984; Butynski, 1985; Struhsaker, 1987
UPPER VOLTA[1]	Extermination completed some decades ago.	Kortlandt, 1965
ZAIRE	Distribution uncertain north of Zaire River, but present in northwest near Congo border; dispersed units in east from Sudan border to south end of Lake Tanganyika; threatened by hunting in all known areas; exports via Congo and Zambia; nominally protected at Garamba (4,920 km^2), Virunga (7,800 km^2), Kahuzi-Biega (6,000 km^2), but no effectively protected habitat exists.	Emlen and Schaller, 1960; Kortlandt, 1965; Rahm, 1965, 1966, 1967; Hendrichs, 1977; Tanno, 1987

Note: The main sources cited above comprise only a small portion of the material examined in compiling this table. See Bibliography for additional sources. Supplementary information was also obtained from several general reports containing data on chimpanzee conservation (Simon and Warland, 1968; Anonymous, 1978b; Teleki and Baldwin, 1979a; Wolfheim, 1983; Mack and Mittermeier, 1984; MacKinnon and MacKinnon, 1986).
1. Now known as Burkina Faso.

Projections from studies at specific sites. At the second and less exact level, the original figures from specific localities were extrapolated to larger surrounding areas, or potential habitats, where comprehensive ground survey data were absent. In such cases supplementary information on the nature and size of those neighboring areas was obtained from other sources, and the local density was extrapolated to the neighboring area (e.g., works of Boesch [1978] in Ivory Coast, McGrew et al. [1981] in Senegal, and Ghiglieri [1984] in Uganda). As the probability of chimpanzees occurring at similar densities in adjacent potential habitats could often be cross-checked, projection on this limited scale involved only minor uncertainty. In nations containing both site studies and area surveys, such as Tanzania and Uganda, the two were compared.

Extrapolation from potential habitat. At the third and most speculative level, it was necessary to estimate populations over large potential habitat areas, assuming that conditions for occupancy were normal. The numbers for Congo, Cameroun, and Zaire were computed in this way. Average density levels from known localities with roughly equivalent habitat conditions were thus applied to larger areas lacking any kind of site or survey work.

On the relevant tables, *known habitat areas* (for which reliable field data exist) and *potential habitat areas* (for which little or no field data are available) are separately indicated. Potential habitat can be further subdivided into *probable habitat areas,* where at least incidental observations by professionals confirm the presence of some chimpanzees, and *possible habitat areas,* where no reliable field information of any kind exists at present even though conditions appear suitable for occupancy.

The three main habitat types used in producing national population estimates – dry savanna-woodland, mosaic grassland-woodland-forest, and humid canopy rain forest – correspond to the Afrotropical Realm classifications developed for the International Union for the Conservation of Nature (IUCN) system of ranking biogeographical provinces (Anonymous, 1985). More refined environmental evaluations were achieved for specific localities or areas by reference to the work of MacKinnon and MacKinnon (1986) and others. With most chimpanzee research occurring in protected areas such as reserves and parks, where some ecological research has also been done, it was feasible to determine the habitat types and dimensions of such sites with some accuracy (Anonymous, 1987a). Field survey and trade data were checked against various general reports addressing broad chimpanzee conservation problems (Kortlandt, 1965; Simon and Warland, 1968; Harrisson, 1971; Suzuki, 1971; Anonymous, 1978b, 1988a; Wolfheim, 1983; Mack and Mittermeier, 1984; Oates, 1985).

Density Estimates

Based on this combination of sources, a total of somewhat less than 1,000,000 square kilometers was identified and mapped as habitat where chimpanzees do occur, probably occur, or possibly occur. Crude density estimates at the most intensively studied sites run from a low of 0.1 to a high of 6.8 chimpanzees per square kilometer, with most localities supporting an average density of considerably less than 1.0 chimpanzee per square kilometer.

In the marginal habitats located at the extreme edge of the continental range of chimpanzees, such as the dry savanna-woodlands of southern Senegal (Baldwin et al., 1982) and southern Mali (Moore, 1985a, 1985b), there are consistently low average densities of about 0.1 chimpanzees per square kilometer. A low density projection is also consistent with the results obtained some time ago by de Bournonville (1967) and Kortlandt (1965) in Guinea, and is even more applicable now due to recently documented population decline (Teleki, 1980; Sugiyama and Soumah, 1988). It is therefore likely that this 0.1 figure is a reliable baseline for projecting potential numbers in unsurveyed dry habitats such as those occurring in Guinea-Bissau.

The somewhat more moist mosaic habitats of grassland-woodland-forests, such as those occurring in western Tanzania (Kano, 1972) and northern Sierra Leone (Harding, 1983, 1984), consistently yield slightly higher densities of about 0.2 chimpanzees per square kilometer, so that figure is more useful in projecting densities for other equivalent habitats in countries such as Nigeria, Ghana, and even parts of southern Sudan.

Finally, rain forest density data from areas as widely dispersed as Liberia (Anderson et al., 1983), Gabon (Tutin and Fernandez, 1984), southeastern Central African Republic (Carroll, 1986a, 1986b; Fay, 1987), and western Uganda (Albrecht, 1976; Butynski, 1985; Struhsaker et al., 1986) indicate that an average figure of 0.3 chimpanzees per square kilometer is representative for many unsurveyed canopy forest blocks, such as those found in southern Cameroun and northern Zaire.

Forest densities do vary, however, and it is essential to keep in mind that density is not exclusively dependent on any single environmental condition (Carroll, 1986a). The highest densities (from 4.0 to 6.8 individuals per square kilometer) were recorded long ago by researchers working in the Budongo Forest Reserve of Uganda (Reynolds and Reynolds, 1965; Sugiyama, 1968; Suzuki, 1971). These densities seem particularly anomalous (Albrecht, 1976) especially now that poaching has become a problem (Anonymous, 1988b, 1989a; Wrangham, pers. comm.). Lower densities vary from 0.4 chimpanzees per square kilometer at Bwindi (Butynski, 1985) to 2.0 at the core of Kibale (Ghiglieri, 1984). The fact that chimpanzee densities in other tropical forest habitats (such as Kibira National Park in Burundi, with a density of 0.5 [Trenchard, 1988]) match the low end of the Ugandan density scale suggests that Uganda's peak density figures are not a reliable reference point for making projections on a continental spectrum. Such variability implies that habitats which initially appear to be grossly similar in configuration can nevertheless differ greatly in chimpanzee carrying capacity, habitat degradation, or hunting pressure (Struhsaker, 1981, 1987; Skorupa, 1983; Ghiglieri, 1984; Van Orsdol, 1986).

The baseline figures of 0.1, 0.2, and 0.3 chimpanzees per square kilometer seem to provide the best options for projecting population numbers across tropical Africa. The total population estimates listed in this report should be viewed with full awareness that combined local pressures may already have pushed chimpanzee numbers below the levels projected, particularly in those areas in western and eastern Africa inhabited by isolated remnant populations.

POPULATION STATUS

The contrast between the chimpanzees' historical distribution (Fig. 1) and current distribution (Fig. 2) is a matter of range depletion and fragmentation, with present range limits at latitudes 13° North and 7° South. Undeniably, severe reductions in *Pan troglodytes* distributions have occurred since historical ranges were mapped in the last century (as described by Yerkes and Yerkes [1929]) and even since the modern mapping efforts of de Bournonville (1967) or Kortlandt and van Zon (1969). Four of the 25 countries once inhabited by chimpanzees have none at all today, five other countries have populations where extermination can be expected soon, and in five more countries the resident populations are so small and dispersed that they are severely at risk. Even in the ten countries presently estimated to contain populations above 1,000 chimpanzees, the range has diminished greatly since the 1950s, a time of unprecedented agricultural land development, natural resource extraction, and human population growth (Goliber, 1985).

Population Estimates

The current status of wild chimpanzees is summarized in Tables 2–4. Only a fraction of the totals appearing in these tables represent localities where systematic fieldwork has been conducted in recent decades. As new information continues to accumulate, adjustments will be necessary to some estimates. For example, according to Sugiyama and Soumah (1988), Guinea, which ranks third in estimated population, probably has less than half the chimpanzees estimated in these tables. New survey work in northern Congo in 1989 confirms the presence of some chimpanzees in the Lac Telle area (Moore, in prep.). And a 1988 resurvey of the Ugalla area of western Tanzania, also by Moore (pers. comm.), indicates that Kano's (1972) density estimate remains valid.

Table 2 presents estimates of chimpanzee numbers in terms of *known habitat areas* and *potential habitat areas,* as well as a combined total for each country. In each case, the base population estimates are followed by projections of the numbers of breeding females likely to be present, inasmuch as that demographic class is the crucial one from a conservation standpoint (Teleki et al., 1976). The proportion of breeding females, calculated at 23% for the average population, was derived from several long-term field studies in different continental locations (Teleki et al., 1976; Baldwin et al., 1982; Goodall, 1983; Hiraiwa-Hasegawa et al., 1984; Sugiyama, 1984). Two sets of totals appear at the bottom of Table 2, the first being less speculative than the second. The second total shows what the real survival status of *Pan troglodytes* might be if the large segments of potentially suitable but unsurveyed habitats in the Congo Basin contain few or no chimpanzees.

Table 3 presents the same estimates in a more condensed form, rounding the original figures to the nearest thousand and summarizing them in terms of subspecies distributions. Two sets of totals also appear at the bottom of this table to illustrate how matters might stand if viable populations occupy only the 18,000 square kilometers of dispersed habitats within northern and eastern Zaire where chimpanzees are currently known to occur.

Table 4 provides a more detailed breakdown of low and high population estimates by region and by nation, and also includes a basic evaluation of survival prospects for each national population. General protective status is rated on a relative scale. The age and accuracy of the data used to generate estimates are also shown as a second measure of reliability.

Three decades of land development have produced such major environmental changes in some parts of Africa that the earliest surveys are becoming outdated. Survey data have a short decay period, and constant monitoring of population status is essential to conservation of chimpanzee populations. Some countries – Liberia, Ivory Coast, Cameroun, Congo, and Zaire – emerge as top priorities for national surveys.

At the most conservative level of projection, Table 3 shows about 17,000 chimpanzees likely occur in known habitat areas while some 175,000 additional chimpanzees may occur in potential habitat areas. As projected in Table 4, there might be as few as 150,000 or as many as 235,000 chimpanzees surviving in tropical Africa today.

Given the extreme fragmentation of the populations, as noted below, even the higher estimates are decidedly meager in the context of species survival potential. We should assume the lower figure in order to leave a greater margin of safety for introducing more effective conservation measures before such extensive damage occurs that it becomes irreversible (Harrisson, 1971). Even the higher estimate does not permit further exploitation of chimpanzees or the destruction of their remaining habitat.

Population Distribution

There has been consensus for some time among field experts that chimpanzee populations are in sharp decline everywhere (Kortlandt, 1965, 1966, 1972; Harrisson, 1971; Anonymous, 1978b, 1988a; Wolfheim, 1983). This section provides detailed information about subspecies populations throughout the entire continent. The picture that emerges is bleak, in some respects far more so than anyone previously expected, and the prognosis for the future is not optimistic for any wild population.

The western subspecies, *P. t. verus,* is most immediately at risk because it is already extinct in at least four nations, nearly so in five others, and rapidly declining in three more (Table 4). Of an original population numbering perhaps 600,000 or more in a vast forested range of nearly two million square kilometers, only about 17,000 chimpanzees remain. In Sierra Leone, for example, the population is estimated to have dropped, mostly within this century, from about 20,000 to only 2,000 ± 500 (Teleki, 1980). In Guinea, where chimpanzees once seemed to be moderately abundant, with estimates of 12,000 to 14,000 in the 1960s (de Bournonville, 1967), there may be only 2,000 to 4,000 still present (Sugiyama and Soumah, 1988). Ivory Coast and Liberia now appear to be the final bastions of survival for the subspecies. Surveys are urgently needed in both countries so that appropriate strategies for protection can be developed.

The eastern subspecies, *P. t. schweinfurthii,* is apparently somewhat more numerous and therefore seemingly less at risk, but extermination is foreseeable in three nations (Table 4). Although 96,000 chimpanzees may survive in the region, survey

Table 2. Estimated numbers of chimpanzees in known habitats and potential habitats within 25 nations of equatorial Africa, encompassing an area of 10.68 million square kilometers.

Region[1]	Nations	Known Habitat			Potential Habitat			Total National Estimates	
		Area (km²)	Estimated chimpanzees	Estimated breeding females	Area (km²)	Estimated chimpanzees	Estimated breeding females	Total estimated chimpanzees[2]	Estimated breeding females
Western	# Guinea	700	200	50	34,000	9,800	2,250	10,000 #	2,300
	‡ Liberia	1,300	400	90	9,000	2,600	600	3,000	690
	Sierra Leone	1,000	300	70	5,600	1,700	390	2,000	460
	Ivory Coast	3,700	750	170	0	0	0	750	170
	Mali	1,000	190	50	6,500	510	110	700	160
	Ghana	600	180	40	1,000	220	50	400	90
	Senegal	300	40	10	1,200	160	40	200	50
	Guinea-Bissau	0	0	0	1,000	100	20	100	20
	Upper Volta[3]	0	0	0	0	0	0	0	0
	Togo	0	0	0	0	0	0	0	0
	Gambia	0	0	0	0	0	0	0	0
	Benin	0	0	0	0	0	0	0	0
Central	‡ Gabon	6,200	2,000	450	200,000	62,000	14,250	64,000	14,700
	‡ Cameroun	7,500	2,400	550	17,500 *	5,600 *	1,300 *	8,000 *	1,850 *
	‡ Congo	0	0	0	13,000 *	4,000 *	900 *	4,000 *	900 *
	Equat. Guinea	100	50	10	13,000	1,950	450	2,000	460
	C. Afr. Republic	3,000	240	50	8,300	660	150	900	200
	‡ Angola (Cabinda)	0	0	0	1,400	400	90	400	90
	Nigeria	0	0	0	1,000	200	50	200	50
Eastern	‡ Zaire	18,000	6,000	1,400	470,000 *	84,000 *	19,300 *	90,000 *	20,700 *
	Uganda	1,000	2,000	450	1,500	2,000	450	4,000	900
	Tanzania	9,000	2,000	460	0	0	0	2,000	460
	Sudan	0	0	0	1,000	300	70	300	70
	Rwanda	0	0	0	300	150	40	150	40
	Burundi	0	0	0	200	100	20	100	20
TOTAL 1		53,400	16,750	3,850	785,500 -470,000	176,450 -84,000	40,530 -19,300	193,200 -84,000	44,380 -19,300
TOTAL 2		53,400	16,750	3,850	315,500	92,450	21,230	109,200	25,080

Notes: # = data subject to revision. The 10,000 total estimate for Guinea may no longer be valid, as Sugiyama and Soumah (1988) suggest a population decrease of at least 50% since the de Bournonville (1967) survey, with perhaps only 3,000 surviving today. * = highly speculative figures based on the assumption that a large portion of the Congo Basin lying west, north, and east of the Zaire River, and including unsurveyed regions of Zaire, Congo, and Cameroun, contains a relatively even distribution of chimpanzees (refer to Total 1). This may be overly optimistic, however, as 60% of the total potential habitat and 46% of the total estimated number of chimpanzees would presumably then be located in Zaire, where in fact there may be few or no chimpanzees outside the known habitat area. Removal of the most highly speculative Zaire figures may therefore provide a more realistic assessment of the present status of chimpanzees in equatorial Africa (refer to Total 2). ‡ = nations in which chimpanzees are heavily hunted for food or ritualistic purposes.

1. Regions correspond to geographical ranges of the three subspecies.
2. See range of high and low estimates in Table 4.
3. Now known as Burkina Faso.

Table 3. Crude numerical estimates of population for the three chimpanzee subspecies in equatorial Africa.

Subspecies	Estimated number of chimpanzees in known habitats		Estimated number of chimpanzees in potential habitats		Total estimated number of chimpanzees in all habitats	Estimated number of breeding female chimpanzees	% of Total chimpanzees Based on Total 1	Total 2	Total known and potential habitat area (km²)	% of Total habitat Based on Total 1	Total 2
P. t. verus (Western Region)	2,000	+	15,000	=	17,000 #	4,000	9	16	67,000	8	18
P. t. troglodytes (Central Region)	5,000	+	74,000	=	79,000	18,000	41	72	271,000	32	74
P. t. schweinfurthii (Eastern Region)	10,000	+	86,000 *	=	96,000 *	22,000 *	50	12	501,000 *	60	8
TOTAL 1	17,000	+	175,000 −84,000	=	192,000 −84,000	44,000 −19,000	100		839,000 −470,000	100	
TOTAL 2	17,000		91,000		108,000	25,000		100	369,000		100

Notes: # = data subject to revision. The total number in the Western Region may be down to 10,000 chimpanzees, because recent survey work by Sugiyama and Soumah (1988) suggests a population decrease in Guinea of at least 50%, with perhaps only 3,000 surviving today. * = highly speculative figures derived by assuming that 470,000 km² of unsurveyed potential habitats in northern and eastern Zaire contain a relatively even distribution of chimpanzees (refer to Total 1). Removal of these highly speculative figures may yield a more realistic assessment of chimpanzee numbers throughout Africa (refer to Total 2). All figures are rounded off to the nearest thousand after computation and collation from original field sources.

Table 4. Current status of chimpanzee populations in three major geographical regions of equatorial Africa.

Region[1]	Country	Estimated Total Population Low	Estimated Total Population High	Survival prospects	Protective status	Data reliability rating	Study period
Western	Benin	0	0	Extermination completed	NA	3	1960-65
	Gambia	0	0	Extermination completed	NA	1	1975-85
	Togo	0	0	Extermination completed	NA	3	1960-65
	Upper Volta[2]	0	0	Extermination completed	NA	3	1960-65
	Guinea-Bissau	0	200	Extermination expected	None	3	1960-65
	Senegal	100	300	Extremely scarce, at risk	Adequate	1	1976-79
	Ghana	300	500	Extermination expected	Nominal	3	1952-76
	Mali	600	800	Extremely scarce, at risk	None	1	1975-84
	Ivory Coast	500	1,000	Highly threatened	Modest	2	1960-80
	Sierra Leone	1,500	2,500	Vulnerably scarce, declining	Nominal	1	1978-81
	Liberia	3,000	4,000	Vulnerably scarce, declining	Nominal	2	1960-83
	Guinea #	8,000	12,000	Moderately abundant locally	None	1	1965-66
	Subtotals	14,000	21,300	57% presumably in Guinea #			
Central	Nigeria	100	300	Extermination expected soon	None	3	1960-75
	Angola (Cabinda)	200	500	Extremely scarce, at risk	None	3	1970-73
	C. Afr. Republic	800	1,000	Vulnerably scarce, declining	None	1	1960-85
	Equat. Guinea	1,000	2,000	Vulnerably scarce, at risk	None	1	1966-68
	Congo *	3,000	5,000	Locally threatened, declining	None	3	1973-74
	Cameroun *	6,000	10,000	Moderately abundant locally	Modest	2	1960-80
	Gabon	51,000	77,000	Still abundant, widespread	Nominal	1	1980-83
	Subtotals	62,100	95,800	80% presumably in Gabon			
Eastern	Burundi	0	200	Extermination expected soon	None	3	1960-76
	Rwanda	100	200	Extermination expected soon	None	3	1960-75
	Sudan	200	400	Extremely scarce, at risk	None	3	1960-65
	Tanzania	1,500	2,500	Moderately abundant locally	Good	1	1960-80
	Uganda	3,000	5,000	Moderately abundant locally	Variable	2	1962-84
	Zaire *	70,000	110,000	Locally abundant, declining	None	2	1960-75
	Subtotals	74,800	118,300	93% presumably in Zaire			
	TOTALS	150,900	235,400	9% Western, 41% Central, 50% Eastern			

Notes: Reliability rating: 1 = High, based on evidence from one or more national/regional field surveys; 2 = Medium, based on estimates extrapolated from local field studies; 3 = Low, based on incidental field observations at multiple sites. # = data subject to revision. Recent survey work by Sugiyama and Soumah (1988) suggests that the low and high estimates for Guinea should be 2,000 and 4,000 respectively. * = highly speculative estimates derived by assuming uniform distributions in potentially available habitats. Totals may be closer to 100,000 and 150,000 if Zaire harbors only 20,000 or even fewer chimpanzees.
1. Regions correspond to geographical ranges of the three subspecies.
2. Now known as Burkina Faso.

data are absent from the 470,000 square kilometers of potential habitat in northern Zaire, where studies of human hunting societies indicate that local chimpanzee populations are heavily persecuted by humans (Harako, 1981; Hart, 1978; Hart and Thomas, 1986; Tanno, 1987). If the average density in Zaire is much lower than the projections offered in Tables 2–4, or if large parts of Zaire contain no chimpanzees even within intact habitat, then the two small, highly studied remnant populations in the national parks of western Tanzania (where protection is still reasonably good despite some pressure from human settlement and some poaching for illegal trade) may be the only stronghold for the subspecies (Hasegawa and Nishida, 1984; Goodall, 1986; Anonymous, 1987b).

The central subspecies, *P. t. troglodytes,* could be the most numerous at this time, given the survey estimates currently available for Gabon (Tutin and Fernandez, 1983, 1984, 1985), but even that population segment faces potentially serious long-range survival problems (Tutin and Fernandez, 1987, 1988). Numbers are very low in two nations, and populations are scarce and highly vulnerable in three others (Table 4). The wildlife resources of neighboring countries have already been substantially depleted (Sabater Pi and Jones, 1967; Spinage, 1980), and care must be taken to avoid a similar outcome in Gabon. Surveys of southern Cameroun, Congo, and the Cabinda area of Angola are particularly needed. Chimpanzees have long been threatened within this central range by domestic subsistence hunting coupled with some commercial taking for export, and may in the future be hard pressed by escalating habitat loss (Sabater Pi and Groves, 1972; Sabater Pi, 1979; Gandini, 1979; Harcourt and Stewart, 1980; Tutin and Fernandez, 1984, 1988; Carroll, 1986a, 1986b; Tanno, 1987).

FACTORS AFFECTING SURVIVAL

All *Pan troglodytes* subspecies are vulnerable to extinction if the factors causing their decline continue to operate at the present pace. Numbers alone cannot illustrate how deeply the various *Pan troglodytes* subspecies are mired in survival problems or reveal the velocity of their retreat toward extinction. Central to a discussion of survival problems is an awareness of the reproductive and social parameters which characterize chimpanzees (see Introduction, this report). Time spans measured in decades, not in months or years, are needed for chimpanzee populations to recover once stability is lost (Teleki et al, 1976).

The current decline in chimpanzee population stems from human pressures. Development of land reduces chimpanzee habitat and fragments their existing populations, increasing the exposure to contagious disease. Fragmentation also makes the chimpanzee more accessible for commercial exploitation in a variety of forms.

Habitat Depletion and Fragmentation

Timbering, mining, farming, and other forms of land development play a significant role in habitat loss. Timber extraction is now a major pressure on forest primates throughout tropical Africa (Gartlan, 1975a, 1975b; Gartlan and Struhsaker, 1972;

Struhsaker, 1987). Logging operations, in both clear-felling and selective-felling forms, often attract agricultural encroachment and commercial hunting of wildlife (Gandini, 1979; Teleki, 1980; Harcourt and Stewart, 1980). Timber extraction and the related human activities can clear a large area of wildlife within a few years, particularly when a commercially valuable target species such as *Pan troglodytes* is present (Teleki, 1980; A. Peal, pers. comm.).

All regional chimpanzee populations are currently undergoing fragmentation into ever smaller and more isolated units (compare Figure 1 and Figure 2), although the process of fragmentation is more advanced in some regions than in others (Jones and Sabater Pi, 1971; Albrecht, 1976; Spinage, 1980; Teleki, 1980). Fragmentation of a regional population restricts gene flow and increases vulnerability to other pressures. This phenomenon may be even more indicative of pending extinction than a general decline in numbers.

Deforestation is most advanced in the western region, where only remnant tracts of primary rain forest remain in all countries. Deforestation is progressing most rapidly in the eastern region, where the remaining forest blocks are being felled at an ever faster pace. In Sierra Leone, for example, only 4% of the original canopy forest still stands, confined mostly to the Gola Forest Reserves, and even that area has been concessioned for felling by the end of the century (Fox, 1968; Clarke, 1969; Cole, 1980). The Golas are also accessible to motorized hunting units crossing the border from Liberia (Teleki, 1980). The Gola chimpanzees, along with other rare fauna, thus face destruction within the century if protective measures are not introduced soon (Oates, 1980; Whitesides, 1985; Merz, 1986; Davies, 1987a, 1987b). Several other forest blocks in the western countries are equally beleaguered by both corporate and local timber-felling operations (Robinson, 1970; Jeffrey, 1970; Dossi et al., 1981; Anadu and Oates, 1982; Lamotte, 1983; Anadu, 1987).

The western subspecies is the most fragmented. Only 2,000 individuals are widely dispersed through the known sites, and perhaps another 15,000 survive in various potential habitat areas (Table 3). The 17,000 estimated total may already be down to 7,000. Only three nations – Guinea, Sierra Leone, and Liberia – have resident populations that definitely exceed 1,000 members, and in each, local ranges are shrinking rapidly due to persistent human persecution (Kortlandt, 1965; Albrecht and Dunnett, 1971; Jeffrey, 1977; Teleki, 1980; Teleki and Baldwin, 1979b, 1981; Robinson and Peal, 1981; Sugiyama and Soumah, 1988). The habitat available within the western region, estimated at about 67,000 square kilometers spread over eight countries, may be shrinking at a rate of 2% or more per annum (Fox, 1968; Verschuren, 1983). Traditionally high levels of hunting for bushmeat and export markets fuel the pace of fragmentation (Anonymous, 1988a).

At the other end of the continent, in Uganda, the closed forest that still covers 3% of the land is shrinking so rapidly that chimpanzees may soon become as scarce as they now are in Rwanda and Burundi (Spinage, 1972; Albrecht, 1976; Verschuren, 1978; Skorupa, 1983; Van Orsdol, 1986; Butynski, 1985; Struhsaker, 1981, 1987). Chimpanzee population distributions in Uganda have changed considerably in recent decades (Stott and Selsor, 1959; Albrecht, 1976). Other eastern region forests inhabited by chimpanzees are also under the pressure of timber-felling operations, and the resident chimpanzees are in peril (Hart and Thomas, 1986; Trenchard, 1988). An estimated 10,000 chimpanzees may occur in known habitats, and perhaps another 86,000 in potential habitat areas located mainly in Zaire (Table 3).

The eastern subspecies is also highly fragmented throughout its 500,000 square kilometer range of potentially suitable habitat dispersed over six countries. Fragmentation is most advanced in Sudan, Uganda, Rwanda, and Burundi (Emlen and Schaller, 1960; Butler, 1966; Eibl et al., 1966; Kock, 1967, 1969; Spinage, 1972; Verschuren, 1978; Struhsaker, 1981, 1987; Skorupa, 1983; Harcourt, 1984; Van Orsdol, 1986; Hart and Thomas, 1986; Tanno, 1987). In Uganda, for instance, forest habitats are dispersed in dozens of small blocks isolated by farmlands (Albrecht, 1976; Butynski, 1985; Struhsaker, 1987), with poaching of chimpanzees occurring in some blocks (Ghiglieri, 1984; Anonymous, 1988b, 1989a). In Tanzania, one of the two populations is being further fragmented by human settlement (Hiraiwa-Hasegawa et al., 1984; Goodall, 1986; Anonymous, 1987b, 1988b; T. Nishida, pers. comm.).

The fragmented populations of the western and eastern subspecies are primarily located in remnant forest and game reserves or in national parks. In many such "protected areas," poaching for meat and live infants is common, as is unauthorized logging, mining, and farming (Petrides, 1965; Riney and Hill, 1967; Happold, 1971; Suzuki, 1971; Jeffrey, 1975; Asibey, 1978; Spinage, 1980; Harcourt and Stewart, 1980; Teleki, 1980, 1986; Teleki and Baldwin, 1981; Teleki and Bangura, 1981; Robinson and Peal, 1981; Dossi et al., 1981; Lamotte, 1983; Wolfheim, 1983; Harcourt, 1984; Ghiglieri, 1984; Van Orsdol, 1986; McKinnon and McKinnon, 1986; Struhsaker, 1987; Verschuren, 1983; Anonymous, 1987b, 1988a).

In the central region, where deforestation has progressed at widely variable rates in different countries, chimpanzee habitats are most severely threatened in Equatorial Guinea and Congo (Sabater Pi and Jones, 1967; Spinage, 1980). Timber-felling may not yet be the chief pressure on forest primates in Gabon and southern Cameroun, but it probably will be soon (Harcourt and Stewart, 1980; Lahm, 1985; Tutin and Fernandez, 1984, 1987, 1988). The main pressure at present stems from a strong local hunting tradition which often targets primates as a preferred food resource (Sabater Pi and Groves, 1972; Sabater Pi, 1979; Gandini, 1979; Harcourt and Stewart, 1980; Tanno, 1987).

The central subspecies is currently less fragmented than other subspecies, at least in the core range area comprising Gabon, southern Cameroun, and southern Central African Republic. Some 5,000 chimpanzees probably inhabit known habitats covering about 17,000 square kilometers of suitable range, and another 74,000 may exist in about 254,000 square kilometers of potential habitat (Table 3). The largest reservoir population is undoubtedly in Gabon, where range fragmentation is just now beginning to occur (Harcourt and Stewart, 1980; Lahm, 1985; Tutin and Fernandez, 1984, 1988). Habitat fragmentation is very evident in countries such as Nigeria, northern Cameroun, Equatorial Guinea, and Congo (Sabater Pi and Jones, 1967; Jones and Sabater Pi, 1971; Gartlan, 1975b; Spinage, 1980; Harcourt and Stewart, 1980; C. Duncan, pers. comm.).

Mining is a major industry in many parts of tropical Africa (Coakley et al., 1984). It is rarely cited as a primary threat to wild chimpanzees, yet large mining operations invariably serve as focal points for human settlement and activity, including intensive hunting to provide bushmeat for the workers and their families. In southern Sierra Leone, all major mining sites are largely devoid of wildlife (Teleki, 1980). In western Liberia, the presence of a large iron mine near the Gola Forest Reserve has contributed to the loss of chimpanzee and other wildlife populations throughout the border area (Teleki, 1980; Oates, 1980). A similar situation exists at Nimba Mountains, on the joint boundaries of Liberia, Guinea, and Ivory Coast (Lamotte, 1983; Verschuren, 1983; Y. Sugiyama, pers. comm.). At both Gola and Nimba, the commercial hunting associated with mining operations is particularly damaging to large mammals because hunters have access to modern firearms and explosives as well as to specialty items such as carbide lamps for night work (Teleki, 1980).

Once deforestation on a major scale combines with hunting, the effect on wildlife populations may be terminal (Teleki, 1980). The present discrepancies between chimpanzee ranges and forest distributions may be a relatively new phenomenon, perhaps associated with failure to compete with escalating human utilization of forest zones (Kortlandt and van Zon, 1969; Kortlandt, 1983). It is clear that the size of the African rain-forest belt has decreased substantially in recent decades due to timber exploitation, agricultural expansion, and other human pressures (Myers, 1979, 1980; Salati and Vose, 1983; MacKinnon and MacKinnon, 1986) and that chimpanzee distributions have been adversely affected everywhere by such changes (Anonymous, 1988a).

There is an alarming uncertainty about which of Africa's surviving forests contain viable chimpanzee populations that could be conserved. The species may be far closer to extinction than is generally realized if northern Zaire, long presumed to be a vital stronghold of *Pan troglodytes,* harbor few or no chimpanzees (Anonymous, 1988a). A related species, *Gorilla gorilla,* is absent there (Anonymous, 1988e).

The available evidence strongly suggests that any type of contact with humans can result in morbidity and mortality increases for the chimpanzees (Teleki et al., 1976). As the habitat becomes more fragmented, the perimeters of the fragments increase, and more contact with humans and their disease organisms occurs.

The most specific information on morbidity has been collected for one small study population in western Tanzania, where substantial losses occurred over a span of 30 years of observation (Goodall, 1983, 1986, pers. comm.). Contagious diseases, parasites, genetic abnormalities, and injuries caused 30 to 40 percent of the deaths in that population. Between 1960 and 1987, at least four major epidemics, including poliomyelitis, influenza, and other respiratory ailments, struck this population, with a high probability that these were contracted from neighboring human settlements. Orphan losses resulting from the deaths of mothers were also high because infants under five years of age are unlikely to survive when deprived of maternal care. Similar events have been confirmed in other long-term studies (Hiraiwa-Hasegawa et al., 1984; Sugiyama, 1984). Population losses accruing from these factors are largely hidden, as only long-term projects can reveal the effects of endemic and introduced diseases.

Commercial Exploitation

Live chimpanzees are exploited for various commercial purposes, chief among which are international trade (biomedicine and entertainment) and local trade (bushmeat market and pet sales). The total number taken per year across Africa is not precisely known, but it is certainly in the thousands. Data relating to international commerce in chimpanzees is available from a variety of sources.[2]

International trade. Since 1975, when the Convention on International Trade in Endangered Species (CITES) was first ratified, the export of live chimpanzees has been authorized by several African nations inhabited by chimpanzees (Cameroun, Congo, Gabon, Guinea, Liberia, Sierra Leone, and Zaire), as well as by some nations lacking endemic chimpanzee populations (Kenya, Malagasy Republic, South Africa, and Togo). According to CITES world trade records, 1,110 chimpanzees entered international commerce in 1975–1986, with only 311 (28%) of those transactions involving captive-bred individuals.

The CITES figures may represent only a fraction of the actual world trade, however, as customs records from some African countries show shipments of large numbers of chimpanzees not listed on CITES records. For example, 406 chimpanzees departed from Sierra Leone between 1975 and 1979, mostly to the United States, without appearing anywhere on CITES records. Other nations, such as Equatorial Guinea and Uganda, exported unregistered chimpanzees in the same period (Teleki, 1988a). The routes and destinations of the chimpanzees taken in Equatorial Guinea for commercial trade have never been confirmed by authorities (Sabater Pi, 1979). Zaire has been a source for illegal shipments to Zambia and probably to Burundi, and a new illicit trade pipeline is developing from Tanzania to Burundi (Teleki, 1987; Anonymous, 1984a, 1987b). A smuggling operation from Uganda to the United Arab Emirates recently was broken by authorities in Kampala (Anonymous, 1988b).

Additional trade records show that Guinea, Liberia, and Sierra Leone have been the source of most chimpanzee exports since the 1940s (Anonymous, 1975; Teleki and Baldwin, 1979a; Teleki, 1980; Wolfheim, 1983; Kavanagh and Bennett, 1984; Mack and Eudey, 1984). So many infant chimpanzees have been shipped from these countries since the 1950s that the majority of the chimpanzees currently held by importing nations are probably *P. t. verus* obtained from this region (Teleki and Baldwin, 1979a, 1981).

Guinea. For more than 70 years, Guinea's chimpanzee population has been subject to pressures from a biomedical station and both direct and indirect export. The long-term effect of these pressures has been devastating (Sugiyama and Soumah, 1988).

Over a span of four decades ending in the late 1950s, a Pasteur Institute biomedical station in Guinea captured many chimpanzees for domestic use and export to Europe (Kortlandt, 1965; Harrisson, 1971). Records indicate that at least 700 chimpanzees were exported, with as many as 3,000 to 4,000 eliminated during capture (Kortlandt, 1965, 1966). In 1950–1956 alone, at least 300 infants were exported, but about 100 of them died in transit (Kortlandt, 1966). Such heavy depredation could not have been sustained for long, considering the thin distributions noted a few years later by de Bournonville (1967) in the same areas. Closing the station in the 1960s fortunately eliminated exports in the 1970s (Harrisson, 1971). Since then, no authorized attempts to resume trade in Guinea are known (Kavanagh and Bennett, 1984), although one export offer was made to overseas buyers in 1983 (Anonymous, 1984b). However, Guinea has continued to be a source for many chimpanzees smuggled into Sierra Leone for re-export overseas by dealers based in Freetown (Teleki and Baldwin, 1979b; Teleki, 1980).

Sierra Leone. In Sierra Leone, the chimpanzee trade has been a major and highly lucrative business for several decades (Kortlandt, 1965, 1966; Wade, 1978; Teleki and Baldwin, 1979b, 1981; Teleki, 1980; Henson, 1983; Redmond, 1986). Indeed, one dealer based in Sierra Leone claims to have supplied 1,500 – 2,000 chimpanzees to the United States (Anonymous, 1988d). Records from the 1950s and 1960s indicate that the annual export rate from Freetown was about 150 infants, with a minimum of 1,500 shipped out during each decade (Kortlandt, 1965; Riney and Hill, 1967). With the wastage estimated conservatively at ten chimpanzees killed for every infant taken, and with mothers being the prime targets, thousands of adult chimpanzees probably died each decade to sustain the trade (Kortlandt, 1965, 1966). Detailed customs records between 1973 and 1979 showed 1,582 infants shipped from Freetown, primarily to the United States (Teleki, 1980). During the decade of the 1970s, the total exports rose to above 200 per year; thus more than 2,000 infants went overseas and many thousands died in the process (Teleki, 1980).

At the present time, trade is the only significant pressure on the Sierra Leone population other than habitat loss (Teleki, 1980). Sierra Leone now has internal laws to restrict trade but continues to export "indigenous" chimpanzees, although most are of Guinea and Liberia origin (Anonymous, 1978a; Teleki, 1980, 1986; McGiffin, 1985). The trade pipeline from Sierra Leone remains open today mainly due to the efforts of biomedical buyers in Europe and Asia (Anonymous, 1976, 1978c, 1979a, 1979b, 1979c, 1988a; Henson, 1983; Sugiyama, 1985b; Cherfas, 1986; Redmond, 1986, 1988; Greisenegger, 1986; Gaski, 1987; Hiroko, 1987; Luoma, 1989). Plans for establishing a major pharmaceutical laboratory have been submitted to Sierra Leone by an Austrian firm (Redmond, 1986).

Liberia. From 1974 to present, the New York Blood Center station, VILAB II, acquired many chimpanzees in Liberia for biomedical use and exported some to the United States (Anonymous, 1975; Teleki and Baldwin, 1979a; Prince et al., 1986). By 1988, the station had 150 live chimpanzees, of which only a fraction were born on location. Many chimpanzees were captured for VILAB by a contracted white hunter, others were captured by local hunters, and some were obtained from pet owners (Prince et al., 1986). The extent to which capture of infants for VILAB involved killing mothers is a matter of contention (A. Prince, pers. comm.; A. Peal, pers. comm.), but it must have been substantial. Two wildlife dealers based in Monrovia, who were at times assisted by the hunter working for VILAB at Robertsfield, were exporting 50 to 100 infants per year to the United States. This continued over a period of several years until Liberian authorities closed down the trade in the late 1970s (Teleki and Baldwin, 1979a; Verschuren, 1983; A. Peal, pers. comm.).

As trade restrictions increase, a new trend toward greater use of wild-born chimpanzees in local biomedical stations is emerging in Africa. Established laboratories in Liberia, Gabon, and Zaire may soon be supplemented by stations in Uganda, Ivory Coast, Nigeria, Tanzania, and probably elsewhere, mainly in connection with malaria and AIDS research (Teleki, 1988b).

Participants in the international marketing of live chimpanzees often downplay the impact of their activities by blaming the decline of wild populations exclusively on habitat shrinkage and local hunting pressure (Anonymous, 1986a). In their view, commercial trade is no less than a practical means of salvaging a valuable resource from wasteful destruction in order to use it for the general benefit of humankind. Both dealers and buyers like to portray themselves as conservationists acting in the best interests of chimpanzee survival (Anonymous, 1986b). Suppliers may insist that they obtain "orphan" infants from tribal hunters or pet owners, benevolently saving them from miserable deaths, while buyers publicly proclaim that trade causes little or no damage to wild populations (Anonymous, 1986a, 1986b; Johnsen, 1987; see also interviews in Redmond, 1986, 1988; Luoma, 1989). These assurances contradict observed facts and promote the approval by authorities of excessively high levels of exploitation.

The harsh reality is that trade in chimpanzees is a generally destructive and often brutal enterprise conducted for profit (Kortlandt, 1965; Harrisson, 1971; Domalain, 1977; Sabater Pi, 1979; Teleki, 1980, 1987; Teleki and Baldwin, 1981; Wolfheim, 1983; Mack and Mittermeier, 1984; Anonymous, 1988a, 1988b). For every live chimpanzee in the trade pipeline, many chimpanzees are killed.

Local trade. Local trade involves both killing for domestic bushmeat and capturing for live pet sales. It is also a link to the export trade because double profits arise for local hunters when chimpanzee mothers are shot for sale at bushmeat markets and their infants are sold to dealers for export overseas (Sabater Pi, 1979; Harcourt and Stewart, 1980; Teleki, 1980, 1987; Tanno 1987). For example, in Liberia adult female chimpanzees are sold as bushmeat for a handsome profit, while their infants, which command a lower value as food due to small body size, are illegally transported to Sierra Leone for sale to an export dealer (Teleki, 1980). Subsistence hunting, by itself, can act as a major pressure on local chimpanzee populations, especially when hunters have access to modern weapons and technology. Liberia may have the highest per capita ownership of firearms in equatorial Africa, with disastrous results for chimpanzees (Verschuren, 1983). Tribal hunting is highly popular and backed by lucrative bushmeat markets in Ghana, Liberia, Equatorial Guinea, and Zaire, to cite but a few examples (Asibey, 1965, 1966; Sabater Pi and Groves, 1972; Jeffrey, 1970, 1977; Sabater Pi, 1979; Harcourt and Stewart, 1980; Teleki, 1980; Teleki and Baldwin, 1981; Robinson and Peal, 1981; Carroll, 1986a, 1986b; Tanno, 1987). In contrast, hunting of chimpanzees for bushmeat is not popular in Sierra Leone. In Gabon and southern Cameroun, the main pressure at present stems from a strong local hunting tradition that often targets primates as a preferred food resource (Sabater Pi and Groves, 1972; Sabater Pi, 1979; Gandini, 1979; Harcourt and Stewart, 1980; Tanno, 1987). In many parts of tropical Africa, wildlife stocks have dropped so dramatically in recent years that hunters may cross international borders that cannot be monitored (Teleki and Baldwin, 1981; Robinson and Peal, 1981; Verschuren, 1983; Tanno, 1987; Teleki 1980, 1986, 1987).

The emergence of a pet market due to high local income levels is one of the by-products of mining activity in Africa. Sale of chimpanzees as pets occurs in both Sierra Leone and Liberia (Teleki, 1980; A. Peal, pers. comm.), and it has also been documented in the copper belt area of northern Zambia (Teleki, 1987). Infant chimpanzees are purchased in Zambia for as much as $500 per pet (D. Siddle, pers. comm.), a price that rivals purchase value in the export trade occurring in other countries (Teleki, 1980). The chimpanzee populations in Zaire's mineral regions may also be vulnerable to exploitation as pets.

Wastage during capture and shipment. Wastage during capture and shipment has always been devastating, but it is particularly so now that population fragmentation is advancing so rapidly. Typically, the infants collected for export are less than two years of age. There is *no possibility* that infants below the age of three years can be captured by any method other than the maiming or killing of their mothers, who are extremely protective of their offspring at such an early stage of development and do not drop or abandon them no matter how extreme the duress (Goodall, 1986). The protective behavior of other group members toward infants has also been definitively established and can include severe attacks by adult males upon hunters attempting to remove the infants from dead mothers (Teleki, 1980; Goodall, 1986, pers. comm.).

During hunting for infants, one or more protective adults, minimally the mothers of the infants, are shot. Many of the infants also die because the shotgun pellets usually spread to hit both mothers and their clinging offspring. Wounds, often infected, are not uncommon among the infants sold on pet and export markets, and buckshot pellets have been removed from many purchased and confiscated chimpanzees (J. Carter, pers. comm.; D. and A. Siddle, pers. comm.; J. Moor-Jankowski, pers. comm.). The hunters may also use pit traps, wire snares, poisoned food, nets, and even dog teams.

The captive infant chimpanzees are commonly trucked from capture sites to village transit points, and then to urban dealerships, in tiny cages or tightly cinched sacks that permit no shift of posture. Often heavy loads are placed over the captives to avoid discovery at checkpoints. Infants are often bound hand and foot with cord or wire, even when they are inside transport cages, and they receive little or no care en route. Starvation and dehydration are routine, as nursing infants cannot consume even the meager solid rations that may be offered them by traders or truck drivers who commonly know nothing about chimpanzee nutrition. Further deaths ensue because dealer holding facilities are usually atrocious in condition and hygiene, and because care is often unavailable at international airports. Finally, infants who survive capture and shipment may die at their destinations from accumulation of physical traumas and emotional stresses.

There are no known exceptions to these capture and transport conditions, as all dealers currently operating in Africa rely on local hunters for a supply of live infants. In the few instances where major capture operations were mounted by white hunters or dealers under reportedly "humane" conditions, as in Sierra Leone and Liberia during the 1970s (F. Sitter, pers. comm.; A. Prince, pers. comm.), the rate of diminution of wild populations was apparently not reduced (Teleki, 1980; Teleki and Baldwin, 1981; A. Peal, pers. comm.). The only recorded case in which a tranquilizer was success-fully used for immobilizing a wild chimpanzee occurred during temporary capture of an adolescent female for medical treatment in a Tanzanian national park (Roy and Cameron, 1972; Roy, 1974).

A study of capture methods used in Equatorial Guinea reported 66 chimpanzees taken for bushmeat and export sales by local hunters: 23 chimpanzees were taken by first shooting the mothers, 27 were snared in ways that caused serious injury or trauma, and 16 were chased to exhaustion and then killed by hunting parties assisted by dogs, with a total of at least 200 chimpanzees dying in the process (Sabater Pi, 1979). Similar techniques have been documented by experts in various parts of Africa (Kortlandt, 1965, 1966; Sabater Pi and Groves, 1972; Hart, 1978; Sabater Pi, 1979; Gandini, 1979; Harcourt and Stewart, 1980; Teleki, 1980, 1987; Teleki and Baldwin, 1981; Teleki and Bangura, 1981; Robinson and Peal, 1981; Stoneley, 1986; Tanno, 1987).

A nationwide study conducted in Sierra Leone in 1979 (Teleki, 1980) suggests that five chimpanzees are killed for every infant captured. This includes the infant's mother, who is often a female of prime breeding age, and other protective adults (Kortlandt, 1965, 1966). However, the full scope of the loss is even more grievous, for it is estimated that of five infants captured, only one reaches the overseas buyer and survives, with the remainder perishing en route (Teleki, 1980). Therefore, for every wild-caught chimpanzee in captivity, as many as 29 animals may have died ([5 captured individuals x 5 chimpanzees killed per individual captured] + 4 captives lost during transport = 29 deaths). Even if mortality is not always so high, it is certain that at least 10 chimpanzees die for every one which reaches its commercial destination.

The killing for local trade and the deaths associated with capture for international trade are aspects of mortality that are not covered in standard demographic analyses. Yet, large numbers are eliminated from the breeding population without ever being recorded. Such hidden mortality is certainly associated with access made possible for hunters by road construction and other kinds of development (Gandini, 1979; Teleki, 1980; Harcourt and Stewart, 1980). Most known localities are today beset by this problem (Robinson, 1970; Lowes, 1970; Suzuki and Suzuki, 1971; Ghiglieri, 1984; Anonymous, 1987b; Y. Sugiyama, pers. comm.; T. Nishida, pers. comm.). The problem is particularly acute for small populations occupying isolated reserves and parks where humans constantly intrude for tree felling, food collecting, game poaching, and other purposes. The Outamba-Kilimi National Park of Sierra Leone, the Sapo National Park of Liberia, the Tai National Park of Ivory Coast, the Kibale Forest Reserve of Uganda, the Gombe and Mahale National Parks of Tanzania, and the Kibira National Park of Burundi – all of which harbor important chimpanzee popula-tions – are cases in point (Teleki, 1980, 1986; Dossi et al., 1981; Skorupa, 1983; Ghiglieri, 1984; Hasegawa and Nishida, 1984; Robinson, 1986; Van Orsdol, 1986; Trenchard, 1988).

SURVIVAL PROSPECTS

While the prospects for chimpanzee survival vary from one locality to another, they are nowhere optimistic. Steps must be taken at every level to improve existing protection for vital habitats and their inhabitants.

Existing Protective Measures

In 1976, the United States upgraded the chimpanzee to Threatened status in its domestic Endangered Species Act (ESA) in order to regulate exploitation that might be detrimental to chimpanzee survival. (The United States was and still may be the world's leading importer of chimpanzees [Anonymous, 1976; Mack and Eudey, 1984]). By 1977, the chimpanzee also appeared in Appendix I of CITES, which meant that the chimpanzee was regarded by member nations as threatened by extinction and in need of strict trade protection (Kavanagh and Bennett, 1984). The International Union for the Conservation of Nature (IUCN) then classified chimpanzees as Vulnerable on the *IUCN Mammal Red Data List*, recognizing that all chimpanzees would become more endangered in the near future if the causal factors of decline continued to operate (Mack and Mittermeier, 1984). By 1985, an IUCN Primate Specialist Group publication, *Action Plan for African Primate Conservation*, rated chimpanzees in a High Priority category due to their vulnerability and their taxonomic uniqueness (Oates, 1985). By early 1988, the western subspecies of chimpanzee was internationally upgraded to Endangered rank by the IUCN, which also emphasized the acute vulnerability of many local populations in other parts of Africa (Anonymous, 1988a). And finally, the United States Department of Interior declared its intent in late 1988 to reclassify wild chimpanzees to Endangered status on the ESA list and, for purposes of trade with the United States, to treat all chimpanzees exported from Africa, captive and wild, as Endangered (Anonymous, 1988g, 1989c).

During this period, progress also occurred on implementing new protective measures within countries inhabited by chimpanzees (Kavanagh and Bennett, 1984; Kavanagh, 1984). Many African nations drafted legislation to protect their wildlife resources. Some made special efforts to limit destruction of chimpanzee populations, although others, such as Sierra Leone and Uganda, continued trade despite enacting domestic laws that banned chimpanzee exports (Henson, 1983; Cherfas, 1986; Redmond, 1986, 1988; Anonymous, 1988b; Luoma, 1989). A number of new national parks, including some that contained important chimpanzee habitats, have been established (Wolfheim, 1983; Teleki, 1986; Anonymous, 1987b). And by 1988, 13 of the 21 nations still inhabited by chimpanzees had joined CITES: the exceptions are Angola, Equatorial Guinea, Gabon, Guinea-Bissau, Ivory Coast, Mali, Sierra Leone, and Uganda.

Some advances in policy restrictions concerning acquisition and use of chimpanzees also appeared on the biomedical front. In 1970, participants at the Second International Conference on Experimental Medicine and Surgery in Primates issued a statement on responsibility for the preservation of threatened species, clearly acknowledging many of the survival problems facing wild chimpanzees (Goldsmith and Moor-Jankowski, 1971). Subsequently, the United Nations Ecosystem Conservation Group and World Health Organization issued a Policy Statement on Use of Primates for Biomedical Purposes that recognized similar issues, including the absence of sustained-yield capture strategies for wild primates, and recommended that Endangered, Vulnerable, and Rare species be used in biomedicine only when they were obtained from self-sustaining captive breeding colonies (Mack and Mittermeier, 1984). By 1978, when the Primate Steering Committee of the National Institutes of Health (NIH) in Washington sponsored a consultation session with field primatologists, it was noted that the chimpanzee was in jeopardy throughout Africa, with only four countries listed as containing abundant populations, and that surveys were urgently needed almost everywhere (Anonymous, 1978a).

International Compliance

Although international concern about the survival of wild chimpanzees has increased during recent years, more needs to be done in Africa and elsewhere to translate legislative advances into concrete improvements that directly benefit chimpanzees. Most parks and reserves within Africa are only nominally protected, and wildlife laws are only selectively enforced (Kavanagh and Bennett, 1984; MacKinnon and MacKinnon, 1986). Some statements, such as the World Health Organization policy, have no legal standing and cannot be enforced.

Placement of chimpanzees on Appendix I of CITES in 1977 has helped regulate, but not necessarily restrict, trade among member nations. Too many exemptions and loopholes exist to ensure strict protection, even among signatory countries. The global volume of trade remains high, with only a portion of the trade registered with CITES (Teleki, 1988a, 1988b; Luoma, 1989). One commonly abused exemption is a CITES provision that allows trade in chimpanzees for breeding and propagation, which often masks other purposes such as biomedical use (Teleki, 1978).

Some CITES parties continue to export and import chimpanzees, in willful noncompliance with international and/or national legislation (Kavanagh, 1984; Gaski, 1987; Redmond, 1988; Luoma, 1989). Nonsignatory countries also continue to supply wild-caught chimpanzees to any interested buyer (Teleki, 1980; Anonymous, 1988b). Biomedical demand for chimpanzees has not abated (Wade, 1978; Johnsen, 1987; King et al., 1988) and sometimes involves irregularities (Cherfas, 1986; Redmond, 1986; Hiroko, 1987; Luoma, 1989). The practice of establishing laboratories in source countries eliminates the publicity risks of international commerce (Henson, 1983; Cherfas, 1986; Redmond, 1986, 1988). American biomedical institutions have recently drawn attention to their interest in bypassing CITES trade restrictions (Teleki, 1988b). However, the United States passed a special law in 1988 to prohibit expenditure of public funds for any biomedical "project that entails the capture or procurement of chimpanzees obtained from the wild" (Anonymous, 1988f).

Some nations, such as Austria[3] and Japan, continue to conduct commercial trade despite emphatic international censure (Sugiyama, 1985b; Greisenegger, 1986; Redmond, 1986, 1988; Gaski, 1987). Other nations have been just as lenient about enforcing laws and treaties restricting chimpanzee trade (Teleki, 1978, 1980). According to some sources, no chimpanzees were imported by the United States after

1974 (Johnsen, 1987; Whitney, 1988). However, at least 88 chimpanzees were imported from Africa for biomedical purposes in 1974 and 1975 (Luoma, 1989). In 1975–1979, 406 chimpanzees were shipped from Sierra Leone to the United States for mostly biomedical purposes, yet they were never registered with CITES; and CITES records show further shipments into the United States occurred in 1982 (Teleki, 1988a; Luoma, 1989). In addition, in 1978 American biomedical buyers attempted, without success, to start a program to purchase hundreds of chimpanzees each year from Sierra Leone in connection with efficacy testing of hepatitis vaccine. Resumption of imports by the United States and other countries, in connection with present and future medical crises, could seriously contribute to the risk of extirpation in many range countries, and even to extinction of the western and eastern subspecies (Anonymous, 1988a).

Conservation Priorities

Habitat protection. The selection of appropriate areas to be protected is crucial to subspecies survival. However, chimpanzee adaptability makes it difficult to identify optimum habitat for the species. Chimpanzee densities in humid tropical forests vary as much as they do anywhere else, so tropical forests should not be assumed to be the best habitat. Some sources suggest that open grassland-woodland-forest may be the most suitable areas in which to maintain chimpanzee lifestyles (Kortlandt, 1972; Teleki, 1977; McGrew et al., 1981). And because deforestation is now clearly a major cause of species decline in all tropical regions (Myers, 1979), and is certainly a major threat to some chimpanzee populations (Jones and Sabater Pi, 1971; Gartlan, 1975a, 1975b; Davies, 1987a, 1987b; Tutin and Fernandez, 1988), it might be advisable to pay greater attention to degradation of suitable open habitats.

Regional priorities. Conservation strategies superior to those now in place must be developed to overcome the pressures that are threatening wild chimpanzees. Because the problems are international in origin and scope, they must be addressed through international cooperation and innovative approaches. Simply intensifying the current strategies along existing lines will probably not suffice.

For the western subspecies, the conservation priorities are an immediate moratorium on commercial trade, coupled with implementation of strict measures to protect local population remnants and their habitats. The target countries for priority action are Guinea, Sierra Leone, Liberia, and Ivory Coast (Teleki and Baldwin, 1981; Robinson and Peal, 1981; Verschuren, 1983; Anonymous, 1984b; McHenry, 1986). Shrinkage of habitat in Guinea must be counterbalanced (Sugiyama and Soumah, 1988). The export ban imposed by Sierra Leone in 1978 must be enforced, and CITES membership encouraged, as that pipeline continues to drain the populations of several neighboring countries (Teleki, 1980, 1986; McGiffin, 1985). Tighter internal controls on market hunting are needed in Liberia, along with establishment of new reserves and parks (Robinson and Peal, 1981). The primary goals for Ivory Coast are better protection for the Tai Forest area and membership in CITES (Boesch, pers. comm.). Because the Tai region may now be the best hope for salvaging the western population, systematic surveys of that area must be undertaken immediately. In Liberia, which has taken steps to set up its first national park, Sapo, surveys are urgently needed in order to develop appropriate protection strategies.

For the eastern subspecies, a top conservation priority is control of habitat degradation through radical improvement of current forest management techniques and better enforcement of laws applying to protected areas. The main target nations are Uganda, Rwanda, and Burundi (Skorupa, 1983; Van Orsdol, 1986; Struhsaker, 1987; Trenchard, 1988). Internal control of domestic hunting within Zaire would probably not be feasible, but the cutting of chimpanzee trade routes to neighboring countries would be definitely beneficial (Hart and Thomas, 1986; Stoneley, 1986; White and Susman, 1986; Anonymous, 1987b, 1988b; Teleki, 1987). Membership in CITES by Uganda is urgently needed. Distributions of chimpanzee populations have been demarcated in Uganda, Rwanda, Burundi, and Tanzania, but a major survey of Zaire is urgently required because virtually no protection exists there for what may be the bulk of the eastern population. A survey of southern Sudan is also needed to define a northern distribution limit for the subspecies (Kortlandt and van Zon, 1969; Kortlandt, 1983; Kock, 1967, 1969).

For the central subspecies, the highest priority is the designation of new protected areas in prime chimpanzee habitats. The main targets are Cameroun, Central African Republic, and Gabon (Harcourt and Stewart, 1980; Carroll, 1986a, 1986b; Fay, 1987; Tutin and Fernandez, 1984; Lahm, 1985). Such action must be taken prior to the concessioning of major forest tracts to timber industries, particularly in Gabon which currently has no national parks (Tutin and Fernandez, 1988). Internal control of market hunting would be advisable, given the traditional tribal preferences for primate bushmeat in this area, but that may not be a practical goal (Sabater Pi and Groves, 1972; Harcourt and Stewart, 1980; Tanno, 1987). Surveys of Congo and southern Cameroun are urgently needed to supplement work completed in Equatorial Guinea, Gabon, and Central African Republic (Sabater Pi and Jones, 1967, Jones and Sabater Pi, 1971; Tutin and Fernandez, 1984; Carroll, 1986a, 1986b; Fay, 1987). CITES membership for Equatorial Guinea and Gabon is also strongly recommended.

International priorities. Chimpanzees could benefit substantially from broad international recognition of their plight in Africa. Reclassification to Endangered status, the highest level of protective ranking available on the *IUCN Red Data List,* should occur in every country (Mack and Mittermeier, 1984). The step taken recently by the United States to add wild chimpanzees in Africa to its Endangered Species List (Anonymous, 1988c, 1989c) sets a fine example for the global community. The assignment of various subspecies to different protective categories, a position presently endorsed by the IUCN (Anonymous, 1988a), is not advisable because neither trade sanctions nor other protective measures can be effectively enforced. This is crucial, given that *Pan troglodytes* subspecies are too similar in external appearance (Reynolds and Luscombe, 1971) to allow for accurate identification of individuals by customs inspectors if shipments are accompanied by bogus trade documents.

Blockage of illegal trafficking in chimpanzee infants should be a top priority for all nations, not only for CITES members. No exemptions to existing trade restrictions should be permitted without an absolute guarantee that species propagation is the sole purpose. Indeed, consideration should be given to imposing an international moratorium on trade in chimpanzees from any source country containing fewer than 5,000 in the wild.

Prospects for Rehabilitation and Captive Breeding

The need to rehabilitate chimpanzees has basically two root causes. First, infants confiscated from illegal traffickers need to be rehabilitated (Anonymous, 1979a, 1979c; Carter, 1981; White and Susman, 1986; D. and S. Siddle, pers. comm.). Rehabilitation is currently the only publicly acceptable response to managing the orphans from this source. Second, unwanted chimpanzees in captivity need to be rehabilitated (Wilson and Elicker, 1976; Pfeiffer and Koebner, 1978; Anonymous, 1978a; Koebner, 1982; Prince, 1985; Prince et al., 1986). This situation is ironic, given a continuing demand by the biomedical industry for new supplies. Proposals to assign orphans to biomedical users are not acceptable because that would simply reinforce the market that causes the problem.

Chimpanzee rehabilitation projects have existed on several continents and in various environmental conditions. Some projects focused on releasing chimpanzees in artificial settings that bore little or no resemblance to native habitats in Africa (Wilson and Elicker, 1976; Pfeiffer and Koebner, 1978). Other projects developed in more natural settings within Africa, some at sites not occupied by wild chimpanzees (Hladik, 1974; Carter, 1978, 1981, 1988; Borner, 1985), and some at sites inhabited by resident populations (Rucks, 1976; Brewer, 1976, 1978). The projects located in Africa commonly involved confiscated chimpanzees, with intensive-care orphanages serving as way stations to full independence at protected sanctuaries (Carter, 1981, 1988; Stoneley, 1986; White and Susman, 1986; Teleki, 1987). Only one project has focused exclusively on "retired" laboratory chimpanzees (Prince, 1985; Prince et al., 1986; Anonymous, 1988d).

Rehabilitation in any circumstance is difficult to achieve because there are many aberrations and inadequacies to overcome (Rijksen and Rijksen, 1979; Harcourt, 1987). Vast human resources are required to maintain rehabilitants for lifespans of 40 years or more. Rehabilitation of chimpanzees, particularly of those used in biomedical experiments, is problematic because there are mental and physical damages to correct in addition to developmental, behavioral, and social anomalies (Wilson and Elicker, 1976; Pfeiffer and Koebner, 1978). The barriers to recovery may be too deeply embedded for individuals who have endured the traumas of capture and transport followed by years of deprivation and suffering in laboratories.

Global biomedical demand to replenish captive laboratory populations with wild-born chimpanzees remains a constant pressure on all wild-chimpanzees (Wolfheim, 1983; Anonymous, 1988a; Luoma, 1989). The presence of some 1,600 captive chimpanzees in North American facilities provides an opportunity to breed enough chimpanzees to satisfy domestic needs and, theoretically, eliminate new demands on wild-born chimpanzees (Seal and Flesness, 1986; Johnsen, 1987). Previous breeding efforts have achieved only a limited success, however, and international trade is expected to continue (Teleki and Baldwin, 1975; Martin, 1981; Stephens, 1987). The failure of the biomedical community to ensure self-sustaining reproductive status for the world laboratory population is due largely to failure in providing suitable living and breeding conditions for most captive chimpanzees (Teleki and Baldwin, 1975; Anonymous, 1978a; Kortlandt, 1978; Martin, 1981; McGrew, 1981; Seal and Flesness, 1986; Goodall, 1987). No international registry of captive-held and captive-born chimpanzees has been established, although a model exists in the International Species Inventory System (ISIS), with voluntary participation.

It remains to be seen whether the new National Chimpanzee Breeding and Research Program (NCBRP) of NIH corrects this failure. According to various documents issued by NIH, the program includes some 350 "successful breeders," chosen from a total of about 1,200 chimpanzees now held by American laboratories. The program includes five institutions, all of which were already producing chimpanzees before September 1986, when the program started. The program is expected to produce 60–100 infants per year, with only half of them scheduled for research. Despite its heavy cost (more than ten million dollars in the first four years), the program will not contribute greatly to national research needs: the same cohort has already been producing 100 infants each year for the preceding five years, and many of the wild-born breeding adults will die in the near future. Meanwhile, some zoos in America have placed colonies on birth control because it is publicly unacceptable to transfer chimpanzees from zoos to laboratories until those laboratories substantially improve conditions of maintenance and care. It is highly improbable that captive breeding will contribute enough chimpanzees in the coming decades to maintain the existing population in the United States.

CONCLUSION

The precipitous decline of chimpanzee populations in Africa is a phenomenon of the present century. The multiple pressures behind the decline vary markedly from region to region, and they are accelerating rapidly due to habitat reduction, local persecution, and commercial exploitation. All wild populations will certainly become more vulnerable if the causal factors of ongoing decline continue operating at existing levels.

Human population increase, combined with escalating rates of natural resource exploitation and land development, is causing major changes in the African landscape. All types of habitats occupied by chimpanzees, not just forest zones, are shrinking at unprecedented rates and causing resident populations to fragment into isolated small units that become increasingly susceptible to catastrophic events such as contagious disease epidemics or intensive commercial exploitation. Competition for scarce natural resources, many of which must be shared with surrounding human groups, further reduces survival options for chimpanzees.

The habitat degradation described above is often cited as the main factor causing rapid decrease of chimpanzees, but human persecution and commercial exploitation are also major problems in many regions. The extermination process has greatly accelerated in recent years with increasing access to modern weaponry (especially in politically unstable parts of Africa) coupled with an escalating demand for bushmeat. One particularly alarming trend is the new pattern of hunting which enables chimpanzees to be taken for both local and international markets simultaneously, with adult carcasses sold as bushmeat and infants sold for export. This double-profit motive encourages slaughter of reproductive females, a habit that no subsistence hunter would normally endorse. The capture methods used to sustain international commerce are particularly inhumane and wasteful. Estimates from many sources indicate that at least ten and perhaps as many as 29 chimpanzees die in the course of successfully supplying one live infant to an overseas buyer. Entire national populations have been depleted by these destructive methods.

Chimpanzee survival problems have not been alleviated so far by legislative measures. The domestic wildlife laws of Africa are neither comprehensive nor consistent, and commonly they are not strictly enforced. Few African countries have enough areas under protection. Most existing reserves and parks contain no sizable breeding units, and the few that do are not efficiently patrolled against poaching and encroachment. International regulations are often less effective than intended, due to the presence of many loopholes and lax enforcement. New protective approaches and actions are urgently needed to avert extinction for the chimpanzee in the critical decades ahead. The Committee for Conservation and Care of Chimpanzees (CCCC),[4] established in 1986, is now preparing a comprehensive Chimpanzee Conservation Action Plan covering the 21 nations where chimpanzees still occur. This plan will be used by World Wildlife Fund–US (WWF–US) as a template for planning projects in Africa.

The conviction remains popular that humankind outranks the chimpanzee. Chimpanzees are classed by law with mice, not men. Yet chimpanzees share close to 99 percent of their genetic endowment with humankind. They also share with us many behavioral, social, and psychological traits once regarded as uniquely human. Among the most significant are self-awareness, compassion for kin, voluntary sharing of resources, cooperative hunting strategy, mourning in response to death, lifetime friendship bonds, long-term memory, capacity for mental anguish, sensitivity to suffering, exceptional learning skills, orphan adoption, artistic skills, use of stone tools, medicinal use of plants, warlike activity, and even cannibalistic tendencies. Such a remarkable species deserves the status of sibling with our species and merits greater respect, more empathy, and far better treatment from humankind than accorded to date. To incarcerate members of this species in social isolation within tiny cages lacking mental stimulation is unconscionable. To exterminate chimpanzees in their homelands would be completely irresponsible.

ENDNOTES

1. The geographical range of each subspecies is treated as a separate continental region, labeled as western, central, and eastern, even though equatorial Africa is not normally partitioned that way in a geopolitical context. The eastern region, for example, covers not only the countries of eastern Africa but also southern Sudan and both northern and eastern Zaire.

2. Assistance in collecting data for this report came from many organizations concerned with chimpanzee survival issues, notably the IUCN/SSC Primate Specialist Group (PSG), the Wildlife Trade Monitoring Unit (WTMU), the Trade Records Analysis of Flora and Fauna in Commerce (TRAFFIC) network, and the International Primate Protection League (IPPL). These organizations were particularly helpful in documenting the scope and impact of commercial trade in live chimpanzees.

3. Some institutions involved in trafficking have adopted an aggressive stance in filing libel lawsuits against public critics (Redmond, 1988; Luoma, 1989). These "intimidation" suits involve millions of dollars in potential damages and have had a "chilling effect" on free speech rights (Gest, 1988). One series of libel actions was commenced in December 1984 in New York by an Austrian pharmaceutical corporation, IMMUNO AG. Earlier, IMMUNO had imported chimpanzees that were confiscated by government authorities and had then tried to establish a station in Sierra Leone, where more wild chimpanzees were to be collected. Subsequently, in 1986, 20 more chimpanzees were imported from Sierra Leone in contravention of CITES, an act ruled as "a violation of the provisions of the Convention and also contrary to its spirit" (Secretariat of CITES, 1988). The suits were finally dismissed in part because the "factual assertions upon which [the] opinions rested were evidently true." IMMUNO nevertheless obtained sizable settlements from all but one of the original defendants "for the obvious reason that the costs of continuing to defend the action were prohibitive" (Anonymous, 1989b). Only one of the defendants "managed to remain in the action to seek a determination on the merits" (Anonymous, 1989d). A strong ruling was issued by the Appellate Division of the New York State Supreme Court against IMMUNO, in part because: "Without exception, the statements at issue were either opinion absolutely privileged under the First Amendment, or statements which the plaintiff utterly failed to show susceptible of being proved false. Indeed, most of the factual statements claimed by the plaintiff to be defamatory were, on the record before us, demonstrably true!" (Anonymous, 1989d).

4. Committee for Conservation and Care of Chimpanzees, 3819 48th Street NW, Washington, DC 20016, USA. Telephone: 202-362-1993.

BIBLIOGRAPHY

Albrecht, H. 1976. Chimpanzees in Uganda. *Oryx* 13:357–361.

Albrecht, H., and S. C. Dunnett. 1971. *Chimpanzees in Western Africa.* Munich: Piper Verlag.

Anadu, P. A. 1987. Prospects for conservation of forest primates in Nigeria. *Primate Conserv.* 8:154–157.

Anadu, P. A., and J. F. Oates. 1982. Forests and primates in southwestern Nigeria: their status, and recommendations for their conservation. Special report, mimeographed, 84pp.

Anderson, J. R., E. A. Williamson, and J. Carter. 1983. Chimpanzees of Sapo Forest, Liberia. *Primates* 24:594–601.

Anonymous. 1975. Chimpanzee collecting in Liberia. *IPPL Newsl.* 2(3):12.

Anonymous. 1976. Chimpanzees among 26 primates designated as endangered or threatened species. U. S. Fish and Wildlife Service *News Release*, Washington, Oct. 29, 1976.

Anonymous. 1978a. Report of the task force on the use and need for chimpanzees. Interagency Primate Steering Committee, NIH, Bethesda.

Anonymous. 1978b. Report of Consultation on Wild Chimpanzee *(Pan troglodytes)* Populations. Interagency Primate Steering Committee, NIH, Bethesda. Also in *IPPL Newsl.* 5(2):14.

Anonymous. 1978c. Sierra Leone clamps down on chimp trade. *IUCN Bull.* 9(9):53–56.

Anonymous. 1978d. Sierra Leone's bright spot. *IUCN Bull.* 9(5):25–28.

Anonymous. 1978e. Merck Sharp and Dohme applies to import 125 chimpanzees. *IPPL Newsl.* 5(1):9–10.

Anonymous. 1979a. Chimpanzees are seized at Amsterdam airport. *IUCN Bull.* 10(1):8.

Anonymous. 1979b. Chimps return to Africa. *IUCN Bull.* 10(8/9):78–79.

Anonymous. 1979c. Amsterdam chimps return to West Africa. *IPPL Newsl.* 6(2):6.

Anonymous. 1983. Sierra Leone's chimps endangered by commercial exploitation. *TRAFFIC Bull.* 5(3/4):48.

Anonymous. 1984a. Illegal traffic in chimpanzees. *IPPL Newsl.* 11(3):8.

Anonymous. 1984b. Guinea to export chimpanzees. *IPPL Newsl.* 11(2):19.

Anonymous. 1985. *1985 United Nations List of National Parks and Protected Areas.* Gland: IUCN.

Anonymous. 1986a. Extra: Alles uber unsere Schimpansen. *Wir Informieren* 5:2–16 (Immuno AG Special Publication).

Anonymous. 1986b. Todlich: Das "Affen-Theater" un die Immuno AG. *Bezirksjournal* 10:6–7.

Anonymous. 1987a. *IUCN Directory of Afrotropical Protected Areas.* Gland: IUCN.

Anonymous. 1987b. Gombe chimps smuggled out of Tanzania. *Daily News*, Tanzania, No. 4526, Sept. 3, 1987.

Anonymous. 1988a. Chimpanzee data sheets. *Threatened Primates of Africa: The IUCN Red Data Book.* Gland: IUCN.

Anonymous. 1988b. Chimp racket blown. *New Vision*, Oct. 17, 1988. Uganda.

Anonymous. 1988c. Notice of finding on petition to reclassify chimpanzee. *U.S. Federal Register* 53(249):52452.

Anonymous. 1988d. Brutal Kinship. National Geographic Explorer series. Aired on cable television Oct. 30, 1988.

Anonymous. 1988e. Gorilla data sheets. *Threatened Primates of Africa: The IUCN Red Data Book.* Gland: IUCN.

Anonymous. 1988f. Public Law 100–436 - Sept. 20, 1988, Sec. 218 (a)(1), National Institutes of Health. Department of Health and Human Services Appropriations Act, 1989.

Anonymous. 1988g. Proposed Endangered Status for chimpanzee and pygmy chimpanzee. *U.S. Federal Register* 54(36):8152–8157.

Anonymous. 1989a. Chimps intercepted at airport. *New Vision*, Jan. 10, 1989. Uganda.

Anonymous. 1989b. Editor wins appeal of libel claim. *New York Law Journal*, January 18, 1989.

Anonymous. 1989c. Chimpanzees proposed for reclassification to Endangered. *End. Sp. Tech. Bull.* 14(3):1–2.

Anonymous. 1989d. Decision of the day (Appellate Division, First Department): Defamation action, plaintiff's burden of proving falsity. *New York Law Journal:* Jan. 30, 1989.

Asibey, E. O. A. 1965. Utilization of wildlife in Ghana. FAO Conference, Kampala. Special report, mimeographed, 25pp.

_____. 1966. Why not bushmeat too? *Ghana Farmer* 10:165–170.

_____. 1978. Primate conservation in Ghana. In D. J. Chivers and W. Lane-Petter, eds., *Conservation. Recent Advances in Primatology,* vol. 2. London: Academic Press.

Badrian, A., and N. Badrian. 1977. Pygmy chimpanzees. *Oryx* 13:463–468.

Baldwin, L. A., and G. Teleki. 1973. Field research on chimpanzees and gorillas: an historical, geographical, and bibliographical listing. *Primates* 14:315–330.

Baldwin, P. J., W. C. McGrew, and C. E. G. Tutin. 1982. Wide-ranging chimpanzees at Mt. Asserik, Senegal. *Internatl. J. Primat.* 3(4):367–385.

Bassus, W. 1975. Der Wildtierbesstand, sein Schutz und seine Bejagung in der Volksrepublik Kongo. *Arch. Natursch. Landschaftsforsch.* 15(4):247–263.

Boesch, C. 1978. Nouvelles observations sur les chimpanzes de la foret de Tai (Cote-D'Ivoire). *La Terre et la Vie* 32:195–201.

Boesch, C., and H. Boesch. 1981. Sex differences in the use of natural hammers by wild chimpanzees: a preliminary report. *J. Human Evol.* 10:585–593.

Booth, A. H. 1956. The distribution of primates in the Gold Coast. *J. West Afr. Sci. Assoc.* 2:122–133.

Borner, M. 1985. The rehabilitated chimpanzees of Rubondo Island. *Oryx* 19(3):151–154.

Bothma, J. du P. 1975. Conservation status of the large mammals of southern Africa. *Biol. Conserv.* 7(2):87–95.

Bourliere, F., E. Minner, and R. Vuattoux. 1974. Les grands mammifers de la region de Lamto, Cote d'Ivoire. *Mammalia* 38(3):433–447.

de Bournonville, D. 1967. Contribution a l'etude du chimpanze en Republique de Guinee. *Bull. Instit. Fond. Afr. Noire* 29A:1188–1269.

Brewer, S. 1976. Chimpanzee rehabilitation. *IPPL Special Report,* 10pp.

_____. 1978. *The Forest Dwellers.* London: Collins.

Butler, H. 1966. Some notes on the distribution of primates in Sudan. *Folia Primat.* 4:416–423.

Butynski, T. M. 1985. Primates and their conservation in the impenetrable (Bwindi) forest, Uganda. *Primate Conserv.* 6:68–72.

Carroll, R. W. 1986a. The status, distribution, and density of the lowland gorilla *(Gorilla gorilla gorilla* Savage and Wyman), forest elephant *(Loxodonta africana cyclotis),* and associated dense forest fauna in southwestern Central African Republic. Special report, mimeographed, 94pp. New Haven: Yale University.

_____. 1986b. Status of the lowland gorilla and other wildlife in the Dzanga-Sangha region of southwestern Central African Republic. *Primate Conserv.* 7:38–41.

Carter, J. 1978. Chimpanzee rehabilitation. Unpubl. report, mimeographed, 13pp. Wildlife Conservation Department, Banjul, The Gambia.

_____. 1981. A journey to freedom. *Smithsonian* April:90–101.

_____. 1988. Freed from keepers and cages, chimps come of age on Baboon Island. *Smithsonian* June:36–49.

Cherfas, J. 1986. Drugs firm accused over chimpanzee cages. *New Scientist* August 21:16.

Clark, C. B. 1977. A preliminary report on weaning among chimpanzees of the Gombe National Park, Tanzania. In S. Chevalier-Skolnikoff and F. E. Poirier, eds., *Primate BioSocial Development.* New York: Garland.

Clarke, J. I. 1969. *Sierra Leone in Maps.* New York: Africana.

Coakley, G. J., and Staff. 1984. *Mineral Industries of Africa.* Bureau of Mines, Department of the Interior, Washington.

Cole, N. H. A. 1980. The Gola Forest of Sierra Leone: a remnant primary tropical rain-forest in need of conservation. *Environ. Conserv.* 7(1):33–40.

Cousins, D. 1978. Gorillas – a survey. *Oryx* 14:254–258.

Curry-Lindahl, K. 1956. Ecological studies on mammals, birds, reptiles and amphibians in the eastern Belgian Congo. *Ann. Mus. Roy. Congo Belge,* vol. 1.

_____. 1974. Conservation problems and progress in equatorial Africa. *Environ. Conserv.* 1(2):111–122.

Davies, A. G. 1987a. Conservation of primates in the Gola Forest Reserves, Sierra Leone. *Primate Conserv.* 8:151–153.

_____. 1987b. *The Gola Forest Reserves, Sierra Leone: Wildlife Conservation and Forest Management.* Gland: IUCN.

Domalain, J. Y. 1977. *The Animal Connection: The Confessions of an Ex-Wild Animal Trafficker.* New York: William Morrow.

Dossi, H., J. L. Guillaumet, and M. Hadley. 1981. The Tai Forest: land use problems in a tropical forest. *Ambio* 10:120–125.

Dupuy, A. R. 1971. Statut actuel des primates au Senegal. *Bull. Instit. Fond. Afr. Noire* 33A:467–478.

Dupuy, A. R., and J. Verschuren. 1977. Wildlife and parks in Senegal. *Oryx* 14:36–46.

Eibl, A., U. H. Rahm, and G. Mathys. 1966. Les mammiferes et leurs tiques dans la foret du Ruggege (Republique Rwandaise). *Acta Trop.* 23:223–263.

Emlen, J. T., and G. B. Schaller. 1960. Distribution and status of the mountain gorilla *(Gorilla gorilla beringei)* – 1959. *Zoologica* 45(1):41-52.

Fay, J. M. 1987. Partial completion of a census of the lowland gorilla *(Gorilla g. gorilla* Savage and Wyman) in southwestern Central African Republic. Special report, mimeographed, 38pp. St. Louis: Washington University.

Fox, J. E. D. 1968. Exploitation of the Gola Forest. *J. West Afr. Sci. Assoc.* 13:185–210.

Galdikas, B. M. F., and G. Teleki. 1981. Variations in subsistence activities of female and male pongids: new perspectives on the origins of hominid labor division. *Current Anthrop.* 22(3):241–256.

Gandini, G. 1979. Problems of ape conservation in Gabon. *IPPL Newsl.* 6(3):12.

Gartlan, J. S. 1975a. The African forests and problems of conservation. In S. Kondo, M. Kawai, and S. Kawamura, eds., *Proceedings of the Symposia of the Fifth Congress of the International Primatological Society, Nagoya, Japan, 1984.* Tokyo: Japan Science Press.

_____. 1975b. The African coastal rain forest and its primates: threatened resources. In G. Bermant and D. G. Lindburg, eds., *Primate Utilization and Conservation.* New York: John Wiley.

Gartlan, J. S., and T. T. Struhsaker. 1972. Polyspecific associations and niche separation of rain-forest anthropoids in Cameroon, West Africa. *J. Zool. London* 168:221–266.

Gaski, A. 1987. Chimpanzees cause controversy. *TRAFFIC(USA)* 7(2/3):25.

Gest, T. 1988. A chilling flurry of lawsuits. *U.S. News and World Report,* May 23. 104(20):64–65.

Ghiglieri, M. P. 1984. *The Chimpanzees of Kibale Forest: A Field Study of Ecology and Social Structure.* New York: Columbia University Press.

Goldsmith, E. I., and J. Moor-Jankowski. 1971. Scientists' responsibility for preservation of threatened species: Statement. *Medical Primatology 1970.* Basel: S. Karger.

Goliber, T. J. 1985. Sub-Saharan Africa: population pressures on development. *Pop. Bull.* 40(1):1–46.

Goodall, J. 1983. Population dynamics during a fifteen-year period in one community of free-living chimpanzees in the Gombe National Park, Tanzania. *Zeitschr. Tierpsych.* 61:1–60.

_____. 1986. *The Chimpanzees of Gombe: Patterns of Behavior.* Boston: Harvard University Press.

_____. 1987. A plea for the chimpanzees. *Amer. Scientist* 75:574–577. Reprinted from *New York Times Magazine,* May 17, 1987.

Greisenegger, I. 1986. Tarzans Geschaefte: Wirbel um die Papiere fur den Schimansenimport der Immuno – Von Affenschmuggel bis Freunderlwirtschaft steht alles zur Diskussion. *Profil* 43:60–62.

Haddow, A. J. 1958. Chimpanzees. *Uganda Wildlife* 1(3):18–20.

Hall, P. 1976. Priorities for wildlife conservation in northeastern Nigeria. *Nigerian Field* 41(3):194–204.

Happold, D. C. D. 1971. A Nigerian high forest reserve. In D. C. D. Happold, ed., *Wildlife Conservation in West Africa.* Gland: IUCN.

Harako, R. 1981. The cultural ecology of hunting behavior among Mbuti Pygmies in the Ituri Forest, Zaire. In R. S. O. Harding and G. Teleki, eds., *Omnivorous Primates.* New York: Columbia University Press.

Harcourt, A. H. 1984. Conservation of the Virunga gorillas. *Primate Conserv.* 4:36–37.

_____. 1987. Options for unwanted or confiscated primates. *Primate Conserv.* 8:111–113.

Harcourt, A. H., and K. J. Stewart. 1980. Gorilla-eaters of Gabon. *Oryx* 15(3)248–251.

Harding, R. S. O. 1983. Large mammals of the Kilimi area, Sierra Leone. Special report, mimeographed, 28pp. Philadelphia: University of Pennsylvania.

_____. 1984. Primates of the Kilimi area, northwest Sierra Leone. *Folia Primat.* 42:96–114.

Harrisson, B. 1971. *Conservation of Nonhuman Primates in 1970.* Basel: S. Karger.

Hart, J. 1978. From subsistence to market: a case study of the Mbuti net hunters. *Human Ecol.* 6:323–353.

Hart, J., and S. Thomas. 1986. The Ituri Forest of Zaire: primate diversity and prospects for conservation. *Primate Conserv.* 7:42–43.

Hartmann, R. 1886. *Anthropoid Apes.* New York: D. Appleton.

Hasegawa, T., and T. Nishida. 1984. Progress report on Mahale National Park. *Primate Conserv.* 4:37–38.

Hendrichs, H. 1977. Untersuchung zur Saugertierfauna in einem palaeotropischen und einem neotropischen Bergregenwald-gebiet. *Saugertierk. Mitteil.* 25(3):214–224.

Henson, N. 1983. Loophole may allow trade in African chimps. *New Scientist* October 20:165.

Hill, W. C. O. 1969a. The nomenclature, taxonomy and distribution of chimpanzees. In G. H. Bourne, ed., *The Chimpanzee,* vol. 1. Basel: S. Karger.

_____. 1969b. The discovery of the chimpanzee. In G. H. Bourne, ed., *The Chimpanzee,* vol. 1. Basel: S. Karger.

Hiraiwa-Hasegawa, M., T. Hasegawa, and T. Nishida. 1984. Demographic study of a large-sized unit-group of chimpanzees in the Mahale Mountains, Tanzania: a preliminary report. *Primates* 25(4):401–413.

Hiroko, K. 1987. Wild chimpanzees: keeping for research a serious issue (English translation). *Yomiuri Shimbun* July 19:1.

Hladik, C. M. 1974. La vie d'un groupe de chimpanzes dans la foret du Gabon. *Science et Nature* 121:5–14.

Hladik, C. M., and G. Viroben. 1974. L'alimentation proteique du chimpanze dans son environment forestier natural. *C. R. Acad. Sci. Paris* 279:1475–1478.

Horn, A. D. 1979. The taxonomic status of the bonobo chimpanzee. *Amer. J. Phys. Anthrop.* 51:273–282.

Itani, J. 1979. Distribution and adaptation of chimpanzees in an arid area. In D. A. Hamburg and E. R. McCown, eds., *The Great Apes.* Menlo Park: Benjamin/Cummings.

Jahnke, H. E. 1974. Conservation and utilization of wildlife in Uganda: a study of environmental economics. *Forsch. Afr., Studienstelle J.F.O., Instit. Wirtsch.* vol. 54.

Jeffrey, S. M. 1970. Ghana's forest wildlife in danger. *Oryx* 4:50–52.

_____. 1974. Primates of the dry high forest of Ghana. *Nigerian Field* 39:117–127.

_____. 1975. Ghana's new forest national park. *Oryx* 13:34–36.

_____. 1977. How Liberia uses wildlife. *Oryx* 14:168–173.

Johns, A. D. 1985. Selective logging and wildlife conservation in tropical rain forest: problems and recommendations. *Biol. Conserv.* 31:355–375.

Johnsen, D. O. 1987. The need for using chimpanzees in research. *Lab Animal* July/August, 19–23.

Jones, C., and J. Sabater Pi. 1971. Comparative ecology of *Gorilla gorilla* (Savage and Wyman) and *Pan troglodytes* (Blumenbach) in Rio Muni, West Africa. *Bibl. Primat.* 13:1–96.

Kabongo, K. M. 1984. Will the pygmy chimpanzee be threatened with extinction like the elephant and the white rhinoceros in Zaire? In R. L. Susman, ed., *The Pygmy Chimpanzee: Evolutionary Biology and Behavior.* New York: Plenum.

Kano, T. 1972. Distribution and adaptation of the chimpanzees on the eastern shore of Lake Tanganyika. *Kyoto Univ. Afr. Stud.* 7:37–129.

_____. 1984. Distribution of pygmy chimpanzees *(Pan paniscus)* in the central Zaire basin. *Folia Primat.* 43:36–52.

Kavanagh, M. 1984. A review of the international primate trade. In D. Mack and R. A. Mittermeier, eds., *The International Primate Trade*, vol. 1. Washington: TRAFFIC(USA).

Kavanagh, M., and E. Bennett. 1984. A synopsis of legislation and the primate trade in habitat and user countries. In D. Mack and R. A. Mittermeier, eds., *The International Primate Trade*, vol. 1. Washington: TRAFFIC(USA).

King, F. A., C. J. Yarbrough, D. C. Anderson, T. P. Gordon, and K. G. Gould. 1988. Primates. *Science* June 10. 240:1475–1481.

Kock, D. 1967. Die Verbreitung des Schimpansen, *Pan troglodytes schweinfurthii* (Giglioli, 1872), im Sudan. *Zeitschr. Saugertierk.* 32:250–255.

_____. 1969. Die Verbreitung der Primaten im Sudan. *Zeitschr. Saugertierk.* 34:193–215.

Koebner, L. 1982. Surrogate human. *Science 82* 3(6):32–39.

Kortlandt, A. 1962. Chimpanzees in the wild. *Sci. Amer.* 206(5):128–138.

_____. 1963. Bipedal armed fighting in chimpanzees. *Proc. XVI Internatl. Congr. Zool. Washington,* vol. 3, pp. 64–65.

_____. 1965. Some results of a pilot study on chimpanzee ecology. Unpubl. report, mimeographed, 59pp. University of Amsterdam.

_____. 1966. Chimpanzee ecology and laboratory management. *Lab. Prim. Newsl.* 5:1–11.

_____. 1972. *New Perspectives in Ape and Human Evolution.* Stichting voor Psychobiologie, Amsterdam.

_____. 1976. Chimpanzee habitats, locomotion, and hand use in the wild. Unpubl. report, mimeographed, 74pp.

_____. 1978. Can enough primates be supplied, and how? Unpubl. report, mimeographed, 11pp.

_____. 1983. Marginal habitats of chimpanzees. *J. Human Evol.* 12:231–278.

_____. 1986. The use of stone tools by wild-living chimpanzees and earliest hominids. *J. Human Evol.* 15:77–132.

Kortlandt, A., and E. Holzhaus. 1987. New data on the use of stone tools by chimpanzees in Guinea and Liberia. *Primates* 28(4):473–496.

Kortlandt, A., and J. C. J. van Zon. 1969. The present state of research on the dehumanization hypothesis of African ape evolution. *Proc. 2nd Internatl. Congr. Primat.*, vol. 3. Basel: S. Karger.

Lahm, S. A. 1985. Mandrill ecology and the status of Gabon's rainforests. *Primate Conserv.* 6:32–33.

Lamotte, M. 1983. The undermining of Mount Nimba. *Ambio* 12(3/4):174–179.

Lowes, R. H. G. 1970. Destruction in Sierra Leone. *Oryx* 10:309–310.

Luoma, J. R. 1989. The chimp connection. *Animal Kingdom* 92(1):38–51.

Mack, D., and A. Eudey. 1984. A review of the U. S. primate trade. In D. Mack and R. A. Mittermeier, eds., *The International Primate Trade*, vol. 1. Washington: TRAFFIC(USA).

Mack, D., and R. A. Mittermeier, eds. 1984. *The International Primate Trade*, vol. 1. Washington: TRAFFIC(USA).

MacKinnon, J. 1976. Mountain gorillas and bonobos. *Oryx* 13(4):372–382.

MacKinnon, J., and K. MacKinnon. 1986. *Review of the Protected Areas System in the Afrotropical Realm.* Gland: IUCN.

MacKinnon, J., K. MacKinnon, G. Child, and J. Thorsell. 1986. *Managing Protected Areas in the Tropics.* Gland: IUCN.

Martin, D. E. 1981. Breeding great apes in captivity. In C. E. Graham, ed., *Reproductive Biology of the Great Apes*. New York: Academic Press.

McGiffin, H. L. 1985. History of primate conservation in Sierra Leone. *IPPL Newsl.* 12(3):3–5.

McGinnis, P. R. 1979. Sexual behavior in free-living chimpanzees: consort relationships. In D. A. Hamburg and E. R. McCown, eds., *The Great Apes.* Menlo Park: Benjamin/ Cummings.

McGrew, W. C. 1981. Social and cognitive capabilities of non-human primates: lessons from the wild to captivity. *Int. J. Stud. Anim. Prob.* 2(3):138–149.

_____. 1983. Animal foods in the diets of wild chimpanzees *(Pan troglodytes):* why cross-cultural variation? *J. Ethol.* 1:46–61.

McGrew, W. C., and C. E. G. Tutin. 1978. Evidence for a social custom in wild chimpanzees. *Man* 13:234–251.

McGrew, W. C., P. J. Baldwin, and C. E. G. Tutin. 1981. Chimpanzees in a hot, dry and open habitat: Mt. Asserik, Senegal, West Africa. *J. Human Evol.* 10:227–244.

McHenry, T. J. P. 1986. Report to the government of the Republic of Liberia on wildlife and national parks legislation. Special report, mimeographed, 52pp. Rome: FAO.

Merz, G. 1986. The status of the forest elephant *Loxodonta africana cyclotis,* Matschie, 1900, in the Gola Forest Reserves, Sierra Leone. *Biol. Conserv.* 36:83–94.

Monath, T. P., and G. E. Kemp. 1973. Importance of non-human primates in yellow fever epidemiology in Nigeria. *Trop. Geogr. Med.* 25:28–38.

Moore, J. 1985a. Chimpanzees in Mali. *WWF Mon. Report,* 155–158.

_____. 1985b. Chimpanzee survey in Mali, West Africa. *Primate Conserv.* 6:59–63.

Myers, N. 1979. *The Sinking Ark.* New York: Pergamon Press.

_____. 1980. The present status and future prospects of tropical moist forests. *Environ. Conserv.* 7:101–114.

Napier, J. R., and P. H. Napier. 1967. *A Handbook of Living Primates.* New York: Academic Press.

Nishida, T. 1979. The social structure of chimpanzees of the Mahale Mountains. In D. A. Hamburg and E. R. McCown, eds., *The Great Apes.* Menlo Park: Benjamin/Cummings.

Nishida, T., and K. Kawanaka. 1972. Inter-unit-group relationships among wild chimpanzees of the Mahali Mountains. *Kyoto Univ. Afr. Stud.* 7:131–169.

Nishida, T., M. Hiraiwa-Hasegawa, T. Hasegawa, and Y. Takahata. 1985. Group extinction and female transfer in wild chimpanzees in the Mahale National Park, Tanzania. *Zeitschr. Tierpsychol.* 67:284–301.

Oates, J. F. 1980. Report on a pilot study of *Colobus verus* and other forest monkeys in southern Sierra Leone, with comments on conservation problems. Unpubl. report, mimeographed, 25pp. Hunter College, New York.

_____. 1985. *Action Plan for African Primate Conservation: 1986–1990.* IUCN/SSC PSG Special Report, Washington.

Petrides, G. 1965. Advisory report on wildlife and national parks in Nigeria, 1962. *Amer. Comm. Internatl. Wildl. Protec. Spec. Publ.* 18.

Pfeiffer, A. J., and L. J. Koebner. 1978. The resocialization of single-caged chimpanzees and the establishment of an island colony. *J. Med. Primat.* 7:70–81.

Phillipson, J. R. 1978. Wildlife conservation and management in Sierra Leone. Special report, mimeographed, 188pp. British Council, Oxford University.

Prince, A. M. 1985. Rehabilitation and release program for chimpanzees. *Primate Conserv.* 5:33.

Prince, A. M., B. Brotman, A. Hannah, M. Donnelly, K. Hentschel, and H. Roth. 1986. Rehabilitation and release into the wild of chimpanzees used in medical research. Special report, mimeographed, 31pp. New York Blood Center, New York.

Rahm, U. 1965. Distribution et ecologie de quelques mammiferes de l'est du Congo. *Zool. Afr.* 1(1):149–166.

_____. 1966. Les mammiferes de la foret equatoriale de l'est du Congo. *Ann. Mus. Roy. Afr. Centr. Tervuren Sci. Zool.* 14:39–121.

_____. 1967. Observations during chimpanzee captures in the Congo. In D. Starck, R. Schneider, and H. J. Kuhn, eds., *Neue Ergebnisse in Primatologie*. Stuttgart: Gustav Fischer Verlag.

Redmond, I. 1986. Law of the jungle. *BBC Wildlife* June, 300–301.

_____. 1988. Aren't chimps God's children too? *BBC Wildlife* 6(4):187–191.

Reynolds, V. 1967. *The Apes*. New York: E. P. Dutton.

Reynolds, V., and G. Luscombe. 1971. On the existence of currently described subspecies in the chimpanzee *(Pan troglodytes)*. *Folia Primat*. 14:129–138.

Reynolds, V., and F. Reynolds. 1965. Chimpanzees of Budongo Forest. In I. DeVore, ed., *Primate Behavior*. New York: Holt-Rinehart-Winston.

Rijksen, H., and A. Rijksen. 1979. Rehabilitation: a new approach is needed. *Tigerpaper* 6(1):16–18.

Riney, T., and P. Hill. 1967. Conservation and management of African wildlife. Special report, mimeographed. Rome: FAO.

Robinson, P. T. 1970. The status of the pygmy hippo and other wildlife in West Africa. Thesis. Michigan State Univ.

_____. 1971. Wildlife trends in Liberia and Sierra Leone. *Oryx* 11:117–122.

_____. 1986. The proposed Sapo National Park in Liberia: a field survey of prospects and problems. *Nat. Geogr. Soc. Res. Reports* 21:425–435.

Robinson, P. T., and A. Peal. 1981. Liberia's wildlife: the time for decision. *Zoonooz* 54(10):7–21.

Roy, A. D. 1974. Rhinophycomycosis enteromorphae occurring in a chimpanzee in the wild in East Africa. *Amer. J. Trop. Med. Hygiene* 23:935.

Roy, A. D., and H. M. Cameron. 1972. Rhinophycomycosis enteromorphae occurring in a chimpanzee in the wild in East Africa. *Amer. J. Trop. Med. Hygiene* 21:234–237.

Rucks, M. G. 1976. Notes on the problems of primate conservation in Bia National Park. Special report, mimeographed, 13pp. Department of Game and Wildlife, Accra, Ghana.

Sabater Pi, J. 1978. *El chimpance y los origenes de la cultura*. Promocion Cultural, Barcelona.

_____. 1979. Chimpanzees and human predation in Rio Muni. *IPPL Newsl*. 6(2):8.

Sabater Pi, J., and C. Groves. 1972. The importance of the higher primates in the diet of the Fang of Rio Muni. *Man* 7:239–243.

Sabater Pi, J., and C. Jones. 1967. Notes on the distribution and ecology of the higher primates of Rio Muni, West Africa. *Tulane Stud. Zool*. 14:101–109.

Salati, E., and P. B. Vose. 1983. Depletion of tropical rain forests. *Ambio* 12(2):67–71.

Sayer, J. A. 1977. Conservation of large mammals in the Republic of Mali. *Biol. Conserv*. 12:245–263.

Seal, U. S., and N. R. Flesness. 1986. Captive chimpanzee populations: past, present, and future. In K. Benirschke, ed., *Primates: The Road to Self-Sustaining Populations*. New York: Springer-Verlag.

Secretariat of CITES. 1988. Notification to the parties, No. 461. Jan. 20, 1988. Lausanne: CITES.

Simon, N., and M. A. G. Warland. 1968. Preliminary status survey of *Pan troglodytes* (Blumenbach 1799). Gland: IUCN.

Skorupa, J. 1983. Effects of selective logging on primates in the Kibale Forest, Uganda. *Primate Conserv*. 3:24.

Spinage, C. A. 1972. The ecology and problems of the Volcano National Park, Rwanda. *Biol. Conserv*. 4(3):194–204.

_____. 1980. Parks and reserves in Congo Brazzaville. *Oryx* 15(3):292–295.

Stephens, M. 1987. NIH's chimpanzee-breeding program: bad news for primates. *Humane Society News* 32(2):22–23.

Stoneley, J. 1986. *Starting Again.* Videotape report on the Chimfunshi Wildlife Orphanage, Zambia.

Stott, K., and C. J. Selsor. 1959. Chimpanzees in western Uganda. *Oryx* 5:108–115.

Struhsaker, T. T. 1981. Forest and primate conservation in East Africa. *Afr. J. Ecol.* 19:99–114.

_____. 1987. Forestry issues and conservation in Uganda. *Biol. Conserv.* 39:209–234.

Struhsaker, T. T., and P. Hunkeler. 1971. Evidence of tool-using by chimpanzees in the Ivory Coast. *Folia Primat.* 15:212–219.

Struhsaker, T. T., P. C. Howard, and A. Kisubi. 1986. Conservation of tropical forest wildlife in western Uganda. WWF Project Annual Report, mimeographed, 19pp. Washington: WWF.

Sugiyama, Y. 1968. Social organization of chimpanzees in the Budongo Forest. *Primates* 9:225–258.

_____. 1973. The social structure of wild chimpanzees: a review of field studies. In R. P. Michael and J. H. Crook, eds., *Comparative Ecology and Behavior of Primates.* London: Academic Press.

_____. 1979. Social structure and dynamics of wild chimpanzees at Bossou, Guinea. *Primates* 20:323–339.

_____. 1984. Population dynamics of wild chimpanzees at Bossou, Guinea, 1976–1983. *Primates* 25:391–400.

_____. 1985a. The brush-stick of chimpanzees found in southwest Cameroon and their cultural characteristics. *Primates* 26(4):361–374.

_____. 1985b. Import of chimpanzees and animal experimentation (in Japanese). *Kagaku* 55:127–130.

Sugiyama, Y., and A. G. Soumah. 1988. Preliminary survey of the distribution and population of chimpanzees in the Republic of Guinea. *Primates* 29(4):569–574.

Susman, R. L., and K. M. Kabongo. 1984. Update on the pygmy chimpanzee in Zaire. *Primate Conserv.* 4:34–36.

Susman, R. L., N. Badrian, A. Badrian, and N. T. Handler. 1981. Pygmy chimpanzee in peril. *Oryx* 16(2):179–183.

Suzuki, A. 1969. An ecological study of chimpanzees in a savanna woodland. *Primates* 10:103–148.

_____. 1971. On the problems of conservation of the chimpanzees in East Africa and the preservation of their environment. *Primates* 12:415–418.

Suzuki, A., and T. Suzuki. 1971. Requiem for a chimp: report on the threat to the chimpanzees of the Budongo Forest. *Africana* 4:23

Tanno, T. 1987. On the distribution of chimpanzees and gorillas and hunting pressure upon them in the lower Ubangi River area. Special report, mimeographed, 7pp. Japan: Hirosaki University.

Teleki, G. 1977. Spatial and temporal dimensions of routine activities performed by chimpanzees in Gombe National Park, Tanzania: an ethological study of adaptive strategy. Dissertation, Pennsylvania State University.

_____. 1978. A summary of evidence pertaining to the application of Merck, Sharp and Dohme Research Laboratories for a permit to import 125 chimpanzees *(Pan troglodytes)* from Sierra Leone. Unpubl. report prepared for the Endangered Species Office, Department of the Interior, Washington, March 16, 1978.

_____. 1980. Hunting and trapping wildlife in Sierra Leone: aspects of exploitation and exportation. Special report, mimeographed, 85pp. Washington: WWF.

_____. 1981. The omnivorous diet and eclectic feeding habits of chimpanzees in Gombe National Park, Tanzania. In R. S. O. Harding and G. Teleki, eds., *Omnivorous Primates.* New York: Columbia University Press.

_____. 1986. Outamba-Kilimi National Park: a provisional plan for management and development. Special report, mimeographed, 206pp. Washington: WWF.

_____. 1987. A visit with Sheila and David Siddle and their sixteen chimpanzees. *IPPL Newsl.* 14(3):3–7.

_____. 1988a. Preliminary investigation of the international chimpanzee trade. Special report, mimeographed, 6pp. Washington: CCCC.

_____. 1988b. Circumvention of CITES restrictions on chimpanzee trade. *FRAME News* 19:4–5.

Teleki, G., and L. A. Baldwin. 1975. Breeding programs aim to keep this a planet of the apes. *Smithsonian* 5:76–81.

_____. 1979a. The status of the chimpanzee *(Pan troglodytes* and *Pan paniscus)* in equatorial Africa: aspects of population destruction, exploitation and conservation. Special report, mimeographed, 80pp. Washington: IUCN/SSC PSG.

_____. 1979b. The need for a systematic survey of chimpanzee populations in West Africa, focusing initially on the Republic of Sierra Leone. Special report, mimeographed, 15pp. Washington: IUCN/SSC PSG.

_____. 1980. Disaster for chimpanzees. *Oryx* 15(4):317–318.

_____. 1981. Sierra Leone's wildlife legacy: options for survival. *Zoonooz* 54(10):21–27.

Teleki, G., and I. Bangura. 1981. Outamba-Kilimi National Park: cornerstone for conservation. *Zoonooz* 54(10):28–31.

Teleki, G., E. E. Hunt, and J. H. Pfifferling. 1976. Demographic observations (1963–1973) on the chimpanzees of Gombe National Park, Tanzania. *J. Human Evol.* 5:559–598.

Trenchard, P. C. 1988. Ecology and conservation of the Kibira National Park, Burundi. Special report, mimeographed, 51pp. Washington: Peace Corps.

Tutin, C. E. G., and M. Fernandez. 1983. Gorilla and chimpanzee census in Gabon. *Primate Conserv.* 3:22–23.

_____. 1984. Nationwide census of gorilla *(Gorilla g. gorilla)* and chimpanzee *(Pan t. troglodytes)* populations in Gabon. *Amer. J. Primat.* 6:313–336.

_____. 1985. Foods consumed by sympatric populations of *Gorilla g. gorilla* and *Pan t. troglodytes* in Gabon: some preliminary data. *Internatl. J. Primat.* 6(1):27–43.

_____. 1987. Gabon: a fragile sanctuary. *Primate Conserv.* 8:160–161.

_____. 1988. Status of populations of chimpanzees *(P. t. troglodytes)* in Gabon. Comments submitted to U. S. Fish and Wildlife Service, Washington.

Tutin, C. E. G., and P. R. McGinnis. 1981. Chimpanzee reproduction in the wild. In C. E. Graham, ed., *Reproductive Biology of the Great Apes.* New York: Academic Press.

Van Orsdol, K. G. 1986. Agricultural encroachment in Uganda's Kibale Forest. *Oryx* 20:115–117.

Verschuren, J. 1975. Wildlife in Zaire. *Oryx* 13:25–33,149–163.

_____. 1978. Burundi and wildlife: problems of an overcrowded country. *Oryx* 14(3):237–240.

_____. 1983. Conservation of tropical rain forest in Liberia. Special report, mimeographed, 51pp. Gland: IUCN.

Wade, N. 1978. New vaccine may bring man and chimpanzee into tragic conflict. *Science* 200:1027–1030.

White, F. J., and R. L. Susman. 1986. Program for the conservation and protection of contraband pygmy and common chimpanzees in the Republic of Zaire. *Primate Conserv.* 7:59–60.

Whitesides, G. H. 1985. Nut cracking by wild chimpanzees in Sierra Leone, West Africa. *Primates* 26:91–94.

Whitney, R. A. 1988. NIH will not obtain chimpanzees from the wild for AIDS research. Press statement issued by the NIH Office of Animal Care and Use.

Wilkinson, A. F. 1974. Areas to preserve in Sierra Leone. *Oryx* 12:596–597.

Wilson, M. L., and J. G. Elicker. 1976. Establishment, maintenance, and behavior of free-ranging chimpanzees on Ossabaw Island, Georgia, U.S.A. *Primates* 17(4):451–473.

Wolfheim, J. 1983. *Primates of the World: Distribution, Abundance and Conservation.* Seattle: University of Washington Press.

Wrangham, R. W. 1975. The behavioural ecology of chimpanzees in Gombe National Park, Tanzania. Dissertation, Cambridge University.

_____. 1977. Feeding behaviour of chimpanzees in Gombe National Park, Tanzania. In T. H. Clutton-Brock, ed., *Primate Ecology.* London: Academic Press.

_____. 1979. Sex differences in chimpanzee dispersion. In D. A. Hamburg and E. R. McCown, eds., *The Great Apes.* Menlo Park: Benjamin/Cummings.

Yerkes, R. M. 1943. *Chimpanzees: A Laboratory Colony.* New Haven: Yale University Press.

Yerkes, R. M., and A. W. Yerkes. 1929. *The Great Apes.* New Haven: Yale University Press.

van Zon, J. C. J., and J. van Orshoven. 1967. Enkele resultaten van de zesde Nederlandse chimpansee-expeditie. *Vakbl. Biol.* 47:161–166.

DEMOGRAPHY OF CHIMPANZEES
IN CAPTIVITY

Katherine Latinen

Concerns about the conservation of wild populations of primates have intensified in recent years because of man's heightened worldwide impact on the flora and fauna of tropical forests. Simultaneously, managers of captive animals have focused on breeding programs for species in their care. Today, managers of both field and captive populations face similar challenges of species preservation and management of small populations. The purposes of this presentation are to describe current efforts in the management of the chimpanzee *(Pan troglodytes)* in North American zoos, to identify demographic and genetic concerns, and to briefly describe husbandry efforts in the captive chimpanzee population.

In 1986, permission by the American Association of Zoological Parks and Aquariums was granted to the Detroit Zoo to establish a regional studbook for chimpanzees residing in North American zoos and exhibit facilities. The need for such a management effort was based on the current legal status of the species, uncertain population levels in the wild, the threatened and vulnerable status of the wild population, and the insecure status of the captive population. At the present time we are collecting data on individual animals from current and past holders of chimpanzees. Utilizing information collected by the International Species Inventory System (ISIS) in Minnesota, along with the updated information, the first edition of the studbook is to be published in 1989.

A studbook is a registry of founders (wild-born animals), their offspring, and all animals currently living in a captive population. The purpose of a studbook is to document vital information for each animal, including birth and death dates, country of origin if wild born, locations throughout life, sire, dam, and other pertinent background information such as social history. This material can then be analyzed to assess the demographic and genetic characteristics of the population. From this analysis, recommendations can be made to improve the management of the population by reducing inbreeding coefficients, equalizing founder representation, or decreasing the rate of loss of genetic variability. If left unregulated in small populations, these factors could jeopardize the overall health and continued existence of that captive population. Studbooks have been recognized as valuable management tools since the establishment of the first studbook for wisent (European bison, *Bison bonasus*) in 1932 (Mohr, 1968).

As of 30 June 1986, 87 British, European, and North American facilities reporting to ISIS held 1,870 chimpanzees (902 males, 959 females, 9 unsexed animals). Of this total population, 408 chimpanzees are known to be wild born, 951 captive born, and 511 of unknown origin. This population is found in breeding, exhibit, and biomedical facilities. Of this worldwide population, 1,676 chimpanzees reside in 73 North American facilities (828 males, 839 females, 9 unsexed). Approximately one-fifth, or 307 chimpanzees (117 males, 185 females, 5 unsexed), are found in 52 zoo and exhibit facilities (ISIS, 1986). These 307 are the animals with which this studbook will be concerned.

DEMOGRAPHIC CONCERNS

The age distribution of these 307 chimpanzees residing in zoo and exhibit facilities is an unstable one. Twenty-five percent of this population (76 animals) are between the ages of 0–10 years, 43% (131 animals) between 11–20 years, and 32% (96 animals) are older than 20 years. The optimal age structure of a self-sustaining population that is constant in size is one in which each age class has an equal or larger number of animals than any older age class (Foose, 1980). There are several factors which contribute to the unbalanced age structure in our captive chimpanzee population, the primary being the lack of breeding in both wild- and captive-born animals.

The earliest reported successful chimpanzee births in North America occurred in 1915 (Bourne, 1971) and 1926 (Fritz and Fritz, 1983). There have been isolated successes in continued breeding programs as evidenced by four generations of captive breeding at Yerkes Regional Primate Center (Martin, 1981) and three generations at the San Diego Zoo (Pournelle, 1961).

Recent data, however, indicate that current levels of chimpanzee breeding are insufficient to sustain the captive population. Blood (1982) reported that in the United States as of 1979, of the then current population of 1,273 chimpanzees, only 180 wild-born females and 83 wild-born males had produced offspring.

Breeding success in the captive-born population is even lower than that of the wild-born population. From 1926 to 1973 there were 411 births in zoo and nonzoo facilities. Only 8 males and 27 females of these captive-born chimpanzees were breeding successfully in 1982 (Fritz and Fritz, 1983).

Seal and Flesness (1986) have also indicated that there is low recruitment of captive-born male chimpanzees. In 1986, of 80 captive-born male chimps older than 15 years of age, 13 had sired 25 surviving offspring. While 80% of the wild-born males have been incorporated into the breeding population, only 16% of the captive-born males are breeding (Seal and Flesness, 1986).

Additionally, family sizes reflect the dependence on few breeding males (Seal and Flesness, 1986). Seal and Flesness have demonstrated that of 113 male chimpanzees which had sired 566 young, 12 of these breeders or approximately 11% have produced 47% of the progeny (263 births). The disproportionate contribution of a few breeding males means that the overall population will have to be larger in order to counteract the resulting increased loss of genetic diversity and the over-representation of specific bloodlines.

CAPTIVE HUSBANDRY

Chimpanzees have been maintained in European zoological collections since the early 18th century (Maple, 1979). Obstacles to their successful captive management historically have been inadequate nutrition and the chimpanzees' susceptibility to human diseases, climate irregularities, and stress (Maple, 1979). Today these basic problems are better understood, and the major questions of captive-chimpanzee management now concern the social and psychological needs of the species as well as the lack of reproduction, especially in captive-born generations.

In recent years, increased emphasis has been placed on providing stimulating environments for all species of great apes in zoos. In order to improve zoo facilities, studies have been conducted to determine the effect of exhibit designs on the rate of abnormal and undesirable behaviors (Maple, 1979). Many zoos are eliminating inadequate conditions and providing both outdoor and indoor facilities with appropriate "furniture" for increased environmental diversity and physical and psychological stimulation. Improvements will continue to be made as the psychological and social needs of chimpanzees are better understood. Three important aspects of our captive-management programs play pivotal roles in efforts at establishing a self-sustaining *ex situ* population.

Restrictive social conditions from birth to five years of age generally result in decreased and abnormal social repertoires in adults (Davenport, 1979; Fritz and Fritz, 1983; Fritz, 1986; Maple, 1979; Reynolds and Reynolds, 1961). Restrictive conditions include lack of maternal contact and socialization through the first six months to two years of life, and solitary living conditions with inadequate physical and mental stimuli. Early social experiences impact on intellectual, social, and sexual development (Davenport, 1979; Fritz and Fritz, 1985) and, ultimately, on successful reproduction as adults. The period of maternal contact must therefore be extended as long as possible.

Hand-rearing techniques for infants have been shown to have significant effects on physical, social, and sexual development (Davenport, 1979; Fritz and Fritz, 1985). When consideration is given to the needs of neonate and infant chimpanzees for tactile, cutaneous, kinesthetic, and vestibular stimulation, normal social development will occur (Fritz and Fritz, 1982, 1985). Imprinting by chimpanzees on humans may occur between 6 and 12 months of age if infants are not introduced to and reared with peers (Fritz and Fritz, 1982, 1985). If hand-rearing is necessary, techniques such as those described by the Primate Foundation of Arizona, which result in normal adult social behaviors (Fritz and Fritz, 1985), should be adopted.

Resocialization of chimpanzees exhibiting abnormal social and sexual behaviors has received increased attention, and significant progress has been made in some breeding facilities (Fritz, 1986; Fritz and Fritz, in press). The rate for converting asocial animals into socially adjusted colony members is approximately 98%, and the rate of conversion to successful breeders is 50% (Fritz, 1986). Conditions facilitating natural rearing of progeny, improved hand-rearing techniques, and resocialization programs must receive increased attention if our goal is to increase the rate of successful reproduction in the captive chimpanzee population.

The zoo chimpanzee population is not breeding at a rate sufficient to maintain a self-sustaining population. Four factors can be identified as contributing to this problem:

1. The low rate of reproduction among animals in the wild-born population may be due to adults who are unsocialized because they were removed from the wild during the critical social development period and raised in a restrictive environment.

2. The lack of reproduction in the captive-born population may be due to rejection by natural mothers, management practices of early removal, and subsequent inadequate hand-rearing techniques. It may also be due to lack of recruitment of captive-born males by managers of breeding programs.

3. There is an intentional limitation on breeding due to the difficulty in placing surplus progeny, especially males, and the difficulty in exchanging animals between established colonies.

4. The infant mortality rate for captive chimpanzees (total North American population) has a significant negative impact on the growth rate of the population (Seal et al., 1985). The death rate for chimpanzees in the first year of life is 21%, with half of the first-year deaths occurring on the day of birth (Seal et al., 1985). Offspring of primiparous females have higher rates of death in the first two years than do offspring of multiparous females (Seal et al., 1985). Birth rank also affects infant mortality rates, with the first three ranks having higher frequencies of day-of-birth deaths (Seal et al., 1985). Occurrences of infant deaths on the day of birth (day-0) increase the risk for recurrence of infant deaths in subsequent pregnancies (Seal et al., 1985). Although infant/fetal mortality rates of 38% and year-1 rates of 32% have been reported for chimpanzees in the wild (Hiraiwa-Hasegawa et al., 1984), in a closed captive population (no new genetic input), infant mortality rates as high as these can have a great impact on the growth rate of the population.

GENETIC CONCERNS

Regarding the genetic status of the captive population, one important issue remains unanswered. Generally three subspecies of the chimpanzee *(Pan troglodytes)* are recognized (Napier and Napier, 1967): *P. t. troglodytes* Blumenbach, 1779; *P. t. verus* Schwarz, 1934; *P. t. schweinfurthii* Giglioli, 1872. The subspecies are defined by physical characteristics, which appear to change during development, and by geographic range. The physical characteristics include pelage, amount of balding, and cutaneous pigmentation. Collection-site information is unavailable for most captive chimpanzees (ISIS, 1986). Crossbreeding of subspecies of *P. troglodytes* is probably occurring in captivity because 284 out of a total of 307 chimpanzees in zoos are identified only as *P. troglodytes* (ISIS, 1986).

For the long-term genetic management of the chimpanzee, it is necessary to decide whether the captive population should be managed on a species or subspecies basis. Genetic research and input from field biologists are needed on this issue.

In summary, significant gains have been made in the captive husbandry of chimpanzees; however, the population is not self-sustaining. Further investigations are needed into improved husbandry and management methods. Implementation of improved techniques in breeding programs will hopefully lead to a self-sustaining population. As the wild-born population gets older, chimpanzee managers are faced with the problems of aging animals and associated infertility. Reduced fertility with increasing age is an expected condition in any population, but it is a serious problem for chimpanzees since the captive-born population is not breeding sufficiently.

There is hope for the future of the captive chimpanzee population if continued attention is paid and action taken by the managers of captive populations. To illustrate this, a regional studbook has already been established for the British population (Badham, 1986). The North American regional studbook will be available in the upcoming year. The National Institutes of Health have made significant efforts in the last decade at establishing a self-sustaining breeding population of chimpanzees. The management and breeding techniques of the biomedical facilities also make a significant contribution to breeding programs worldwide.

In the area of field conservation efforts, the IUCN/SSC Primate Specialist Group recently published the Action Plan for African Primate Conservation: 1986–1990 (Oates, 1986). That report marked the need for coordinated conservation efforts for the *in situ* populations of threatened primate species. The organization of the symposium "Understanding Chimpanzees" and the conservation session reflected the shared concern for the precarious situation of the wild population of chimpanzees. However, continued mutual support is needed among field biologists, breeding facilities, biomedical facilities, and zoos in order to guarantee the continued survival of both our wild and captive populations of chimpanzees.

REFERENCES

Badham, M. 1986. *Survey of Greater Anthropoid Apes in the British Isles and Ireland – Studbook of Chimpanzees.* Atherstone: East Midland Zoological Society/Twycross Zoo.

Blood, B. D. 1982. Prospects for a self-sustaining captive chimpanzee breeding program. In A. B. Chiarelli and R. S. Corruccini, eds., *Advanced Views in Primate Biology,* pp. 143–146. Berlin: Springer-Verlag.

Bourne, G. H. 1971. *The Ape People.* New York: G. P. Putnam's Sons.

Davenport, R. K. 1979. Some behavioral disturbances of great apes in captivity. In D. A. Hamburg and E. R. McCown, eds., *The Great Apes,* pp. 341–357. Menlo Park, Calif.: Benjamin/Cummings.

Foose, T. 1980. Demographic management of endangered species in captivity. *International Zoo Yearbook* 20:154–166.

Fritz, J. 1986. Resocialization of asocial chimpanzees. In K. Bernirschke, ed., *Primates: The Road to Self-Sustaining Populations,* pp. 351–359. New York: Springer-Verlag.

Fritz, J., and P. Fritz. 1982. Great ape hand-rearing with a goal of normalcy and a reproductive continuum. *AAZV Proceedings,* pp. 27–31.

_____. 1983. Captive chimpanzee population crisis. *AAZPA Regional Proceedings,* pp. 79–83.

_____. 1985. The hand-rearing unit: management decisions that may affect chimpanzee development. In C. E. Graham and J. A. Bowen, eds., *Clinical Management of Great Apes,* pp. 1–34. New York: Alan R. Liss.

_____. In press. *Resocialization of Asocial Captive Chimpanzees: A Prerequisite to a Humane Solution of the Captive Housing Crisis.*

Hiraiwa-Hasegawa, M., T. Hasegawa, and T. Nishida. 1984. Demographic study of a large-sized unit-group of chimpanzees in the Mahale Mountains, Tanzania: a preliminary report. *Primates* 25(4):401–413.

ISIS. 1986. SDR abstract – *Pan troglodytes* ssp. June 30, 1986.

Maple, T. L. 1979. Great apes in captivity: the good, the bad, and the ugly. In J. Erwin, T. L. Maple, and G. Mitchell, eds., *Captivity and Behavior,* pp. 239–272. New York: Van Nostrand Reinhold.

Martin, D. E. 1981. Breeding great apes in captivity. In C. Graham, ed., *Reproductive Biology of Great Apes,* pp. 343–373. New York: Academic Press.

Mohr, E. 1968. Studbooks for wild animals in captivity. *International Zoo Yearbook* 8:160–166.

Napier, J. R., and P. H. Napier. 1967. *A Handbook of Living Primates.* New York: Academic Press.

Oates, J. F. 1986. Action plan for African primate conservation: 1986–1990. IUCN/SSC Primate Specialist Group.

Pournelle, G. H. 1961. Three generations of chimpanzees at San Diego Zoo. *International Zoo Yearbook* 2:84.

Reynolds, V., and F. Reynolds. 1961. The natural environment and behavior of chimpanzees and suggestions for their care in zoos. *International Zoo Yearbook* 2:141–144.

Seal, U. S., and N. R. Flesness. 1986. Captive chimpanzee populations – past, present, and future. In K. Bernirschke, ed., *Primates: The Road to Self-Sustaining Populations,* pp. 46–55. New York: Springer-Verlag.

Seal, U. S., N. Flesness, and T. Foose. 1985. Neonatal and infant mortality in captive-born great apes. In C. E. Graham and J. A. Bowen, eds., *Clinical Management of Great Apes,* pp. 193–203. New York: Alan R. Liss.

AREA STATUS REPORT: TANZANIA

Jane Goodall

Tanzania, for many years, has had a very good record so far as conservation is concerned. Fully 25% of the land surface is under protected status of some sort or another – national parks, game reserves, conservation areas, and forest reserves.

For many years chimpanzees were only protected in the Gombe National Park (originally the Gombe Stream Game Reserve). During the past few years, however, the Mahale Mountains National Park has been gazetted. This is the area where, since 1965, Toshisada Nishida and other Japanese primatologists have been studying chimpanzees.

The Gombe National Park supports approximately 150 chimpanzees in an area of roughly 30 square miles. When I first arrived at Gombe, I could climb up to the rift escarpment (which forms the eastern boundary) and look eastward, away from Lake Tanganyika, across wild chimpanzee country stretching away in the distance. Gradually, however, cultivation has crept closer and closer to the boundaries, and today Gombe National Park is, for all intents and purposes, an island.

Up until 1976 we sometimes saw chimpanzees in patches of forest between the park and Kigoma (12 miles south). Today, however, these remnant groups have gone, and we no longer see even their nests. And there are no chimpanzees left to the north of the park. Thus there is no possibility of genetic exchange except between the three social groups (communities) living in the park. This is cause for grave concern in the future.

Recently I discovered that there is some hunting of chimpanzees around the periphery of the park. So far this does not pose a serious threat since the culprits are young boys, armed with spears, who are principally concerned with catching bushbuck and bushpigs. However, they enjoy hunting baboons which they then feed to their hunting dogs. And, I am told, the greatest sport is to hunt chimpanzees. Usually they are unsuccessful.

Ten years ago, when a young chimpanzee was found dead close to the village that lies just outside the northern boundary, there was a different attitude towards these

360

apes. The people said that they had never before seen a chimpanzee really close up. They were amazed at the similarities between the body of this youngster and their own. They were particularly impressed by his eyelashes, his ears, and his hands and fingernails. Although a primarily Muslim community, they decided to give the chimpanzee a proper burial. I have not yet been able to find out the attitude of the older and more responsible people in the villages towards the spearing of chimpanzees by their young men.

The Gombe chimpanzees are also under pressure from tourism. Fortunately this is not yet heavy; nevertheless, many people come to see the chimpanzees, particularly during the summer months. And, as yet, there is no control over the number of tourists allowed into the park. There is no question that some of the chimpanzees are tense, even fearful, when they encounter a party of tourists. I believe that the ranging patterns of some of the shyer females have been affected. However, plans are underway, with full cooperation from National Parks, for setting up tourist facilities in the north and south of the park, so as to relieve the pressure on the central study group. Some tour groups are cooperating in this venture.

It is especially important to build additional ranger stations in the north and south of the park to control illegal hunting. Villagers also enter the park to cut poles for building. (Note: During my brief tour of Western Germany in November 1986, I was able to raise most of the money necessary for building one of these ranger stations.)

Towards the end of 1986 (in October) I discovered that two infant chimpanzees were being kept, illegally, in a small village just outside Kigoma. It is unclear where they came from, although it seems likely that they originated from the Mahale Mountains National Park. More recently I have been told, by several sources, that young chimpanzees are occasionally smuggled across the border and into Burundi. From there they are exported, presumably for medical research. I have also heard that adult chimpanzees are occasionally sold to people from Zaire for food.

Very alarming, in view of the panic caused by AIDS (Acquired Immune Deficiency Syndrome), is information that a certain expatriate, a Scandinavian, is trying to persuade the Tanzanian government to lift the ban on the export of live chimpanzees. He has been heard to say, "You have a potential gold mine – two parks full of chimpanzees." Needless to say, I am doing my best to ensure that those responsible do not follow his advice.

The chimpanzee population of Tanzania, in the past, spread along the shores of Lake Tanganyika and was linked with chimpanzees in Burundi and Rwanda. In Burundi there is a tiny remnant group left, but soon this will be gone. In some areas, in the past, chimpanzees extended quite far inland, east of the lake. But even back in the mid-70s, the chimpanzees in some of these forests were dying because the fig trees were being poisoned so that more timber trees could be grown and forested by the local people. Many of those little inland pockets are gone now.

In conclusion, in Tanzania today we are left with approximately 700 chimpanzees in Mahale National Park, 150 in Gombe, and perhaps another 1,000 in other parts of southwestern Tanzania, as estimated by Japanese primatologists.

CONSERVATION STATUS OF
PAN PANISCUS

*Richard K. Malenky, Nancy Thompson-Handler,
and Randall L. Susman*

In this report, Central Africa is defined geographically as Zaire, The Central African Republic (C.A.R.), and The People's Republic of the Congo (Brazzaville). Ecologically the area is dominated by lowland rain forests associated with the drainage of the Zaire and Ubangi Rivers. Field studies of chimpanzees in this area began only recently. Consequently, our knowledge of chimpanzee ecology, behavior, and conservation status is less advanced than that of chimpanzees in other parts of Africa such as the Gombe Stream Reserve or the Mahale Mountains of Tanzania.

In addition, it is difficult to extrapolate the conservation status of Central African apes using data from other areas. Central Africa is unique ecologically in that both species of chimpanzee *(Pan troglodytes* and *Pan paniscus)* are found there. They are allopatric in their distribution, and the Zaire River is a natural boundary that separates them (Schwarz, 1929; Coolidge, 1933; Napier and Napier, 1967; van den Audenaerde, 1984). *Pan troglodytes* occurs north of the river, while *Pan paniscus* (bonobo) is found exclusively to the south of the river.

What follows is primarily (1) a summary of what is known about the current distribution of *Pan paniscus* and (2) some of the factors that are most likely to threaten its continued existence in the wild. The most notable aspect of the present discussion is the paucity of reliable information on the status of great apes (and primates in general) in Central Africa. For example, it is not even possible at present to report on the status of *Pan troglodytes* to the north and west of the Zaire River due to a total lack of reliable survey data (IUCN, 1988; Oates, 1985). Carroll is the only researcher presently studying great apes in Central Africa north of the Zaire River. His major focus is the lowland gorilla, but he has noted the presence of "sparse" populations of chimpanzees in southwestern C.A.R. (Carroll, in prep.). This lack of data is even more regrettable considering the fact that the Central Zaire Basin contains roughly 70% of the rain forest that remains on the African continent today.

In theory, the range of *Pan paniscus* extends to the east beyond Ikela, west to Lake Tumba, and south to the extent of the rain forest itself (Fig. 1). Efforts to estimate the range and distribution of bonobos began shortly after they were first recognized as a distinct species. Coolidge (1933) published a distribution map based on information available from museum collections (principally at Tervuren and Berlin) (Fig. 1). The bulk of the data came from the collections of the Royal Museum of the Congo in Brussels, material that was compiled by Dr. H. Schouteden, director of the museum at that time (see van den Audenaerde, 1984). The first in-country survey of existing populations was carried out in 1955 by Vandebroek as an adjunct to his collecting mission to the Belgian Congo. Vandebroek increased the number of known bonobo localities and added significantly to the documented range of *Pan paniscus* (Fig. 1). However, no further data on *Pan paniscus* in the wild were collected until the early 1970s when Nishida went to Zaire and initiated the most recent series of field studies (Nishida, 1972a, 1972b). Horn also conducted a survey in the early 1970s on the west bank of Lac Tumba (Horn, 1980). In the mid-to-late 1970s, the area north of Befale around the Maringa and Lomako Rivers was surveyed by the Badrians (Badrian and Badrian, 1977). However, Kano has conducted the most extensive surveys to date. On two separate expeditions, he has surveyed numerous areas between the Lopori River in the north and the headwaters of the Lomela River in the south (Kano, 1980).

The Badrians surveyed most extensively in the Zone of Befale (Fig. 2). They had

Figure 1. Map of estimated range of *Pan paniscus* (adapted from Susman et al., 1981).

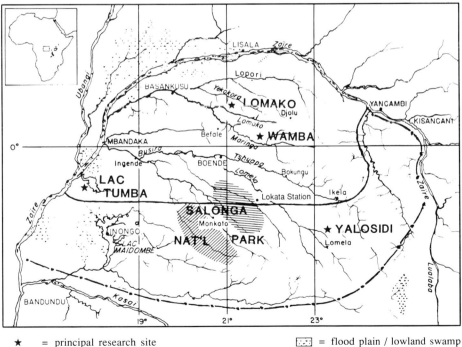

★ = principal research site
——— = extent of range (after Coolidge, 1933)
— • — = extent of range (after van den Audenaerde, 1984)

▨▨ = flood plain / lowland swamp
▨▨ = national park

originally intended to work in Salonga National Park, which is one of the largest parks in Africa and was established in part to protect the bonobo. However, in Boende they were informed that Kano and Nishida had recently surveyed the park and were unsuccessful in finding any populations of bonobos there. Therefore, the Badrians opted to search north of Boende. In 1974 they took up temporary residence in the Lomako Forest, where they were able to conduct an eight-month study of *Pan paniscus* (Badrian and Badrian, 1977). Surveys conducted in and around Befale led them to the conclusion that bonobo populations have significantly diminished in recent years due to human activity, and those that remained in 1977 probably existed only in isolated pockets (Badrian and Badrian, 1977).

Kano's surveys of *Pan paniscus* included areas between Yalosidi and Wamba as well as Befale and Salonga National Park (for greater detail see Kano, 1984). Although Kano's efforts are admired by all who have worked in Zaire, the information he collected is not as comprehensive as he would have liked due to the extreme difficulty of travel and logistics in Zaire (Kano, pers. comm.). The roads Kano suffered were (and are) in serious disrepair, and it is difficult to obtain adequate supplies (indeed some of Kano's travels were on bicycle after his vehicle broke down).

All of the researchers mentioned above were looking for study sites in which to conduct long-term studies of the behavioral ecology of bonobos. They were not there

Figure 2. A detailed map of the Zone of Befale.

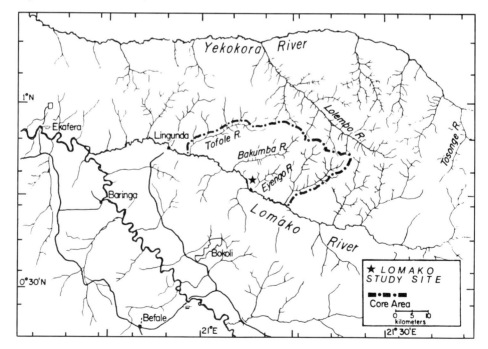

specifically to do survey work. However, all agree that the distribution of *Pan paniscus* within the Central Zaire Basin is fragmented. Only a proper on-site survey will establish the bonobo's true distribution and number. In order to correct this situation, Kano, Nishida, and Kuroda are planning a more extensive survey of this species throughout its entire range. They are now looking for funds to begin that work in the near future.

EXISTING THREATS TO *PAN PANISCUS* IN ZAIRE
Hunting

The consequences of our ignorance regarding the locations of viable populations of bonobos are worsened by the pressures that threaten remaining populations. Hunting, which varies from one place to the next, is certainly one of these pressures. There have been no systematic studies that deal with hunting pressures over the whole range of the species. As a result, much of the available information is anecdotal. For example, in our study area, the Lomako Forest, there are taboos against hunting bonobos. In other areas nearby, this is not the case (Kano, 1984; Horn, 1980; Nishida, 1972b), and people do hunt them for food. In addition, there is hunting in many areas to secure young bonobos for sale as gifts or pets. Mainly adult females are killed, and the young are taken and sold within the country. Such animals are found for sale even in the capital city of Kinshasa. Infant bonobos are also found for sale on large river boats that travel along the Zaire River and some of its tributaries (Susman et al., 1981; Mubalamata, 1984). Dr. Kano (pers. comm.) reports that recently at Wamba one male and one adult female were killed by poachers, three infants were taken for sale, and three other adult females were missing from the study population (possibly also killed by poachers). We are certain that bonobos are similarly hunted in the Lomako Forest. This is especially the case when researchers are not present at the study site.

Habitat Destruction

Another potential threat to bonobos is habitat destruction from commercial logging. There are eight logging concessions that have been leased for commercial operations between the Lopori and Tshuapa Rivers (Fig. 3). Together these tracts cover approximately 25% of the total range of the species. Most importantly, they comprise the central core of the range of *Pan paniscus*. This is precisely the area most likely to harbor the remaining populations of *Pan paniscus* since it is more distant from the major centers of commercial activity along the Zaire River.

Approximately 50,000 hectares of forest (including the Lomako Forest study site) are presently protected from commercial logging (Fig. 3). However, unless there are significant populations of bonobos in Salonga National Park, this protected area is the only haven that exists for *Pan paniscus* throughout its entire range.

Figure 3. Map of concessions granted to logging companies within the range of *Pan paniscus* (adapted from Susman et al., 1981).

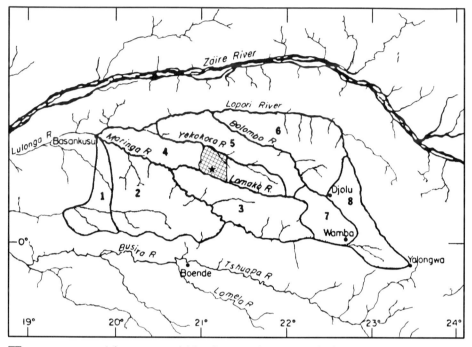

◰ = area protected from commercial logging   ★ = research site

CONCLUSION

It is clear that reliable survey data for both species of chimpanzee are urgently needed to determine population sizes and distribution. The identification of viable populations in undisturbed forest is vital to the design and implementation of future conservation efforts. Anthropologists, primatologists, and behavioral ecologists have expressed deep interest in the study of these animals. However, in order to increase the chance for success in protecting remaining chimpanzee populations, a greater commitment to matters of conservation must be made in the near future by more members of the professional community.

POSTSCRIPT

Some new information on the distribution of *Pan paniscus* has come to our attention since the initial writing of this chapter. A survey team under the aegis of Carsten Bresch of the University of Freiburg, West Germany, observed groups of bonobos in two areas (Bresch, pers. comm.). Their first sighting of a group of 20 animals occurred within a two-hour walk of the Lokata Station of Salonga National Park (Fig. 1). Park rangers there reported the presence of at least three other groups within the boundaries of the northern sector of the park. Bresch's team also located bonobos near an abandoned lumber camp in the region of Ekafera (Fig. 2). The animals were sighted in recently cut forest that borders a lumber road about 5 km from the camp. Additional bonobos were heard, and signs of recent scavenging were noted near the village of Tofili (in the region of Ekafera, Fig. 2). A more detailed survey of Salonga National Park sponsored by the World Wildlife Fund–Belgium and the Institute Zairois pour la Conservation de la Nature (IZCN) has confirmed the presence of significant numbers of bonobos within the boundaries of the northern sector of the park. This huge national park has been declared a World Heritage Site by the International Union for the Conservation of Nature (IUCN), and plans for establishing a major research center there are now underway (J. Sayer, pers. comm.).

The government of Zaire has also shown positive interest and action in fostering the conservation of previously known populations of bonobos. As of 1987, the area around Wamba, where Japanese researchers have studied several communities of bonobos since the early 1970s, is officially protected (Kuroda, pers. comm.). Further surveys within the area containing the Lomako Forest are planned by the IZCN to determine boundaries for a protected area for the population of bonobos there (Lanjouw, pers. comm.).

ACKNOWLEDGMENTS

We wish to thank Drs. Toshisada Nishida, Takayoshi Kano, and Suehisa Kuroda (Kyoto University), who generously provided information beyond that presented in their publications. We are grateful to the organizers of the symposium "Understanding Chimpanzees," especially Paul Heltne and Linda Marquardt, for making it possible for us to collaborate with these and other researchers. The Lomako Forest Pygmy Chimpanzee Project would not exist today if it were not for the pioneering work of Noel and Alison Badrian. Dr. Carsten Bresch (University of Freiburg), Jeffrey Sayer (IUCN), and Annette Lanjouw (Zaire Gorilla Project) have been extremely helpful in keeping us updated on the latest developments within Zaire. We also wish to extend our appreciation to the government of Zaire for its continued interest in the welfare of *Pan paniscus* and the support it has given to foreign researchers. We wish to thank the Departement de Recherches Scientifique and the Institute Zairois pour la Conservation de la Nature for their support in our research efforts. Figures for this chapter were drawn by Luci Betti.

The authors also wish to acknowledge the following institutions for their support during the course of our most recent research: The National Science Foundation, The L.S.B. Leakey Foundation, The New York Zoological Society, and the American Association of University Women.

REFERENCES

van den Audenaerde, D. F. E. T. 1984. The Tervuren Museum and the pygmy chimpanzee. In R. Susman, ed., *The Pygmy Chimpanzee: Evolutionary Biology and Behavior,* pp. 3–11. New York: Plenum Press.

Badrian, A., and N. Badrian. 1977. Pygmy chimpanzees. *Oryx* 13(5):463–468.

Carroll, R. W. In prep. Relative density, range extension, food preferences, and conservation potential of the lowland gorilla in the Dzanga-Sangha region of southwestern Central African Republic.

Coolidge, H. J. 1933. *Pan paniscus*, pygmy chimpanzee from south of the Congo River. *Amer. J. Phys. Anthropol.* 17(1):1–57.

Horn, A. 1980. Some observations of the ecology of the bonobo chimpanzee *(Pan paniscus* Schwarz, 1929) near Lac Tumba, Zaire. *Folia Primatol.* 34:145–169.

IUCN. 1988. *Threatened Primates of Africa. IUCN Red Data Book,* pp. 106-122. Gland, Switz.: IUCN.

Kano, T. 1980. Social behavior of wild pygmy chimpanzees *(Pan paniscus)* of Wamba: a preliminary report. *Journal of Human Evolution* 9:243–260.

_____. 1984. Distribution of pygmy chimpanzees *(Pan paniscus)* in the Central Zaire Basin. *Folia Primatol.* 43:36–52.

Mubalamata, K. 1984. Will the pygmy chimpanzee be threatened with extinction as are the elephant and the white rhinoceros in Zaire? In R. Susman, ed., *The Pygmy Chimpanzee: Evolutionary Biology and Behavior,* pp. 415–418. New York: Plenum Press.

Napier, J. R., and P. H. Napier. 1967. *A Handbook of Living Primates: Morphology, Ecology and Behavior of Nonhuman Primates,* p. 238. London and New York: Academic Press.

Nishida, T. 1972a. A preliminary study of pygmy chimpanzees: searching for elya (II). *Monkey* 12:929–934.

_____. 1972b. Preliminary information of the pygmy chimpanzees *(Pan paniscus)* of the Congo basin. *Primates* 13:415–425.

Oates, J. F. (compiler). 1985. IUCN/SSC Primate Specialist Group Action Plan for African Primate Conservation: 1986–1990. UNEP and World Wildlife Fund.

Susman, R. L., N. Badrian, A. Badrian, and N. T. Handler. 1981. Pygmy chimpanzees in peril. *Oryx* 16:179–183.

Schwarz, E. 1929. Das vorkommen des Schimpansen auf den linken Kongo-Ufer. *Rev. Zool. Bot. Afr.* 16:424–426.

5

EVOLUTION
and EPILOGUE

HOMINOID SOCIOBIOLOGY AND
HOMINID SOCIAL EVOLUTION

Michael P. Ghiglieri

INTRODUCTION

Reconstructing human evolution is one goal of field studies of the great apes. Existing models of hominid evolution are of two major types: *referential* models, where real observations (e.g., a gathering economy, a living ape species, etc.) are used as referents for other phenomena less amenable to direct study (e.g., behavioral evolution); and *conceptual* models based on "theories: sets of concepts or variables that are defined, and whose relationships are analytically specified" (e.g., evolutionary theory and behavioral ecology) (Tooby and DeVore, 1987). Tooby and DeVore argued that reconstruction of human evolution *requires* conceptual models to "substantiate the claim that a living species is a good parallel to an extinct one."

Wrangham (1987) identifies a suite of social behaviors shared by the common ancestor of humans, chimpanzees *(Pan troglodytes)*, bonobos *(Pan paniscus)*, and gorillas *(Gorilla gorilla)*. The following essay goes beyond Wrangham's (1987) phylogenetic referential model and considers reproductive function in evolutionary terms. Hence, it attempts to elucidate the *principles* that predict how reproductive problems are solved functionally through individual social adaptations shaped by the processes of natural selection – particularly sexual selection and behavioral ecology. This entire analysis draws heavily on data reviewed in detail by Ghiglieri (1987). The second goal of this essay is to refine a phylogenetic referential model of the suite of social behaviors of the *most recent* common ancestor of hominids, chimpanzees, and bonobos. In this attempt I readily acknowledge a debt to Wrangham (1977, 1979a, 1979b, 1980, 1987), to Itani (1980), and to other contributors to *The Evolution of Human Behavior: Primate Models* (Kinzey, 1987) for ground breaking on several levels.

Wrangham considers all extant African hominoids as a single large evolutionary clade. Thus his final suite of social behaviors for the hominid ancestor excludes the following strategies because they are uncharacteristic of gorillas: strict male retention, territoriality, and fusion-fission sociality (all common to chimpanzees, bonobos, and humans). Apropos of recent cytogenetic evidence, my analysis considers the more recent ancestor of the smaller clade containing chimpanzees, bonobos, and humans to be a more appropriate and revealing model for the hominid ancestor.

REFERENTIAL MODEL OF SOCIAL STRUCTURE FOR THE HOMINID ANCESTOR: Sociobiological Comparisons of the Hominoidea

Social behavior does not fossilize well. Reconstructing the phylogenetic relationships of living hominoids depends to a great extent on methods of comparing living populations (e.g., Sibley and Ahlquist, 1984; Tooby and DeVore, 1987).

Table 1 summarizes basic sociobiological data on all five extant hominoids (Ghiglieri, 1987). The orangutan's male exogamy and female retention, with a very dispersed matriarchal harem system, is similar to the most common primate social system (Wrangham, 1980). Including orangutans in a referential model for the hominid ancestor would reduce commonly shared social traits to a very low number. This conclusion is supported by Wrangham (1987), by new cytogenetic studies (Tanner, 1987; Pilbeam, 1986), and by Sibley and Ahlquist's (1984) DNA-DNA hybridization experiments (see also Lewin, 1988a, 1988b). The cumulative evidence of Galdikas (1984) indicates that orangutan society, in contrast to other great apes, is basically matriarchal, and exogamy is primarily via males. The rarity of female exogamy among nonhuman primates in general (found in perhaps 5% of species) suggests that the shared tendency for female exogamy among African hominoids evolved after the split of the orangutan ancestor from the human-ape stem.

I argue that the social structure of mountain gorillas also departs significantly from the probable suite of social behaviors of the hominid ancestor. My departure from Wrangham (1987) is based on the DNA-DNA hybridization experiments by Sibley and Ahlquist (1984) plus those summarized by Lewin (1988a, 1988b) that indicate the gorilla ancestor split from the ape-human stem approximately 8.0–9.9 mya, whereas the split of hominids from the common ancestor of chimpanzees, bonobos, and humans occurred approximately 6.3–7.7 mya. Even if these estimated dates ultimately are proved incorrect, the $\Delta T_{50}H$ values generated by the DNA-DNA hybridizations between gorillas and humans are significantly greater than the values between chimpanzees and bonobos and humans (see also Pilbeam, 1986).

The second line of evidence is derived from field studies as discussed in detail by Ghiglieri (1987) and summarized in Table 1. While a minority of mountain gorillas exhibit what appears to be incipient male retention in social groups, the most common social group is a one-male harem in which the silverback leader breeds strictly in an individual effort and does not share mates (Harcourt, 1981). By comparison, the male kin-group community system of chimpanzees, bonobos, and humans (see Table 1) shows several differences in sociality. Of the 15 social behaviors listed in 11 categories in Table 1, gorillas differ from chimpanzees in 8. In contrast to chimpanzees, mountain gorillas exhibit frequent male exogamy, a one-male breeding group instead of a multimale system, a lack of fusion-fission sociality, no sharing of mates, no territorial behavior, no cooperative group defense by males, no cooperative competition for mates, and extreme sexual dimorphism.

Michael P. Ghiglieri

Table 1. Summary of social structures of great apes and humans.

SOCIAL CHARACTERISTIC	MOUNTAIN GORILLA	ORANGUTAN
1. Female exogamy	Most common	Rare
2. Male exogamy	Less common, but frequent	Most common
3. Associations between females	Due to common attraction to silverback harem male	Are between female kin (matriarchal)
4. Bonds between adult females	Extremely weak, dissolve with loss of harem male	Weak but strong enough to cause matriarchal home range cluster
5. Type of social group; tendency for fusion-fission sociality	One-male harem, stable, closed. No tendency for fusion-fission sociality	Solitary, independent adults. *Very* slight tendency for fusion-fission among females
6. Mating system; mate sharing	Polygynous, no sharing of mates	Polygynous, mate-sharing facultative
7. Territoriality; defense by males; level of aggression	Neither sex territorial	Males vary; most not territorial; a few fight for dominance
8. Mating competition between males	Intense between individuals, severe fights	Intense between individuals, fights
9. Infanticide	By males against infants sired by rival males	Not seen; may not occur
10. Parental investment by adult males	Slight, consists of protecting infant from rival males	None
11. Sexual dimorphism of adults	Extreme; males are 221% the weight of females	Extreme; males are 237% the weight of females

Note: See Ghiglieri (1987) for sources and discussion.

Table 1. (continued)

CHIMPANZEE	BONOBO	HUMAN
Common	Apparently common	Most common
Rare; none after infancy	None observed	Uncommon
Due to common attraction to males	Due to common attraction males or territory (?)	Due to marriage to husband
Weak, but may include mutual aid, rare adoption of orphans	Weak but strongest among Hominoidea: include plant food sharing, genito-genital rubbing and mutual grooming	Weak, limited cooperation
Multimale kin-group: stable, closed community with multiple females. Fusion-fission sociality	Multimale, stable, closed community with multiple females. Fusion-fission sociality	Multimale kin-group: stable, semi-closed community, multiple females. Fusion-fission sociality
Community polygyny, promiscuous mate sharing	Community polygyny, promiscuous mate sharing	Polygynous, true promiscuity nonexistent
Communal territoriality by male kin-group; expansion by killing unrelated males	Communal territoriality by cooperating males; some fights serious	Communal territoriality by cooperating males in lethal warfare
Lethal between communities, also occurs within	Intense between communities, much milder within	Can be lethal between communities, less so within
By males against infants sired by rival males	Has not been seen	Mostly by males against infants of rival males
Slight, mostly consists of protection of infant from rival males	Slight, but greater than among chimps. Includes food sharing and grooming	Greatest among Hominoidea and substantially closer to parity with females
Moderate; males are 123% the weight of females	Moderate; males are 136% the weight of females	Moderate; males are 120-130% the weight of females

Table 2. The probable suite of social behaviors of the hominid ancestor, a composite referential model based on the social behaviors of chimpanzees, bonobos, and humans.

Elements of social behavior common to chimpanzees, bonobos and humans. (Data from Table 1 and Ghiglieri, 1987)	Elements of social behavior common to gorillas, chimpanzees, bonobos and humans. (Data from Wrangham, 1986: Table 3.7)
1. Female exogamy	1. Female exogamy
2. Bonds between females are weak but do exist (unlike gorillas)	2. No alliance bonds between females
3. Social groups are closed	3. Closed social networks (groups)
4. Social groups are stable, multi-male, multi-female communities	4. Male-dominated intergroup relationships
5. Males are active in territorial defense	5. Males are active in hostile interactions between groups
6. Males see, attack, and in two species, may kill rival males	6. Stalk and attack interactions by males against non-group males
7. Polygynous mating pattern	7. Polygynous mating pattern
8. Males sometimes travel alone	8. Males sometimes travel alone
9. Females commonly travel alone	
10. Fusion-fission sociality common *within* communities	
11. Female associations due primarily to attraction to same male(s)	
12. Male retention (endogamy)	
13. *Communal territoriality* typical	
14. Mating competition between males of same community mild relative to that between communities	
15. Sexual dimorphism is moderate; males cooperate in alliance against rivals	

Notes: Data in the first column are from Table 1 and Ghiglieri (1987). Data in the second column are a similar referential model proposed by Wrangham (1986) based on all four species of the African Hominoidea. See text for sources and discussion.

It is my contention that virtually all of these differences are traceable either directly or indirectly to two important evolutionary developments among the chimpanzee-bonobo-human ancestor: the strict retention of male offspring in the natal social group and female exogamy. As male reproductive strategies go, male retention is something of a quantum evolutionary leap because it sets the stage for the evolution of cooperative behaviors among male kin leading to communal reproductive strategies. Mountain gorillas exhibit this tendency in a very incipient form, which, compared to its absence among orangutans, suggests the ancestor of African Hominoidea may have been slightly preadapted in this direction. It is tempting to conclude that the two million years separating the branching of gorillas and humans was the period during which the adaptation of strict male retention arose.

Column 1 of Table 2 lists as the most probable social profile for the hominid ancestor the social behaviors common to chimpanzees, bonobos, and humans (from Table 1). Clearly, much commonality is lost by Wrangham's (1987) inclusion of gorilla traits (see column 2 of Table 1). The suite of social behaviors of the hominid ancestor based on the chimpanzee-bonobo-human clade contains a complete "sub-suite" (items 4–8, 10, 12–14 of Table 2) of rare social behaviors that clearly are adaptations to enhance reproductive success among males via a communal reproductive strategy. Data from the smaller clade of chimpanzees, bonobos (Ghiglieri, 1987), and humans (Ember, 1978; Murdock, 1967) suggest extreme communal territoriality.

Strict male retention is associated with, and perhaps essential to, the evolution of male communal territoriality. To be reproductively successful against rival males employing individual strategies, such male kin-groups must (1) attract and recruit multiple females, (2) breed with them more successfully, (3) insure that the quality of their territory promotes a higher rate of survival to maturity of their offspring, (4) be more capable of defending their females, offspring, and territory, and (5) maintain solidarity among males of a kin-group to foster mutual support in confrontations with rivals. The social behaviors in column 1 of Table 2 seem designed to insure exactly these functions. Unlike gorillas and orangutans, males of the chimpanzee-bonobo-human clade retain male offspring, live in closed social groups containing multiple females, mate polygynously, restrict their ranging to a communal territory, and are *cooperatively* active in territorial defense. And, apparently, when a neighboring community weakens, the males of some chimpanzee communities make a *concerted* strategic effort to stalk, attack, and kill their rivals (Ghiglieri, 1987, 1988; Goodall, 1986; Goodall et al., 1979; Nishida et al., 1985). Humans do this also; for instance, the Yanomamo Indians of Venezuela accrue 309% greater reproductive success by engaging in mortal fights than by avoiding fights altogether (Chagnon, 1983, 1988). The immense sexual dimorphism in body size among gorillas contrasted with the much more moderate levels among chimpanzees, bonobos, and humans (Table 1) suggests that the latter species have relied for a long time on cooperation among male kin, instead of body size, for success in male-male competition. With regard to the social evolution of the hominid ancestor, this propensity may well have been a critical preadaptation for male cooperation in dangerous scavenging or hunting (Morgan, 1979).

Fusion-fission sociality is common to the chimpanzee-bonobo-human clade. In all three species, adults of either sex travel alone temporarily. Among chimpanzees, increased frequency of solitary travel coincides with seasons of reduced food availability and smaller food patches (Ghiglieri, 1987), as it does also among humans (Harako, 1981; Hayden, 1981; Silberbauer, 1981). The significance of fusion-fission sociality is that *despite* environmental difficulties, individuals continue to socialize in large groups. All three species exhibit sophisticated behaviors to facilitate reunions during fusion. Indeed, Bauer (1980) reported that reunions and social parties among Gombe chimpanzees occurred for social rather than ecological reasons. Bauer also noted that party sizes were significantly larger in the *periphery* of the territory than in the core. Large groups permit successful defense of the territory (and reproductive females) from the males of other communities. As a behavioral adaptation among chimpanzees, fusion-fission sociality is the solution to two problems: (1) male-male competition, which requires a large social group for reproductive advantage, and (2) the distribution and abundance of rare food resources which can only support small groups on a stable basis. Among early hominids, fusion-fission sociality may have been a preadaptation for a division of labor on a daily basis and for diversification of ecological modes (Hayden, 1981).

Because communal breeding strategies by male kin are so extremely rare among nonhuman primates (Wrangham, 1980), the chance that each of the most recent three species from the common ape-human stem evolved these strategies independently seems infinitesimal. The behaviors appear to be homologies and of considerable adaptive value. Hence, very likely the hominid ancestor exhibited this suite of social traits.

TOWARD A STRATEGIC MODEL OF HOMINID EVOLUTION

Among the Hominoidea, individual reproductive strategies – not the influences of predators or food resources – seem to dominate and shape social structure most. Within the evolutionary lineage whose species possess the greatest cognitive abilities (Rumbaugh, 1970), individuals seem to solve their ecological problems so readily as to have surplus cognitive resources remaining to focus on social interactions. Cheney et al. (1986), in fact, argue that natural selection for intelligence may have acted strongest in the social domain: "During primate evolution, group life exerted strong selective pressure on the ability to form complex associations, reason by analogy, make transitive inferences, and predict the behavior of fellow group members."

Perhaps as a by-product of selection for intelligence, and spurred by sexual selection (Darwin, 1871), kin selection (Hamilton, 1964), and reciprocal altruism (Trivers, 1971), *the rare strategy of individual males cooperating* to enhance their reproductive success evolved in the most recent ancestor of the ape-human stem. Note that enhanced success may have been accomplished only by reducing the success of uncooperative males. In the section above, I described the male-retentiveness of the chimpanzee-bonobo-human clade as a reproductive adaptation conferring advantage over males attempting to reproduce individually. Below I note four components of a strategic model which describe how males might gain reproductive advantage by communally defending a territory containing females. (See Tooby and DeVore [1987] for an excellent discussion of necessary criteria for a "strategic" model of hominid evolution.)

1. The first component considers mating systems to be products of individual adaptations to maximize reproductive success. If males can manage to lower their parental investment in their offspring (Trivers, 1972) – laying the greater burden on females and thereby freeing the males to reenter the mating arena, sire additional offspring, and, by doing so, more than make up for any loss of reproductive success incurred by having lowered their parental investment – polygyny will become the dominant reproductive strategy. I believe that this is a necessary first step toward a male-retentive communal system.

2. The second component is kin selection. In essence, the more closely related two individuals are, the greater inclusive fitness in which one individual shares when the other is reproductively successful (Hamilton, 1964). The evolution of cooperation among male apes seems most likely a consequence of inclusive fitness and kin selection.

3. The third component is reciprocal altruism (Trivers, 1971; Axelrod and Hamilton, 1981; Packer, 1977), a relationship between individuals wherein one performs an act at some cost to one's self in order to benefit one's "partner," but does so in expectation that the favor will be repaid in kind when appropriate. This sort of relationship could be critical as a payoff for taking the individual risks involved with communal defense of a group of females against other groups of males, even when such cooperation is already rewarded through inclusive fitness. For reciprocal altruism to work within a complex social network, however, the following conditions may have been precursors: longevity, low dispersal, and high rates of mutual dependence (Trivers, 1971).

4. The fourth component is flexible use of alternate behavioral strategies, depending on ecological and demographic factors. Maynard Smith's (1978, 1982) concept of evolutionarily stable strategies sets standards for measuring the reproductive values of alternate morphologies or behaviors within a population. In populations whose environments change periodically, different traits may alternately become reproductively superior. The very different male reproductive strategies of the extant Hominoidea (Ghiglieri, 1987) illustrate alternate evolutionarily stable strategies. If kin-related males communally defend a territory containing females whom they share as mates, males employing alternate strategies apparently cannot match or exceed them reproductively. And a solitary male chimpanzee apparently cannot compete successfully against the community system.

It is my contention that the underlying genetic and psychological mechanisms required by each of the first three models are operating among males of the chimpanzee-bonobo-human clade, just as they operate among male lions, who, when they lose male kin, readily form alliances with non-kin based on reciprocal altruism (Pusey and Packer, 1983). Maynard Smith's concept of evolutionarily stable strategies offers an hypothetical framework to test this contention.

Clearly hominids show unique adaptations within the chimpanzee-bonobo-human clade. I propose that the hominid ancestor, equipped with the unusual male reproductive strategy of relying on communal defense by male kin, took the additional step of mating in stable, exclusive bonds with one or more females (see also Foley and Lee, 1989). Because such males could be far more sure of paternity, they donated much more parental investment in their recognized offspring than had been the case among their hominoid ancestors. This greater parental investment by male hominids (assuming that females continued their high level of investment) should then have raised the reproductive success of both sexes above that of populations characterized by lower male investment.

Strategic models of primate social structure which hypothesize reproductive advantage require field testing of reproductive success among individual apes and humans who employ *specific ecological or life history strategies* (see Clutton-Brock et al., 1982; Endler, 1986). Among long-lived primates, however, such data are difficult to collect even for females (e.g., Galdikas, 1981; Goodall, 1983; Harcourt et al., 1981). Determination of lifetime reproductive success for samples of males will require entirely new levels of sophistication and dedication in field research (e.g., Mitani, 1985).

REFERENCES

Axelrod, R. T., and W. D. Hamilton. 1981. The evolution of cooperation. *Science* 211:1390–1396.

Bauer, H. R. 1980. Chimpanzee society and social dominance in evolutionary perspective. In D. R. Omark, F. F. Strayer, and D. Freedman, eds., *Dominance Relations: Ethological Perspectives on Human Conflict,* pp. 97–119. New York: Garland.

Chagnon, N. A. 1983. *Yanomamo: The Fierce People.* 3rd ed. New York: Holt, Rinehart & Winston.

———. 1988. Life histories, blood revenge, and warfare in a tribal population. *Science* 239:985–992.

Cheney, D., R. Seyfarth, and B. Smuts, 1986. Social relationships and social cognition in nonhuman primates. *Science* 234:1361–1366.

Clutton-Brock, T. H., F. E. Guiness, and S. D. Albon. 1982. *Red Deer: Behaviour and Ecology of Two Sexes.* Chicago: University of Chicago Press.

Darwin, C. 1871. *Sexual Selection and the Descent of Man.* London: John Murray.

Ember, C. R. 1978. Myths about hunter-gatherers. *Ethnology* 17:439–448.

Endler, J. A. 1986. *Natural Selection in the Wild.* Princeton: Princeton University Press.

Foley, R. A., and P. C. Lee. 1989. Finite social space, evolutionary pathways, and reconstructing hominid behavior. *Science* 243:901–906.

Galdikas, B. M. F. 1981. Orangutan reproduction in the wild. In C. E. Graham, ed., *Reproductive Biology of the Great Apes,* pp. 281–300. New York: Academic Press.

———. 1984. Adult female sociality among wild orangutans at Tanjung Puting Reserve. In M. F. Small, ed., *Female Primates, Studies by Female Primatologists,* pp. 217–235. New York: Alan B. Liss.

Ghiglieri, M. P. 1987. Sociobiology of the great apes and the hominid ancestor. *Journal of Human Evolution* 16(4):319–357.

———. 1988. *East of the Mountains of the Moon.* New York: The Free Press.

Goodall, J. 1983. Population dynamics during a fifteen-year period in one community of free-living chimpanzees in the Gombe National Park, Tanzania. *Z. Tierpsychol.* 61:1–60.

———. 1986. *The Chimpanzees of Gombe.* Cambridge: Harvard University Press.

Goodall, J., A. Bandoro, E. Bergman, C. Busse, H. Matama, E. Mpongo, A. Pierce, and D. Riss. 1979. Intercommunity interactions in the chimpanzee population of the Gombe National Park. In D. A. Hamburg and E. R. McCown, eds., *The Great Apes,* pp. 13–54. Menlo Park, Calif.: Benjamin/Cummings.

Hamilton, W. D. 1964. The genetical theory of social behavior, I, II. *J. Theoret. Biol.* 31:295–311.

Harako, R. 1981. The cultural ecology of hunting behavior among Mbuti pygmies in the Ituri Forest, Zaire. In R. S. O. Harding and G. Teleki, eds., *Omnivorous Primates Gathering and Hunting in Human Evolution,* pp. 499–555. New York: Columbia University Press.

Harcourt, A. H. 1981. Intermale competition and the reproductive behavior of the great apes. In C. E. Graham, ed., *Reproductive Biology of the Great Apes,* pp. 301–318. New York: Academic Press.

Harcourt, A. H., J. Stewart, and D. Fossey. 1981. Gorilla reproduction in the wild. In C. E. Graham, ed., *Reproductive Biology of the Great Apes,* pp. 265–279. New York: Academic Press.

Hayden, B. 1981. Subsistence and ecological adaptations of modern hunter-gatherers. In R. S. O. Harding and G. Teleki, eds., *Omnivorous Primates Gathering and Hunting in Human Evolution,* pp. 344–421. New York: Columbia University Press.

Itani, J. 1980. Social structure of African great apes. *J. Reprod. Fert. Suppl.* 28:33–41.

Kinzey, W. G., ed. 1987. *The Evolution of Human Behavior: Primate Models.* New York: SUNY Press.

Lewin, R. 1988a. Conflict over DNA results. *Science* 241:1598–1600.

_____. 1988b. DNA clock conflict continues. *Science* 241:1756–1759.

Maynard Smith, J. 1978. The evolution of social behavior. *Scientific American* 239(3):176–192.

_____. 1982. *Evolution and the Theory of Games.* Cambridge: Cambridge University Press.

Mitani, J. C. 1985. Sexual selection and adult male orangutan calls. *Anim. Behav.* 33:272–283.

Morgan, C. J. 1979. Eskimo hunting groups, social kinship, and the possibility of kin selection in humans. *Ethology and Sociobiology* 1(1):83–86.

Murdock, G. P. 1967. Ethnographic atlas: a summary. *Ethnology* 6:109–236.

Nishida, T., M. Hiraiwa-Hasegawa, T. Hasegawa, and Y. Takahata. 1985. Group extinction and female transfer in wild chimpanzees in the Mahale National Park, Tanzania. *Z. Tierpsychol.* 67:284–301.

Packer, C. 1977. Reciprocal altruism in *Papio anubis. Nature* 265:441–443.

Pilbeam, D. 1986. Distinguished lecture: hominoid evolution and hominoid origins. *American Anthropologist* 88(2):295–312.

Pusey, A. E., and C. Packer. 1983. The once and future kings. *Natural History* 92(8):54–63.

Rumbaugh, D. M. 1970. Learning skills of anthropoids. In L. A. Rosenblum, ed., *Primate Behavior,* pp. 1–70. London: Academic Press.

Sibley, C. G., and J. E. Ahlquist. 1984. The phylogeny of the hominoid primates, as indicated by DNA-DNA hybridization. *Journal of Molecular Evolution* 20:2–15.

Silberbauer, G. 1981. Hunter/gatherers of the Central Kalahari. In R. S. O. Harding and G. Teleki, eds., *Omnivorous Primates: Gathering and Hunting in Human Evolution,* pp. 455–498. New York: Columbia University Press.

Tanner, N. M. 1987. The chimpanzee model revisited and the gathering hypothesis. In W. G. Kinzey, ed., *The Evolution of Human Behavior: Primate Models,* pp. 3–27. New York: SUNY Press.

Tooby, J., and I. DeVore. 1987. The reconstruction of hominid evolution through strategic modelling. In W. G. Kinzey, ed., *The Evolution of Human Behavior: Primate Models,* pp. 183–237. New York: SUNY Press.

Trivers, R. L. 1971. The evolution of reciprocal altruism. *Quart. Rev. Biol.* 46:35–57.

_____. 1972. Parental investment and sexual selection. In B. Campbell, ed., *Sexual Selection and the Descent of Man: 1871–1971,* pp. 136–179. Chicago: Aldine.

Wrangham, R. W. 1977. Feeding behaviour of chimpanzees in Gombe National Park, Tanzania. In T. H. Clutton-Brock, ed., *Primate Ecology,* pp. 504–538. London: Academic Press.

_____. 1979a. On the evolution of ape social systems. *Social Science Information* 18:335–368.

_____. 1979b. Sex differences in chimpanzee dispersion. In D. A. Hamburg and E. R. McCown, eds., *The Great Apes,* pp. 481–490. Menlo Park, Calif: Benjamin/Cummings.

_____. 1980. An ecological model of female-bonded primate groups. *Behaviour* 75:262–299.

_____. 1987. The significance of African apes for reconstructing human social evolution. In W. G. Kinzey, ed., *The Evolution of Human Behavior: Primate Models,* pp. 51–71. New York: SUNY Press.

EPILOGUE:
UNDERSTANDING CHIMPANZEES AND BONOBOS, UNDERSTANDING OURSELVES

Paul G. Heltne

The growing body of research on chimpanzees and bonobos constitutes one of the most exciting episodes in human exploration of the world around us. As the preceding chapters reveal, humankind has at least two sibling species which I believe must now be recognized as sentient.[1] Yet this recognition is accompanied by the stark fact that between 1990 and 2000, a large portion of the remaining habitat of chimpanzees and bonobos will be destroyed unless we mount concerted conservation efforts. That is, a significant part of the biological and cultural diversity of the large-brained species most nearly related to us could be wiped out by human action during the coming decade.

As studies of chimpanzees and bonobos progress, we will learn more about the similarities and differences between these species and our own. But in the questions investigators are now asking, one senses a shift back toward the concerns of Wolfgang Köhler – the shift is from "How do chimpanzees and bonobos differ from us?" to "What, indeed, are the capacities of these extraordinary creatures?" And in learning about chimpanzees and bonobos, we humans will continue to discover immensely important information about ourselves.

We humans have traditionally accorded to ourselves a whole range of characteristics which we thought so unique or so advanced in degree of development as to constitute essentially qualitative differences between us and other organisms. With pride we counted emotions, strong family relationships, helpfulness, tool use and fabrication, forethought, insight, language, the overwhelming ability to modify our environments, and the importance of individual history as being among our distinguishing characteristics. With dismay we recognized that warfare, murder, purposeful cannibalism, and our ability to deplete and destroy our resources were among our most peculiar features. We saw that the continuous sexual receptivity of the human female knit the familial fabric, and for a long time we thought this characteristic uniquely human. Yet as a result of intensive and long-term study of *Pan troglodytes* (chimpanzees) and *Pan paniscus* (bonobos), we now know that humans are thoroughly a part of a continuum. This idea unfolds dramatically in the chapters of this book.[2]

If we reflect, we realize that it requires free time to be human in the ways enumerated above. Thus it is important to know whether chimpanzees have time for activities beyond subsistence. In this volume, Basuta shows that chimpanzees do not forage according to simple optimality rules. This suggests that chimpanzees choose their foods carefully and that they have the leisure to pursue their food selection and other activities.[3]

Within this leisurely milieu, chimpanzee families have a rich emotional life, as Goodall clearly demonstrates. Older and younger sisters and brothers engage in learning play. Older siblings nurture the younger and, following the death of their mother, see them through depression. Even an older chimpanzee may become so depressed on the loss of its mother that, in the absence of a strong relationship with another chimpanzee, it dies. In some very special cases, an unrelated individual may adopt and care for an infant or behave as an "aunt," greatly increasing the infant's chance of survival.

It is within this dynamic and nurturing emotional context that young chimpanzees learn to use and fashion tools: stick tools and leaf tools (Goodall), stone tools and wooden tools (Kortlandt). They learn to participate in the hunt (Goodall) and learn to recognize and use plants with medicinal properties (Wrangham and Goodall). These activities are extraordinary in themselves, and when the stories of the individual actors are recounted in detail, the activities can only be perceived as insightful and purposeful.

Chimpanzees and bonobos are long-lived animals, and the importance of individual histories and group histories has been realized only through long-term studies in the natural environment. Sustained research is needed to reveal the development of behavioral differences between sexes (Hiraiwa-Hasegawa). Goodall shows that deprivation or maltreatment of females during rearing can lead to deficient, negligent, and even harmful maternal behavior at maturity. It is very important to note that a brief cross-sectional study would not recognize such maternal behavior as aberrant nor reveal its origins – such a study could only view this behavior as part of the range of usual.

The quarter century of research at Mahale (Nishida)[4] has formed the basis for detailed studies of group dynamics. Because of this continuous research at Mahale, we know that groups deteriorate and that individual lives are dramatically altered as individuals meld into another group (Nishida, Hasegawa). One of the most crucial current questions concerns the demography of wild populations, especially those in small isolated groups such as in Bossou, Guinea (Sugiyama), whose survival can only be described as tenuous. Gathering such demographic data is a long-term project.

At both Mahale and Gombe, researchers have observed purposeful murder and cannibalism and, at Gombe, intertroop warfare. Long-term studies are essential for uncovering these infrequent events. Similarly, if we are to comprehend the complex vocal communications that chimpanzees use in their native habitats (Boehm), protracted data gathering is necessary. Wherever studies are pursued over the course of several years, new levels of cultural diversity within and between populations have been revealed (McGrew). Thus it is clear that only extended behavioral studies can put observations on chimpanzees and bonobos into perspective and reveal their significance (R. Fouts).

Remarkable contrasts exist between chimpanzees and bonobos (de Waal). Among bonobos, sexual behaviors are very important in group processes (Kano). Bonobos show delay of infant physical development, which has strong implications for behavioral development (Kuroda). Such retardation may be an important hetero-chronic component in human evolution. Female sociality among bonobos (White) may also point in the direction of human social structure. Clearly, bonobos merit long-term studies. It may be that the lesser-studied bonobos show even greater similarities to humans than do chimpanzees. As we work at understanding chimpanzees and bonobos, we may find some answers to intriguing questions about human evolution (Ghiglieri).

In the language laboratories, studies of chimpanzees and bonobos show that they can learn to communicate with us, in our languages, with alacrity. Chimpanzees make statements, produce requests, ask questions, joke, tease, lie, commiserate, generalize, answer questions, respond to commands, refer to things not present, recall, predict, and indicate state of mind and emotion in American Sign Language (Gardner and Gardner) and in computer language (Rumbaugh). Chimpanzees avidly teach sign language to younger animals using highly effective techniques similar to the ones by which they were taught, they use sign language among themselves during play and in other exchanges, and they even talk to themselves in sign language, as human children do (D. Fouts).

The $2\frac{1}{2}$-year-old bonobo Kanzi astonished the researchers caring for him (Savage-Rumbaugh et al.). He had watched his foster mother being trained in the Yerkish computer language. When she was returned to the colony for breeding, the young Kanzi, without prompting and without training, began to communicate correctly through the computer. Matsuzawa taught the chimpanzee Ai to count, to signal the correct count by using the appropriate arabic numeral, and to relate a rather large series of symbols to objects in her environment. She learned all 26 English letters and how to reconstruct tripartite symbols from their elements. Ai performed the reconstruction more rapidly and more accurately than a graduate student tested for comparison.

All these results indicate that we need to concentrate even more on learning about bonobos and chimpanzees as themselves rather than simply as analogs for us (Menzel, Rumbaugh). Chimpanzees and bonobos have extraordinary capacity that has only begun to be elucidated. Therefore, at a minimum, behavioral testing programs for them must be an interesting challenge to their sophisticated mental equipment.

In light of our growing understanding, we have a very special responsibility for the health and living conditions of chimpanzees and bonobos in captivity (Latinen). Small cages with sterile environments simply can no longer be permitted in research settings or in zoos. Captive environments should be spacious and full of interesting objects; they should facilitate locomotion and encourage social interaction. We must share records of captive chimpanzees and bonobos through a central data bank so that their well-being can be assured and their reproduction optimized, eliminating the need for exploitation of wild populations.

In their African homelands, the bonobo and each of the three subspecies of the chimpanzee require the most sophisticated conservation efforts.[5] The plight of these species is aptly and compellingly related by Teleki, Malenky et al., McGrew, and Goodall. Deforestation has split chimpanzee populations into widely separated, often small, remnant populations. These groups are highly susceptible to human disease and to hunting for meat or export. Even in the national parks and reserves, chimpanzees and bonobos are subject to a variety of challenges. Many areas urgently require surveys to establish or update demographic information. Current ranges of the eastern and western chimpanzee subspecies are drastically reduced and extremely fragmented. Although little is known about the central chimpanzee subspecies, the safest assumption is that they are as threatened as chimpanzees are elsewhere (Teleki). Solutions to the problems of conservation must be creatively pursued on every level from local to global. To do otherwise is to destroy these species.

A corollary of the endangered status of chimpanzees and bonobos, and of their similarity to humans, is that only under circumstances of gravest need should these species be used for biomedical experimentation: that is, only when we expect very great benefit (preferably for the chimpanzees or bonobos as well as humans) and only when chimpanzees or bonobos alone can fulfill the necessary experimental objective.

In conclusion, we now glimpse the rich possibilities of learning from, about, and with our sibling species. It is imperative that, in the year 2000, we be able to look back, not in bitter irony at a grievously lost opportunity, but in prevailing hope and a continuing commitment to the surviving chimpanzee and bonobo populations. Let us be able to say that we cared, that we committed the full range of human ingenuity so that large populations of chimpanzees and bonobos could retain stable resource bases and secure social communities in which to rear their young, transmit their culture, and work out their specific evolutionary destinies.

ENDNOTES

1. The work on gorillas, not covered in this volume, may suggest a third sentient species. Other apes and the monkeys have not been adequately tested at this time.

2. The authors first shared preliminary versions of these chapters in November 1986 at a symposium organized by The Chicago Academy of Sciences to celebrate 25 years of field research in Tanzania by teams of Japanese researchers and by Jane Goodall and her collaborators. The symposium also celebrated the appearance of Jane's major book *The Chimpanzees of Gombe*.

The symposium had several outcomes. The first was a growing realization that chimpanzees and bonobos are our sibling species. The second was a realization of the importance of extended visits between field sites. The third outcome was a determination to repeat the conference in five years to review the progress of research and conservation efforts. The fourth was the conviction that *Pan* was endangered in most, and quite possibly all, of its species' ranges. Among the public and professionals at the symposium there was a ground swell to establish an action arm that would pursue the cause of chimpanzee care and conservation. By the end of the symposium, the participants in the scientific and public sessions had contributed more than nine thousand dollars to the formation of a Committee for Conservation and Care of Chimpanzees (CCCC) to be staffed by Geza Teleki. Those wishing to join or desiring more information about CCCC may write to CCCC at 3819 48th Street NW, Washington, DC 20016. At the time of this writing, the United States has changed the designation of wild-born chimpanzees from Threatened to Endangered on its Endangered Species List. For purposes of trade with the United States, all chimpanzees in Africa are treated by the federal government as Endangered. This makes it more difficult to import chimpanzees into the United States. The CCCC played an instrumental role in this legislation. It continues to work for conservation measures protecting chimpanzees and bonobos in Africa and to press for legislation mandating improved conditions for captive chimpanzees.

3. As with chimpanzees, natural history observations strongly suggest the same conclusion for bonobos and for other species of primates.

4. That studies at Mahale and Gombe have continued for nearly 30 years is a piece of scientific good fortune. The possibilities of comparison between two populations, fairly close together, and studied in somewhat different ways, are enormous and will repay researchers for decades more. It is to be hoped that the principal researchers from each of the Tanzanian sites will soon be able to exchange visits. As of this writing, very few persons have conducted observations at both of the field stations.

Unfortunately for many researchers, the bulk of the work at Mahale is reported only in Japanese. The entire field of chimpanzee studies would be tremendously advanced by the publication of this material in translation.

5. The opportunity to study the interaction between chimpanzees and lowland gorillas, which exists in several locations, should be pursued with the utmost vigor before deforestation and other pressures threaten or destroy this special situation.

CONTRIBUTORS

Christopher Boehm
   Department of Sociology,
      Anthropology, and Philosophy
   Northern Kentucky University
   Highland Heights, Kentucky 41076

Deborah H. Fouts
   Friends of Washoe
   Department of Psychology
   Central Washington University
   Ellensburg, Washington 98926

Roger S. Fouts
   Friends of Washoe
   Department of Psychology
   Central Washington University
   Ellensburg, Washington 98926

Beatrix T. Gardner
   Department of Psychology
   University of Nevada-Reno
   Reno, Nevada 89557

R. Allen Gardner
   Department of Psychology
   University of Nevada-Reno
   Reno, Nevada 89557

Michael P. Ghiglieri
   Department of Anthropology
   Northern Arizona State University
   NAU Box 15200
   Flagstaff, Arizona 86011

Jane Goodall
   Gombe Stream Research Centre
   Director's Office
   P. O. Box 727
   Dar es Salaam, Tanzania

Toshikazu Hasegawa
   Department of Psychology
   Teikyo University
   Otsuka, Hachioji
   Tokyo 192-03, Japan

Paul G. Heltne
   Director
   The Chicago Academy of Sciences
   2001 North Clark Street
   Chicago, Illinois 60614

Mariko Hiraiwa-Hasegawa
   Department of Anthropology
   University of Tokyo
   Hongo Bunkyo-ku
   Tokyo 113, Japan

William D. Hopkins
   Department of Psychology
      and Language Research Center
   Georgia State University
   University Plaza
   Atlanta, Georgia 30303

G. Isabirye-Basuta
   Department of Zoology
   Makerere University
   P. O. Box 7062
   Kampala, Uganda

Takayoshi Kano
   Department of Sociology
   Primate Research Institute
   Kyoto University
   Kanrin, Inuyama
   Aichi 484, Japan

Adriaan Kortlandt
   88, Woodstock Road
   Oxford OX2 7ND
   United Kingdom

Suehisa Kuroda
   Laboratory of Physical
      Anthropology
   Faculty of Science
   Kyoto University
   Sakyo-ku, Kyoto, Japan 606

Katherine Latinen
  Curator / Tropics
  Minnesota Zoological Garden
  12101 Johnny Cake Ridge Road
  Apple Valley, Minnesota 55124

Richard K. Malenky
  Department of Ecology and Evolution
  SUNY at Stony Brook
  Stony Brook, New York 11794-8081

Linda A. Marquardt
  The Chicago Academy of Sciences
  2001 North Clark Street
  Chicago, Illinois 60614

Tetsuro Matsuzawa
  Primate Research Institute
  Kyoto University
  Inuyama, Aichi 484, Japan

William C. McGrew
  Department of Psychology
  University of Stirling
  Stirling FK9 4 LA, Scotland

Emil W. Menzel, Jr.
  Department of Psychology
  SUNY at Stony Brook
  Stony Brook, New York 11794

Toshisada Nishida
  Department of Zoology
  Faculty of Science
  Kyoto University
  Kitashirakawa-Oiwakecho
  Sakyo-ku, Kyoto, Japan 606

Mary Ann Romski
  Department of Communication and
    Language Research Center
  Georgia State University
  University Plaza
  Atlanta, Georgia 30303

Duane M. Rumbaugh
  Language Research Center and
    Department of Psychology
  Georgia State University
  Atlanta, Georgia 30303
        and
  Yerkes Regional Primate
    Research Center
  Emory University
  Atlanta, Georgia 30322

Sue Savage-Rumbaugh
  Language Research Center and
    Department of Biology
  Georgia State University
  Atlanta, Georgia 30303
        and
  Yerkes Regional Primate
    Research Center
  Emory University
  Atlanta, Georgia 30322

Rose A. Sevcik
  Department of Psychology
    and Language Research Center
  Georgia State University
  University Plaza
  Atlanta, Georgia 30303

Yukimaru Sugiyama
  Primate Research Institute
  Kyoto University
  Inuyama, Aichi 484, Japan

Randall L. Susman
  Department of Anatomical Sciences
  School of Medicine
  SUNY at Stony Brook
  Stony Brook, New York 11794-8081

Geza Teleki
  Committee for Conservation
    and Care of Chimpanzees
  3819 48th Street, NW
  Washington, DC 20016

Nancy Thompson-Handler
  Department of Anthropology
  Yale University
  New Haven, Connecticut 06520

Frans B. M. de Waal
  Wisconsin Regional Primate
    Research Center
  University of Wisconsin
  1223 Capitol Court
  Madison, Wisconsin 53715-1299

Frances J. White
  Department of Biological Anthropology
    and Anatomy
  Duke University
  Durham, North Carolina 27706

Richard W. Wrangham
  Department of Anthropology
  University of Michigan
  Ann Arbor, Michigan 48109

NAME INDEX

SUBJECT INDEX

Design	The Chicago Academy of Sciences in collaboration with Diane Hutchinson, designer
Typography	The Chicago Academy of Sciences using Aldus PageMaker® 3.01 and Microsoft ® Word 3.01 on an Apple® Macintosh™ SE, output by Black Dot Graphics on Linotronic® 300
Graphics Production	Howell For Graphics
Text	10/13 Times Roman